SC-FDMA for Mobile Communications

SC-FDMA for Mobile Communications

Fathi E. Abd El-Samie
Faisal S. Al-kamali
Azzam Y. Al-nahari
Moawad I. Dessouky

CRC Press
Taylor & Francis Group
Boca Raton London New York

CRC Press is an imprint of the
Taylor & Francis Group, an **informa** business

CRC Press
Taylor & Francis Group
6000 Broken Sound Parkway NW, Suite 300
Boca Raton, FL 33487-2742

First issued in paperback 2017

© 2014 by Taylor & Francis Group, LLC
CRC Press is an imprint of Taylor & Francis Group, an Informa business

No claim to original U.S. Government works

ISBN-13: 978-1-4665-1071-5 (hbk)
ISBN-13: 978-1-138-19994-1 (pbk)

Contents

Preface

The single-carrier frequency division multiple access (SC-FDMA) system is a well-known system that has recently become a preferred choice for mobile uplink channels. This is attributed to its advantages such as the low peak-to-average power ratio (PAPR) and the use of frequency domain equalizers. Low PAPR allows the system to relax the specifications of linearity in the power amplifier of the mobile terminal, which reduces cost and power consumption. Moreover, it has a similar throughput performance and essentially the same overall complexity as the orthogonal frequency division multiple access (OFDMA) system. Due to these advantages, SC-FDMA has been chosen as the uplink transmission method in the long-term evolution (LTE) system.

However, the SC-FDMA system suffers from several problems such as link performance loss in a frequency-selective channel when high-order modulation techniques are used. In addition, the presence of carrier frequency offsets (CFOs) between the transmitter and the receiver results in a loss of orthogonality among subcarriers and an intercarrier interference (ICI). CFOs also introduce multiple access interference (MAI) and degrade the bit error rate (BER) performance in the SC-FDMA system. Moreover, even though the SC-FDMA transmitted signals are characterized by low signal envelope fluctuations, the performance degradation due to nonlinear amplification may substantially affect the link performance of the system. As a result, there is a need to enhance the performance of the SC-FDMA system. This book deals with these problems, and its main objective is to enhance the performance of the SC-FDMA system.

The book presents an improved discrete cosine transform (DCT)-based SC-FDMA system. Simulation results show that the DCT-based SC-FDMA system provides better BER performance than the DFT-based SC-FDMA and OFDMA systems, while the complexity of the receiver is slightly increased. Moreover, it was concluded that the PAPR of the DCT-based SC-FDMA system is lower than that of the OFDMA system.

In addition, a new transceiver scheme for the SC-FDMA system is introduced and studied. Simulation results illustrate that the proposed transceiver scheme provides better performance than the conventional schemes and it is robust to the channel estimation errors. It was concluded that the immunity of the proposed scheme to the nonlinear amplification and the noise enhancement problems is higher than that of the conventional scheme.

The problem of CFOs is investigated and treated for the single-input single-output (SISO) SC-FDMA system. A new low-complexity equalization scheme, which jointly performs the equalization and CFO compensation in the SISO SC-FDMA system, is presented in this book. The mathematical expression of this equalizer is derived by taking into account the MAI and the noise. A low-complexity implementation of this equalization scheme using a banded matrix approximation is also presented. From the obtained simulation results, this equalization scheme is able to enhance the performance of the SC-FDMA system, even in the presence of estimation errors.

Furthermore, the problem of CFOs is investigated and treated for the multiple-input multiple-output (MIMO) SC-FDMA system. Three equalization schemes for the MIMO SC-FDMA system in the presence and the absence of CFOs are presented. First, a low-complexity regularized zero forcing (LRZF) equalization scheme is introduced. This simplifies the matrix inversion process by performing it in two steps. In the first step, the interantenna interference (IAI) is cancelled. In the second step, the intersymbol interference (ISI) is mitigated. A regularization term is added in the second step of the matrix inversion to avoid noise enhancement. The LRZF scheme is also developed for the MIMO SC-FDMA system in the presence of CFOs. A solution has been derived to jointly perform the equalization and CFO compensation. Finally, an equalization scheme for SISO SC-FDMA system is developed for the MIMO SC-FDMA system in the presence of CFOs by taking into account the MAI and the noise. Computer simulations confirm that the discussed equalization schemes are able to mitigate the effects of CFOs and the multipath channel, even in the presence of estimation errors. It has been deduced that the performance of each of the discussed equalizers outperforms that of the conventional schemes.

Cooperative communication with SC-FDMA is also considered in this book. A background on the cooperative communication and cooperative diversity is presented. A distributed space–time coding scheme is also presented and its performance is evaluated. Distributed SFC for broadband relay channels is also studied and space–time/frequency coding schemes for SC-FDMA systems are described. In addition, relay selection schemes for improving the physical layer security are presented.

Finally, MATLAB® codes for all simulation experiments are included in Appendices F and G at the end of the book.

MATLAB® is a registered trademark of The MathWorks, Inc. For product information, please contact:

The MathWorks, Inc.
3 Apple Hill Drive
Natick, MA 01760-2098 USA
Tel: 508-647-7000
Fax: 508-647-7001
E-mail: info@mathworks.com
Web: www.mathworks.com

Authors

 **Fathi E. Abd El-Samie** received his BSc (Honors), MSc, and PhD from Menoufia University, Menouf, Egypt, in 1998, 2001, and 2005, respectively. Since 2005, he has been a teaching staff member with the Department of Electronics and Electrical Communications, Faculty of Electronic Engineering, Menoufia University. He currently serves as a researcher at KACST-TIC in Radio Frequency and Photonics for the e-Society (RFTONICs), King Saud University. He is a coauthor of about 200 papers in international conference proceedings and journals and of 4 textbooks. His research interests include image enhancement, image restoration, image interpolation, super-resolution reconstruction of images, data hiding, multimedia communications, medical image processing, optical signal processing, and digital communications. Dr. Abd El-Samie received the Most Cited Paper Award from the *Digital Signal Processing* journal in 2008.

Faisal S. Al-kamali received his BSc in electronics and communications engineering from the Faculty of Engineering, Baghdad University, Baghdad, Iraq, in 2001. He received his MSc and PhD in communication engineering from the Faculty of Electronic Engineering, Menoufia University, Menouf, Egypt, in 2008 and 2011, respectively. He joined the teaching staff of the Department of Electrical Engineering, Faculty of Engineering and Architecture, Ibb University, Ibb, Yemen, in 2011. He is a coauthor of several papers in international conferences and journals. His research interests include CDMA systems, OFDMA systems, single-carrier FDMA (SC-FDMA) system, MIMO systems, interference cancellation, synchronization, channel equalization, and channel estimation.

Azzam Y. Al-nahary received his BSc in electronics and communications engineering from the University of Technology, Baghdad, Iraq. He received his MSc and PhD from Menoufia University, Egypt, in 2008 and 2011, respectively. He was also a postdoctoral fellow in the Department of Electrical and Information Technology, Lund University, Sweden. He currently serves as an assistant professor in the Department of Electrical Engineering, Ibb University, Yemen. His research interests include MIMO systems, OFDM, cooperative communications, and physical layer security.

Moawad I. Dessouky received his BSc (Honors) and MSc from the Faculty of Electronic Engineering, Menoufia University, Menouf, Egypt, in 1976 and 1981, respectively, and his PhD from McMaster University, Canada, in 1986. He joined the teaching staff of the Department of Electronics and Electrical Communications, Faculty of Electronic Engineering, Menoufia University, Menouf, Egypt, in 1986. He has published more than 200 scientific papers in national and international conference proceedings and journals. He currently serves as the vice dean of the Faculty of Electronic Engineering, Menoufia University. Dr. Dessouky received the Most Cited Paper Award from *Digital Signal Processing* journal in 2008. His research interests include spectral estimation techniques, image enhancement, image restoration, super-resolution reconstruction of images, satellite communications, and spread spectrum techniques.

1

INTRODUCTION

1.1 Motivations for Single-Carrier Frequency Division Multiple Access

The significant expansion seen in mobile and cellular technologies over the last two decades is a direct result of the increasing demand for high data rate transmissions. So, in recent years, the existing and the incoming wireless mobile communication systems are occupying more and more transmission bandwidths than the conventional ones to support broadband multimedia applications with high data rates for mobile users [1]. In addition, wireless mobile technologies are also moving rapidly toward small and low cost devices. However, broadband wireless channels suffer from a severe frequency-selective fading, which causes ISI [2,3]. As the bit rate increases, the problem of ISI becomes more serious. Conventional equalization in the time domain has become impractical, because it requires one or more transversal filters with a number of taps covering the maximum channel impulse response length [4,5].

Orthogonal frequency division multiplexing (OFDM) and OFDMA systems have received a lot of attention in the last few years due to their abilities to overcome the frequency-selective fading impairment by transmitting data over narrower subbands in parallel [6–8]. However, they have several inherent disadvantages such as the high Peak-to-average power ratio (PAPR) and the sensitivity to CFOs [9–13]. To solve the problems encountered in the uplink of these systems, much attention has been directed recently to another system, namely the single-carrier with frequency domain equalization (SC-FDE) system [2–5], because it has a lower PAPR. To provide multiple access in broadband wireless networks, the SC-FDE system has been naturally combined with frequency division multiple access (FDMA), where different orthogonal subcarriers are allocated to different user equipments, and the new system is referred to as the SC-FDMA system [10,14,15].

1

Recently, the SC-FDMA system, which is the main topic of this book, has attracted much attention due to its low PAPR. It is recognized as a close relative to the OFDMA system, since it takes advantage of the OFDMA system in combination with DFT spreading prior to the OFDMA modulation stage. The main advantages of the SC-FDMA system are that the envelope fluctuations are less pronounced and the power efficiency is higher than that of the OFDMA system [9,14]. Moreover, the SC-FDMA system has a similar throughput performance and essentially the same overall complexity as the OFDMA system [9,14]. Because of these advantages, the SC-FDMA system has been adopted by the third generation partnership project (3GPP) for uplink transmission in the technology standardized for LTE of cellular systems [16], and it is the physical access scheme in the uplink of the LTE-advanced [17]. The implementation of the SC-FDMA with the systems equipped with more transmitting and receiving antennas and cooperative systems is a very promising way to achieve large spectrum efficiency and capacity of mobile communication systems. Furthermore, one of the key features of the LTE-advanced is the application of the MIMO technique for uplink transmission [17].

However, even though the SC-FDMA transmitted signal is characterized by lower signal envelope fluctuations, the performance degradation due to the nonlinear amplification may substantially affect the link performance of the system. In addition, the SC-FDMA system suffers from the link performance loss in frequency-selective channels, when high-order modulation techniques are used [1]. Moreover, the orthogonality of the SC-FDMA system relies on the condition that the transmitter and receiver operate with exactly the same frequency reference. If this is not the case, the perfect orthogonality of the subcarriers is lost causing ICI and MAI [18]. Frequency errors typically arise from a mismatch between the reference frequencies of the transmitter and the receiver local oscillators. On the other hand, due to the importance of using low-cost components in the mobile terminal, local oscillator frequency drifts are usually greater than those in the base station and are typically dependent on temperature changes and voltage variations. These differences from the reference frequencies are widely referred to as CFOs. As a result, there is a need to enhance the link performance of the SC-FDMA system, which is the main objective of this book.

Up to now, the DFT only is used to implement the SC-FDMA system. This motivated us to apply other sinusoidal transforms for the SC-FDMA system such as the DCT. This book will refer to the DFT-based SC-FDMA system as DFT-SC-FDMA and to the DCT-based SC-FDMA system as DCT-SC-FDMA.

1.2 Evolution of Cellular Wireless Communications

The concept of cellular wireless communications is to divide large zones into small cells to provide radio coverage over a wider area than the area served by a single cell. This concept was developed by researchers at Bell Laboratories during the 1950s and 1960s [19]. The first cellular system was created by Nippon Telephone and Telegraph (NTT) in Japan in 1979. From then on, the cellular wireless communication has evolved. The first generation (1G) of cellular wireless communication systems utilized analog communication techniques, and it was mainly built on frequency modulation and FDMA. Digital communication techniques appeared in the second generation (2G) systems, and the spectrum efficiency was improved obviously. Time division multiple access (TDMA) and code division multiple access (CDMA) have been utilized as the main multiple access schemes. The two most widely accepted 2G systems were global system for mobile (GSM) and interim standard (IS-95).

The third generation (3G) systems were designed to solve the problems of the 2G systems and to provide high quality and high capacity in data communication. International Mobile Telecommunications 2000 (IMT-2000) was the global standard for 3G wireless communications, defined by a set of interdependent International Telecommunication Union (ITU) recommendations. IMT-2000 provided a framework for worldwide wireless access by linking the diverse system–based networks. The most important 3G standards are the European and Japanese Wideband-CDMA (WCDMA), the American CDMA2000, and the Chinese time-division synchronous CDMA.

IMT-2000 provided higher transmission rates; a minimum speed of 2 Mbps for stationary or walking users and 384 kbps in a moving vehicle, whereas 2G systems provided only speeds ranging from 9.6 to 28.8 kbps. After that the initial standardization in both WCDMA and CDMA2000 has evolved into 3.5G [9]. Currently, 3GPP LTE is considered as the prominent path to the next generation of cellular systems beyond 3G. The ITU has recently issued requirements for IMT-Advanced, which constitute the official definition of the fourth generation (4G) [20]. The ITU recommends operation in up to 100 MHz radio channels and a peak spectral efficiency of 15 bps/Hz, resulting in a theoretical throughput rate of 1.5 Gbps.

1.3 Mobile Radio Channel

In mobile wireless communications, the transmitted signal is subject to various impairments caused by the transmission medium combined with the mobility of transmitters and/or receivers. Path loss is an attenuation of the signal strength with the distance between the transmitter and the receiver antenna. The frequency reuse technique in cellular systems is based on the physical phenomenon of path loss. Unlike the transmission in free space, transmission in practical channels, where propagation takes place in atmosphere and near the ground, is affected by terrain contours. As the mobile moves, slow variations in mean envelope over a small region appear due to the variations in large-scale terrain characteristics, such as hills, forests, and clumps of buildings. The variations resulting from shadowing are often described by a log-normal distribution [21]. Power control techniques are often used to combat the slow variations in the mean-received envelope due to path loss and shadowing.

Compared to the large-scale fading due to the shadowing, multipath fading, often called fast fading, refers to the small-scale fast fluctuations of the received signal envelope resulting from the multipath effect and/or receiver movement. Multipath fading results in the constructive or destructive addition of arriving plane wave components and manifests itself as large variations in amplitude and phase of the composite-received signal in time [22]. When the channel exhibits a deep fade, fading causes a very low instantaneous signal-to-noise ratio (SNR).

1.3.1 Slow and Fast Fading

The distinction between slow and fast fading is important for the mathematical modeling of fading channels and for the performance evaluation of communication systems operating over these channels. This notion is related to the coherence time (T_{ch}) of the channel, which measures the period of time over which the fading process is correlated (or equivalently, the period of time after which the correlation function of two samples of the channel response taken at the same frequency but different time instants drops below a certain predetermined threshold). The coherence time is also related to the channel Doppler spread f_d by [22]:

$$T_{ch} \approx \frac{1}{f_d} \qquad (1.1)$$

The fading is said to be slow if the symbol time duration (T) is smaller than the channel coherence time; otherwise, it is considered to be fast. In slow fading, a particular fading level affects several successive symbols, which leads to burst errors, whereas in fast fading, the fading decorrelates from symbol to symbol.

1.3.2 Frequency-Flat and Frequency-Selective Fading

Frequency selectivity is also an important characteristic of fading channels. If all the spectral components of the transmitted signal are affected in a similar manner, the fading is said to be frequency-nonselective or equivalently frequency-flat. This is the case in narrowband systems, in which the transmitted signal bandwidth is much smaller than the channel coherence bandwidth. This bandwidth measures the frequency range over which the fading process is correlated and is defined as the frequency bandwidth over which the correlation function of two samples of the channel response taken at the same time but different frequencies falls below a suitable value. In addition, the coherence bandwidth is related to the maximum delay spread τ_{max} by [22]:

$$B_{ch} \approx \frac{1}{\tau_{max}} \qquad (1.2)$$

On the other hand, if the spectral components of the transmitted signal are affected by different amplitude gains and phase shifts, the fading is said to be frequency-selective. This applies to wideband systems in which the transmitted bandwidth is wider than the channel coherence bandwidth. The frequency-selective channel can be modeled or represented as a tapped delay line. The baseband channel impulse response can be expressed as follows:

$$h(t) = \sum_{l=0}^{L-1} h(l)\delta(t - \tau(l)) \qquad (1.3)$$

where
  $h(l)$ and $\tau(l)$ represent the complex fading and the propagation delay of the lth path
  L is the number of multipath components of the channel impulse response

In our simulations, we assume block fading, where the path gains stay constant over each block duration.

1.3.3 Channel Equalization

A frequency-selective channel causes problems in single-carrier communication systems, since the signals from different echoes arrive at different time instants to the receiver. This means that the received signal is the sum of signals sent at different time instants and the so-called ISI arises [23]. When the duration of the transmitted symbols is large as compared to the duration of the channel impulse response, the symbols are only slightly disturbed by the channel. However, if the delay spread is no longer small as compared to the symbol duration, then the ISI spans over one or more symbols, and it severely affects the received signal. As a consequence, it is necessary for the optimum receiver to compensate for or reduce the ISI in the received signal. The compensator for the ISI is called an equalizer. Because in a wireless mobile environment the channel impulse response is time-variant, the equalizer must also change or adapt to the time-varying channel characteristics [24–27].

The required number of equalizer coefficients increases with the amount of ISI. Today, there is a growing interest in high data rate mobile communication, i.e., the transmitted symbol duration needs to be shorter and shorter. Accordingly, the implementation complexity of the equalization process becomes too high. It is therefore imperative to use alternative approaches that support high data rate transmission over multipath fading channels. The multicarrier transmission scheme is an effective technique to combat multipath fading in wireless communications, but it has several inherent disadvantages such as large PAPR and sensitivity to CFOs [9].

Recently, much attention has been focused on single-carrier communication systems, which combine the Cyclic Prefix (CP) concept of the multicarrier systems with single-carrier transmission to enable efficient FDE for severe frequency-selective fading channels. FDE provides a simple yet very efficient solution for combating ISI. The main advantage of the FDE lies in its low complexity when compared to the time-domain equalization.

1.4 Multicarrier Communication Systems

A high data rate stream typically faces a problem in having a symbol period much smaller than the channel delay spread, if it is transmitted serially. This generates ISI, which can only be mitigated by means of a complex equalization procedure. In general, the equalization complexity grows with the square of the channel impulse response length. Multicarrier systems have received a lot of attention in the last few years due to their ability to overcome the frequency-selective fading impairments at high data rate applications.

The basic principles of using multicarrier systems have been found in the late 1960s, where Chang in [28] published his elegant theory concerning data transmission of the orthogonal signals over bandwidth-limited multichannel environments. In [28], the basic fundamentals of simultaneous parallel data transmission over bandwidth-limited propagation channels without ISI have been introduced. However, the required high complexity due to the synchronization and modulation issues widely limited the application of such a scheme, and therefore, Weinstein and Ebert [29] proposed a modified system, namely the OFDM system, in which the inverse fast Fourier transform (IFFT)/ fast Fourier transform (FFT) was applied to generate the orthogonal subcarriers. Their scheme reduced the implementation complexity significantly, by taking advantage of the IFFT/FFT.

Multicarrier transmission is based on splitting a high-rate data stream into several parallel low-rate substreams that are transmitted on different frequency channels, i.e., on different subcarriers [30]. The motivations for using this technique are that no complex equalization is required, and a high spectral efficiency can be achieved.

1.4.1 OFDM System

The OFDM system is simply defined as a form of multicarrier systems, where the carrier spacing is carefully selected so that each subcarrier is orthogonal to the other subcarriers. Two signals are orthogonal if their dot product is zero. That is, if you take two signals, multiply them together and if their integral over an interval is zero, then the two signals are orthogonal in that interval.

Orthogonality can be achieved by carefully selecting the subcarrier spacing, such as letting the subcarrier spacing be equal to the reciprocal of the useful symbol period. As the subcarriers are orthogonal, the spectrum of each subcarrier has a null at the center frequency of each of the other subcarriers in the system, as shown in Figure 1.1. This results in no interference between the subcarriers, allowing them to be spaced as close as theoretically possible. Figure 1.2 shows the block diagram of the OFDM system. The input data symbols are

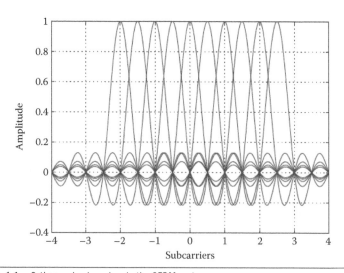

Figure 1.1   Orthogonal subcarriers in the OFDM system.

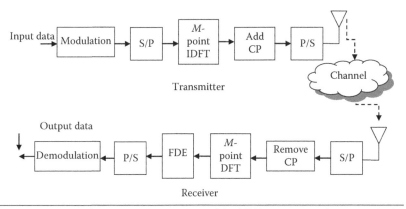

Figure 1.2 Transmitter and receiver structures of the OFDM system over a frequency-selective channel.

modulated. Then, the modulated symbols are serial-to-parallel (S/P) converted and processed by the inverse DFT (IDFT). Finally, the CP is added, and the signal samples are transmitted. The receiver performs the reverse operation of the transmitter. More technical details on the OFDM system are found in [7,30,31]. The major advantages of the OFDM system can be summarized as follows:

- High spectral efficiency, since it uses overlapping orthogonal subcarriers in the frequency domain
- Simple digital realization by using the FFT/IFFT operation
- Different modulation schemes can be used on individual subcarriers, and can be adapted to the transmission conditions on each subcarrier
- Flexible spectrum adaptation

Because of these advantages, the OFDM system has been adopted as a modulation choice by several wireless communication systems such as wireless local area networks (LANs), digital video broadcasting, and WiMAX [7]. Although OFDM has proved itself as a powerful modulation technique, it has its own challenges [9]:

- Due to the narrow spacing and spectral overlap between the subcarriers, the OFDM system has high sensitivity to CFOs and phase errors. These errors lead to a loss of subcarriers orthogonality. Accordingly, a degradation of the global system performance occurs

- Due to the large number of subcarriers, the OFDM system has a large dynamic signal range with a relatively high PAPR. This tends to reduce the power efficiency of the power amplifier (PA)
- Sensitivity to the resolution and dynamic range of the digital-to-analog (D/A) and analog-to-digital (A/D) converters. Since the OFDM system suffers from large envelope fluctuations, a high-resolution D/A converter is required at the transmitter and a high-resolution A/D converter operating with a high dynamic range is required at the receiver side
- Loss in power and spectral efficiency due to the CP insertion
- A need for an adaptive or coded scheme to overcome spectral nulls in the channel. In the presence of a null in the channel, there is no way to recover the data of the subcarriers that are affected by the null unless we use rate adaptation or a coding scheme

1.4.2 OFDMA System

The OFDMA system is a multiuser version of the OFDM system, and all that were previously mentioned about the OFDM system also hold for the OFDMA system. Each user in an OFDMA system is usually given certain subcarriers during a certain time to communicate. Usually, subcarriers are allocated in contiguous groups for simplicity and to reduce the overhead of indicating which subcarriers have been allocated to each user. One of the major problems with an OFDMA system is to synchronize the uplink transmissions, because every user has to transmit his frame so that he avoids interfering with the other users. The OFDMA system for mobile communications was first proposed in [32] based on multicarrier FDMA, where each user is assigned a set of randomly selected subcarriers. Figure 1.3 shows the block diagram of the OFDMA system.

1.4.3 Multicarrier CDMA System

The basic multicarrier CDMA (MC-CDMA) signal is generated by a serial concatenation of classical direct-sequence CDMA (DS-CDMA) and OFDM systems.

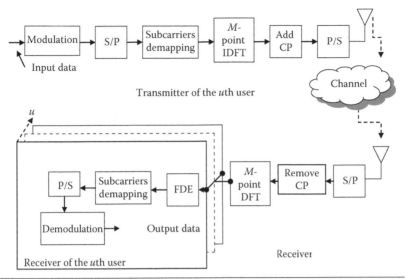

Figure 1.3   Transmitter and receiver structures of the OFDMA system over a frequency-selective channel.

Each chip of the DS-CDMA system is mapped onto different subcarriers. Thus, with the MC-CDMA system, the chips are transmitted in parallel on different subcarriers, in contrast to the serial transmission with the DS-CDMA system. Figure 1.4 shows the block diagram of the MC-CDMA system.

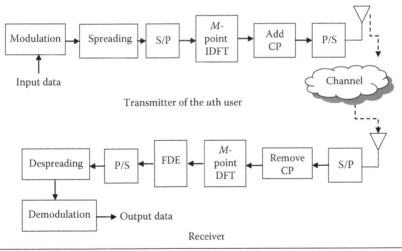

Figure 1.4   Transmitter and receiver structures of the MC-CDMA system over a frequency-selective channel.

1.5 Single-Carrier Communication Systems

Uplink transmissions in mobile communication systems are often limited by the terminal power, among other factors such as time dispersion and interference. Thus, the mobile terminals must transmit data at the lowest possible power but over kilometers of distance. Single-carrier systems have significantly lower PAPRs, and are much more tolerant to PA nonlinearities. These characteristics are important for uplink transmission [1].

1.5.1 SC-FDE System

For broadband multipath channels, conventional time-domain equalizers are impractical because of the complexity (very long channel impulse response in the time domain). FDE is more practical for such channels. The SC-FDE system is another way to fight the frequency-selective fading channel. It delivers a performance similar to that of the OFDM system with essentially the same overall complexity, even for channel delays [2]. Figure 1.5 shows the block diagram of the SC-FDE system.

At the transmitter of the SC-FDE system, we add a CP, which is a copy of the last part of the block, to the input data at the beginning of each block in order to prevent interblock interference (IBI) and also to make the linear convolution with the channel impulse response look like a circular convolution. It should be noted that the circular

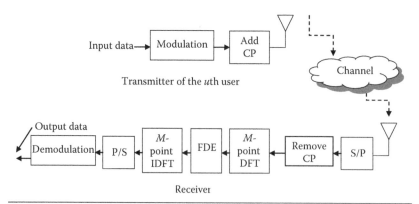

Figure 1.5   Transmitter and receiver structures of the SC-FDE system over a frequency-selective channel.

convolution problem exists for any FDE since multiplication in the frequency domain is equivalent to a circular convolution in the time domain [33]. When the data signal propagates through the channel, it is linearly convolved with the channel impulse response. An equalizer basically attempts to invert the channel impulse response, and thus channel filtering and equalization should have the same type of convolution, either linear or circular convolution. The SC-FDE receiver transforms the received signal into the frequency domain by applying a DFT and performs the equalization process in the frequency domain. After equalization, the signal is brought back to the time domain via the IDFT process and the detection is performed.

Comparing the two systems in Figures 1.2 and 1.5, it is interesting to find the similarity between the OFDM and the SC-FDE systems. Overall, they both use the same functional blocks, and the main difference between them is in the utilization of the DFT and IDFT operations.

In the OFDM system, an IDFT block is placed at the transmitter to multiplex data into parallel subcarriers, and a DFT block is placed at the receiver for FDE, while in the SC-FDE system, both the DFT and IDFT blocks are placed at the receiver for FDE. Thus, one can expect that the two systems have similar link level performance and spectral efficiency. However, there are distinct differences that make the two systems perform differently. At the receiver, the OFDM system makes data detection on a per-subcarrier basis in the frequency domain, whereas the SC-FDE system makes it in the time domain after the additional IDFT operation. Because of this difference, the OFDM system is more sensitive to nulls in the channel spectrum and it requires channel coding or power rate control to overcome this deficiency. Also, the duration of the modulated time symbols is expanded in the case of OFDM with parallel transmission of the data blocks during the elongated time period. In summary, the SC-FDE system has advantages over the OFDM system as follows [9]:

• Lower PAPR due to single-carrier modulation at the transmitter
• Robustness to spectral nulls
• Lower sensitivity to CFOs
• Lower complexity at the transmitter, which is an advantage at the mobile terminal in the cellular uplink communications

1.5.2 DFT-SC-FDMA System

The DFT-SC-FDMA system is an extension of the SC-FDE system to accommodate for multiple access. It is a combination of the FDMA and SC-FDE systems, which has a similar structure and performance to that of the OFDMA system [9]. DFT-SC-FDMA signals are characterized by the significantly lower signal envelope fluctuations, which result in less energy consumption in the PA, hence prolonging the battery life. This feature makes the application of the DFT-SC-FDMA system in the uplink of wireless cellular standards very promising. In [1], it has been shown that reducing the uplink signal peakiness by a few dBs could result in huge improvements in coverage area and range. In particular, a 2.5 dB reduction in signal peakiness in the DFT-SC-FDMA system relative to the OFDMA system approximately doubles the coverage area. This means huge savings in the deployment cost as the DFT-SC-FDMA system would require half the number of base stations to cover a geographic area as that required by the OFDMA system. However, this comparison only holds for coverage-limited situations in, for example, rural areas.

The main difference between the OFDMA and the DFT-SC-FDMA transmitters is the DFT mapper, as we will be seen in Figure 2.2 in Chapter 2. On the other hand, the DFT-SC-FDMA system has much in common with the OFDMA system except for the additional DFT and IDFT blocks at the transmitter and receiver, respectively. It should be noted that the additional DFT/IDFT blocks increase the complexity, especially at the transmitter, and therefore lower fluctuations of the signal envelope are reached by the larger computational processing. During the LTE proposals evaluation phase, a great deal of emphasis was given to the coverage aspects and hence the low signal peakiness was highly desirable. Based on this fact, the DFT-SC-FDMA system was selected to be the multiple access scheme for the LTE uplink. The DFT-SC-FDMA system will be discussed in detail in Chapter 2.

2

DFT-SC-FDMA System

2.1 Introduction

In cellular systems, the wireless communication service in a given geographical area is provided by multiple base stations. The downlink transmissions in cellular systems are one-to-many, while the uplink transmissions are many-to-one. A one-to-many service means that a base station transmits simultaneous signals to multiple users' equipments in its coverage area. This requires that the base station has very high transmission power capability, because the transmission power is shared for transmissions to multiple users' equipments [1]. In contrast, in the uplink, a single user's equipment has all its transmission power available for its uplink transmissions to the base station. In the uplink, the design of an efficient multiple access and multiplexing scheme is more challenging than on the downlink due to the many-to-one nature of the uplink transmissions. Another important requirement for uplink transmissions is the low signal peakiness due to the limited transmission power at the user's equipment [1]. Moreover, wireless communications recently are moving rapidly toward small and low-cost devices. As a result, there is a need for a multiple access scheme to satisfy all these requirements.

OFDMA is a popular high data rate uplink multiple access system, which is currently in use in IEEE 802.11 [34] and IEEE 802.16 [7]. The OFDMA system increases the cell range significantly as compared to the OFDM system that uses TDMA for multiple access. This increase is attributed to the fact that the available transmit power is transmitted only in a fraction of the channel bandwidth, and hence the SNR is improved. However, the OFDMA system suffers from a PAPR problem and this favors single-carrier transmissions. In order to solve the problems encountered in the uplink of the OFDMA system, much attention has been directed recently to another multiple access system, the DFT-SC-FDMA system [1,9], since a single-carrier system with an OFDMA-like multiple access would combine the advantages of the two techniques: the low PAPR and the large coverage area.

This chapter gives an explanation of the DFT-SC-FDMA system. It gives a description for the two methods of subcarriers mapping in the DFT-SC-FDMA system. It also gives a study for the PAPR in the DFT-SC-FDMA system. Finally, the impacts of the radio resources allocation, the input block size, the output block size, and the PA on the performance of the DFT-SC-FDMA system are investigated. Note that through this book, vectors and matrices are represented in boldface.

2.2 Subcarrier Mapping Methods

There are two methods to map the subcarriers among users [9]: the localized mapping method and the distributed mapping method. The former is usually referred to as localized FDMA (LFDMA) scheme, while the latter is usually called distributed FDMA transmission scheme. With the LFDMA transmission scheme, each user's data is transmitted with consecutive subcarriers, while with the distributed FDMA transmission scheme, the user's data is transmitted with distributed subcarriers. Because of the spreading of the information symbol across the entire signal band, the distributed FDMA scheme is more robust to frequency-selective fading. Therefore, it can achieve more frequency diversity. For the LFDMA transmission over a frequency-selective fading channel, the multiuser diversity and the frequency-selective diversity can also be achieved if each user is given subcarriers with favorable transmission characteristics. The distributed FDMA scheme with equidistance between occupied subcarriers in the whole band is called the interleaved FDMA (IFDMA) scheme.

The IFDMA scheme provides a low PAPR, but at the cost of a higher sensitivity to CFOs and phase noise like the OFDMA system [9,18]. The LFDMA scheme is more robust to MAI, but it incurs a higher PAPR than the IFDMA scheme [9]. In this book, we will refer to the localized DFT-SC-FDMA system as DFT-LFDMA, the interleaved DFT-SC-FDMA system as DFT-IFDMA, the localized

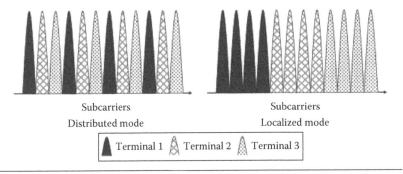

Figure 2.1 Subcarriers mapping schemes for multiple users (3 users, 12 subcarriers, and 4 subcarriers allocated per user). (From Myung, H.G. and Goodman, D.J., *Single Carrier FDMA: A New Air Interface for Long Term Evaluation*, John Wiley & Sons, Chichester, U.K., 2008.)

DCT-SC-FDMA system as DCT-LFDMA, and the interleaved DCT-SC-FDMA system as DCT-IFDMA.

An example of the DFT-SC-FDMA system with 3 users, 12 subcarriers, and 4 subcarriers allocated per user is illustrated in Figure 2.1.

2.3 DFT-SC-FDMA System Model

The block diagram of the DFT-SC-FDMA system is shown in Figure 2.2. One base station and U uplink users are assumed. There are totally M subcarriers and each user is assigned a subset of subcarriers for the uplink transmission. For simplicity, we assume that each user has the same number of subcarriers, N. As shown in Figure 2.2, the DFT-SC-FDMA system has much in common with the OFDMA system except for the additional DFT and IDFT blocks at the transmitter and receiver, respectively. For this reason, the DFT-SC-FDMA system is sometimes referred to as the DFT-spread or DFT-precoded OFDMA system. The transmitter of the DFT-SC-FDMA system uses different subcarriers to transmit information data, as in the OFDMA system. However, the DFT-SC-FDMA system transmits the subcarriers sequentially, rather than in parallel. This approach has the advantage of enabling a low PAPR, which is important to increase cell coverage and to prolong the battery lifetime of mobile terminals.

At the transmitter side, the encoded data is transformed into a multilevel sequence of complex numbers in one of several possible modulation formats. The resulting modulated symbols are grouped

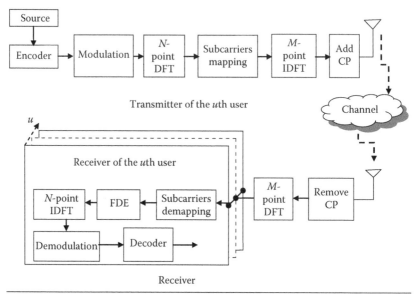

Figure 2.2   Structure of the DFT-SC-FDMA system over a frequency-selective channel.

into blocks, each containing N symbols and the DFT is performed. The signal after the DFT can be expressed as follows:

$$X(k) = \sum_{n=0}^{N-1} x(n)e^{-\frac{j2\pi}{N}nk} \tag{2.1}$$

where

N is the input block size

$\{x(n):n = 0, ..., N-1\}$ represents the modulated data symbols

The outputs are then mapped to M ($M > N$) orthogonal subcarriers followed by the M-point IDFT to convert to a time-domain complex signal sequence. $M = QN$ is the output block size. Q is the maximum number of users that can transmit, simultaneously. Notice that the remaining $(M-N)$ subcarriers may be used by the other users communicating in the cell, thus a promising multiuser access is achieved. The resulting signal after the IDFT can be given as follows:

$$\bar{x}(m) = \frac{1}{M} \sum_{l=0}^{M-1} \bar{X}(l)e^{\frac{j2\pi}{M}ml} \tag{2.2}$$

where $\{\bar{X}(l): l = 0, ..., M-1\}$ represents the frequency-domain samples after the subcarriers mapping scheme. Before transmission through the wireless channel, a CP is appended in front of each block to provide a guard time preventing IBI in the multipath channel.

At the receiver side, the CP is removed from the received signal and the signal is then transformed into the frequency domain via an M-point DFT. After that, the subcarriers demapping and the FDE processes are performed. Finally, the equalized signal is transformed back into the time domain via an N-point IDFT, followed by the demodulation and decoding processes. After removing the CP, the received signal can be written as follows:

$$r = \sum_{u=1}^{U} H_C^u \bar{x}^u + n \qquad (2.3)$$

where

$\bar{x}^u$ is an $M \times 1$ vector representing the block of the transmitted symbols of the uth user

n is the $M \times 1$ vector of zero-mean complex additive white Gaussian noise (AWGN) with variance σ_n^2

H_C^u is an $M \times M$ circulant matrix describing the multipath channel between the uth user and the base station

It is given by

$$H_C^u = \begin{bmatrix} h^u[0] & 0 & . & 0 & h^u[L-1] & . & h^u[1] \\ . & h^u[0] & . & & & & . \\ . & & . & . & & . & h^u[L-1] \\ h^u[L-1] & & & . & . & & 0 \\ 0 & . & & & . & . & . \\ . & . & . & & & . & 0 \\ 0 & . & 0 & h^u[L-1] & . & . & h^u[0] \end{bmatrix}$$

$$(2.4)$$

The circulant matrix H_C^u can be efficiently diagonalized via the DFT (F) and the IDFT (F^{-1}). It can be expressed as follows [4]:

$$H_C^u = F^{-1} \Lambda^u F \qquad (2.5)$$

where Λ^u is an $M \times M$ diagonal matrix containing the DFT of the circulant sequence of H_C^u. Since there is no interference between users, the $N \times 1$ received block for the uth user in the frequency domain is given by

$$R_d = \Lambda_d^u F_N^{-1} x^u + N \tag{2.6}$$

where

Λ_d^u is the $N \times N$ diagonal matrix containing the channel frequency response associated with the N subcarriers allocated to the uth user

x^u is the $N \times 1$ vector containing the data symbols of the uth user

F_N^{-1} is the $N \times N$ DFT matrix

N is the $N \times 1$ frequency-domain noise vector

The equalized symbols in the frequency domain are obtained as follows:

$$\hat{x}^u = W^u R_d \tag{2.7}$$

where W^u is the $N \times N$ FDE matrix of the uth user. The well-known equalization techniques such as the minimum mean square error (MMSE), the decision feedback, and the turbo can be applied in the FDE. Based on the MMSE criterion, the FDE W^u can be expressed as follows [4]:

$$W^u = \left(\Lambda^{u^H} \Lambda^u + \left(\frac{1}{\text{SNR}} \right) I \right)^{-1} \Lambda^{u^H} \tag{2.8}$$

where I is the identity matrix. The main disadvantage of the MMSE equalizer is the need for an accurate estimation of the SNR at the receiver side. This can be performed by transmitting carefully selected pilot sequences and eventually measuring the SNR at the receiver side.

Some properties of the DFT-SC-FDMA system are listed as follows [9]:

- For perfect synchronization, as in the OFDMA system, the DFT-SC-FDMA system can achieve MAI-free transmission by allocating different subcarriers to different users.
- The DFT-SC-FDMA system guarantees orthogonality among users over a multipath channel provided that the length of the CP is longer than the channel impulse response.
- The DFT-SC-FDMA system has lower PAPR than the OFDMA system, thereby providing better coverage and longer terminal talk time.

- The simple single-tap equalizer can be used in the frequency domain for channel equalization, with the zero-forcing (ZF) or MMSE criterion. The computational complexity of the equalizer is independent of the length of the channel impulse response.
- The DFT-SC-FDMA system is more robust to spectral nulls.

However, the DFT-SC-FDMA system has some disadvantages as follows:

- It has a higher complexity than the OFDMA system. At the transmitter side, it requires an additional DFT process with a dynamic DFT size relying on the number of allocated subcarriers. At the receiver side, it requires an additional IDFT process.
- Noise enhancement at the linear receiver occurs. The IDFT block spreads the noise contribution to the faded subcarriers, and it is enhanced by the equalizer over the whole bandwidth.
- Due to the similarity between the DFT-SC-FDMA system and the OFDMA system, the DFT-SC-FDMA system is also sensitive to CFOs, the in-phase/quadrature-phase (I/Q) imbalance, and the phase noise.

2.4 Time-Domain Symbols of the DFT-SC-FDMA System

This section derives the time-domain symbols without pulse shaping for each subcarriers mapping scheme.

2.4.1 Time-Domain Symbols of the DFT-IFDMA System

To derive the time-domain symbols for the DFT-IFDMA system, Equation 2.2 can be written as follows:

$$\bar{x}(m) = \frac{1}{M} \sum_{l=0}^{M-1} \bar{X}(l) e^{\frac{j2\pi}{M} ml} = \frac{1}{Q} x(n) \tag{2.9}$$

where $\bar{X}(l)$ represents the signal after the interleaved subcarriers mapping process. It can be expressed as follows:

$$\bar{X}(l) = \begin{cases} X(l/Q) & l = Q \cdot k & (0 \le k \le N-1) \\ \\ 0 & & \text{otherwise} \end{cases} \tag{2.10}$$

where $0 \le l \le M-1$. The derivation of the time-domain signals for the DFT-SC-FDMA system is found in [35,36]. When the subcarriers allocation starts from the vth subcarrier $(0 < v \le Q-1)$, the time-domain symbols for the DFT-IFDMA system can be described as follows [36]:

$$\bar{x}(m) = \frac{1}{Q} e^{j2\pi \frac{mv}{M}} x(n) \tag{2.11}$$

where $m = N \cdot q + n$, $0 \le q \le Q-1$ and $0 \le n \le N-1$.

In Equation 2.11, there is an additional phase rotation of $e^{j2\pi \frac{mv}{M}}$, when the subcarriers allocation starts from the vth subcarrier instead of the zeroth subcarrier.

2.4.2 Time-Domain Symbols of the DFT-LFDMA System

If $q = 0$, the time-domain symbols of the DFT-LFDMA system can be described as follows [35]:

$$\bar{x}(m) = \frac{1}{M} \sum_{l=0}^{M-1} \bar{X}(l) e^{\frac{j2\pi}{M} ml} = \frac{1}{Q} x(n) \tag{2.12}$$

where $m = Q \cdot n + q$, $0 \le n \le N-1$, and $0 \le q \le Q-1$. For the localized subcarriers mapping method, $\bar{X}(l)$ can be expressed as follows:

$$\bar{X}(l) = \begin{cases} X(l) & (0 \le l \le N-1) \\ \\ 0 & (N \le l \le M-1) \end{cases} \tag{2.13}$$

If $q \ne 0$, the time-domain symbols for the DFT-LFDMA system can be described as follows [35]:

$$\bar{x}(m) = \frac{1}{Q}\left(1 - e^{j2\pi \frac{q}{Q}}\right) \cdot \frac{1}{N} \sum_{p=0}^{N-1} \frac{x(p)}{1 - e^{j2\pi\left\{\frac{(n-p)}{N} + \frac{q}{Q \cdot N}\right\}}} \tag{2.14}$$

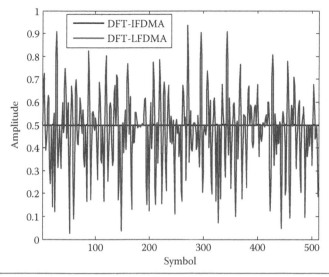

Figure 2.3 Amplitude of the signals in the DFT-SC-FDMA system.

When the radio resource allocation is taken into consideration, Equation 2.14 can be modified as follows [36]:

$$\bar{x}(m) = \frac{1}{Q} e^{j2\pi\frac{qv}{Q}} \left(1 - e^{j2\pi\frac{q}{Q}}\right) \cdot \frac{1}{N} \sum_{p=0}^{N-1} \frac{x(p)}{1 - e^{j2\pi\left\{\frac{(n-p)}{N} + \frac{q}{Q.N}\right\}}} \qquad (2.15)$$

Figure 2.3 shows the amplitude of the signal for each subcarriers mapping scheme for $M = 512$, $N = 128$, and $Q = 4$. It is clear that there are more fluctuations and higher peaks for the DFT-LFDMA system.

2.5 OFDMA vs. DFT-SC-FDMA

Figure 2.4 shows how a series of QPSK symbols is mapped into time and frequency domains by the OFDMA and the DFT-SC-FDMA systems. For clarity, the example here uses only four subcarriers over two symbol periods with the payload data represented by QPSK modulation. Real LTE signals are allocated in units of 12 adjacent subcarriers (180 kHz) called resource blocks [37]. The most obvious difference between the two systems is that the OFDMA system transmits the four QPSK data symbols in parallel, one per

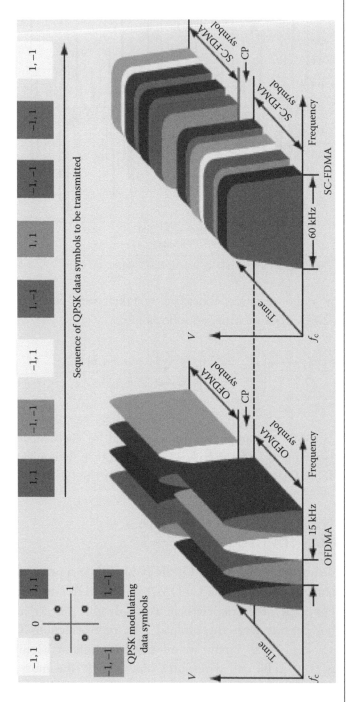

Figure 2.4   How OFDMA and DFT-SC-FDMA transmit a sequence of QPSK data symbols. (From Rumney, M., *Agilent Meas. J.*, 1, January 2008.)

subcarrier, while the DFT-SC-FDMA system transmits the four QPSK data symbols in series at four times the rate, with each data symbol occupying a 60 kHz bandwidth. Visually, the OFDMA signal is clearly multicarrier with one data symbol per subcarrier, but the DFT-SC-FDMA signal appears to be more like a single carrier, which explains the "SC" in its name, with each data symbol being represented by one wide signal.

2.6 Power Amplifier

To determine the effect of the PAPR on the performance of the DFT-SC-FDMA system, the nonlinear PA model must be defined. The PA is generally one of the most significant cost components of user terminals, and the relationship between the PA cost and the maximum power rating is an important technology issue. It should be as power-efficient as possible to increase the operation time of the device. PAs used in radio transmitters have nonlinear characteristics, which can cause significant distortion to the signals, whose instantaneous power fluctuations come too close to the PA output saturation power. The nonlinear characteristics of the PA cause out-of-band and in-band distortions, which usually degrade the transmission quality, seriously. Normally, PAs are operated with a certain power back-off, which can be defined as the ratio of the maximum saturation output power to the average output power. The larger the back-off is, the less the nonlinear distortion will be. However, for a given transmitted power, a larger power back-off lowers the PA efficiency and increases the overall power consumption and battery drain [38,39].

PAs are divided into classes according to the biasing used. A class-A amplifier is defined as an amplifier that is biased so that the current flows for the full time on either the positive or negative half cycle of the input signal. The class-A amplifier is the most linear of all amplifier types, but the maximum efficiency of the amplifier is limited to 50%. This poor efficiency causes high power consumption, which leads to warming in physical devices [40,41]. To achieve a better efficiency, the amplifier can be biased so that the current flows only half the time on either the positive or negative half cycle of the input signal. An amplifier biased like this is called a class-B amplifier.

High demands on linearity make class-B amplifiers unsuitable for a system with high PAPR. In practice, the amplifier is a compromise between classes A and B, and is called a class-AB amplifier [39].

In our simulations, the Rapp model [38] will be used to evaluate the effect of the PA. It is used to model the solid-state PA behavior. The complex envelope of the input signal into the amplifier can be represented as follows:

$$V_{in}(n) = |V_{in}(n)| e^{j\phi(n)} \tag{2.16}$$

where

$|V_{in}(n)|$ is the amplitude of the input signal

$\phi(n)$ is the phase of the input signal

The transmitted signal at the output of the amplifier may be expressed as follows [38]:

$$V_{out}(n) = \frac{|V_{in}(n)|}{\left[1 + \left(\frac{|V_{in}(n)|}{v_{sat}}\right)^{2p}\right]^{1/2p}} e^{j\phi(n)} \tag{2.17}$$

where

$V_{out}(n)$ is the complex output signal

v_{sat} is the output saturation level

The parameter p, often called the "knee factor," controls the smoothness of the characteristics. We will take $p = 2$.

The PA can be described by either the input back-off (IBO) or the output back-off (OBO). They are defined relative to the peak power P_{max} of the PA as follows:

$$IBO = 10 \log_{10}\left(\frac{P_{max}}{P_{av,in}}\right) \quad \text{and} \quad OBO = 10 \log_{10}\left(\frac{P_{max}}{P_{av,out}}\right) \tag{2.18}$$

where

$P_{av,in}$ is the mean power of the signal at the input of the PA

$P_{av,out}$ is the mean power of the signal at the output of the PA

2.7 Peak Power Problem

2.7.1 Sensitivity to Nonlinear Amplification

When transmitting data from the mobile terminal to the network, a PA is required to boost the outgoing signal to a level high enough to be picked up by the network. A signal with large peaks results in severe clipping effects and nonlinear distortion, if this signal is amplified by a PA, which has nonlinear characteristics. In order to limit these effects, it is desirable for the PA to operate in its linear region. On the other hand, the PA has to operate with a large IBO from its peak power. However, applying a large IBO value for the amplifier leads to a low power efficiency, and therefore to a high power consumption, which is not acceptable for mobile wireless communication systems. To prevent this, specific PAPR reduction techniques that positively contribute to the spectral efficiency of the transmission are commonly applied at the transmitter.

2.7.2 Sensitivity to A/D and D/A Resolutions

A high PAPR requires high-resolution A/D and D/A converters, since the dynamic range of the signal is proportional to the PAPR. High-resolution D/A and A/D conversion requires additional complexity, cost, and power burden on the system. The resolution of the A/D converter is defined as the number of bits used to represent the analog input signal. Using a high-resolution A/D converter reduces the quantization error. For the D/A converter, incrementing the code applied to the converter produces smaller step sizes in the analog output.

2.7.3 Peak-to-Average Power Ratio

The PAPR is defined as the ratio between the peak power and the average power of the transmitted signal. A high PAPR means that we need a power back-off to operate in the linear region of the PA. This reduces the power efficiency of the amplifier and results in a lower mean output power for a given peak power-rated device.

We can express the theoretical relationship between the PAPR (dB) and the transmit power efficiency as follows [40,41]:

$$\eta = \eta_{max} \times 10^{-\frac{PAPR}{20}} \qquad (2.19)$$

where

η is the power efficiency

η_{max} is the maximum power efficiency

For class-A PA, η_{max} is 50% and for class-B PA, η_{max} is 78.5% [40]. Figure 2.5 shows the relationship between the PAPR and the efficiency graphically. It is evident from the figure that the high PAPR degrades the transmit power efficiency.

In [42], Wulich and Goldfeld showed that the amplitude of a single carrier–modulated signal does not have a Gaussian distribution, and that it is difficult to analytically derive the exact form of the distribution. So, the exact solution of the PAPR is impossible [42]. As an alternative, they explained a way to derive an upper bound for the complementary distribution of the instantaneous power using the Chernoff bound.

In [41], the author has derived the upper bound for the complementary distribution of the instantaneous power using the Chernoff

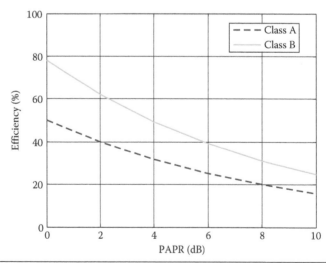

Figure 2.5 A theoretical relationship between PAPR and transmit power efficiency for ideal class-A and class-B amplifiers.

bound for the DFT-IFDMA and the DFT-LFDMA signals. He found that the peak power of the DFT-SC-FDMA signal for both subcarriers mapping schemes is indeed lower than that of the OFDM signal.

So we will adopt the numerical analysis in order to precisely characterize the peak power characteristics of the DFT-SC-FDMA and the DCT-SC-FDMA signals. The PAPR can be expressed as follows:

$$\text{PAPR (dB)} = 10 \log_{10} \left(\frac{\max\left(|x(m)|^2\right)}{\dfrac{1}{M} \displaystyle\sum_{m=0}^{M-1} |x(m)|^2} \right) \qquad (2.20)$$

where $x(m)$ is the mth symbol of the transmitted signal. To evaluate the PAPR characteristics, it is important to compute the complementary cumulative distribution function (CCDF), which is the probability that the PAPR is higher than a certain PAPR value.

2.8 Pulse-Shaping Filters

Pulse-shaping filters are used to restrict the bandwidth of the transmitted signals, while minimizing the likelihood of decoding errors [22]. Two pulse-shaping filters will be considered: the raised-cosine (RC) and the root raised-cosine (RRC) filters. The impulse response of an RC filter is [22]

$$h(t) = \text{sinc}\left(\pi \frac{t}{T}\right) \frac{\cos\left(\pi \gamma \frac{t}{T}\right)}{1 - 4\left(\gamma \frac{t}{T}\right)^2} \qquad (2.21)$$

where

γ is the roll-off factor, which ranges between 0 and 1

T is the symbol period

The RRC filter is an implementation of a low-pass Nyquist filter, i.e., one that has the property of vestigial symmetry. The impulse response of this filter is given by [22]:

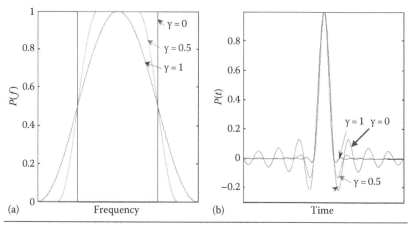

Figure 2.6 (a) Frequency-domain and (b) time-domain representations of the RC filter. (From Myung, H., Single carrier orthogonal multiple access technique for broadband wireless communications, PhD thesis, Polytechnic University, Brooklyn, NY, January 2007.)

$$h(t) = \frac{\sin\left(\pi\frac{t}{T}(1-\gamma)\right) + 4\gamma\frac{t}{T}\cos\left(\pi\frac{t}{T}(1+\gamma)\right)}{\pi\frac{t}{T}\left(1-(4\gamma\frac{t}{T})^2\right)} \qquad (2.22)$$

Figure 2.6 shows the RC filter, graphically, in the frequency and time domains. The roll-off factor γ changes from 0 to 1, and it controls the amount of out-of-band radiation. It is clear that $\gamma = 0$ generates no out-of-band radiation and as γ increases the out-of-band radiation increases.

When $\gamma = 1$, the nonzero portion of the spectrum is a pure raised cosine. In the time domain, the pulse-shaping filter has higher side lobes, when γ is close to 0 and this increases the peak power for the transmitted signal after pulse shaping. We further investigate the effect of pulse shaping on the peak power characteristics of the DFT-SC-FDMA signals described in Section 2.9.

2.9 Simulation Examples

The performance of the DFT-SC-FDMA system has been evaluated by simulations. For comparison purposes, the OFDMA system has also been simulated.

Table 2.1 Simulation Parameters

Simulation method	Monte Carlo
System bandwidth	5 MHz
Modulation type	QPSK and 16-QAM
CP length	20 samples
M	512
N	128
Subcarriers spacing	9.765625 kHz
Channel coding	Convolutional code with rate = 1/2
Subcarriers mapping technique	Localized and interleaved
Channel model	Vehicular-A outdoor channel
Channel estimation	Perfect and MMSE channel estimator
Equalization	MMSE

2.9.1 Simulation Parameters

A DFT-SC-FDMA system with 4 users has been assumed. A convolutional code with memory 7 and octal generator polynomial (133,171) has been chosen as the channel code. It is the same as that used in [2]. The fading was modeled as being quasi-static. The channel model used for simulations is the vehicular-A channel [43]. Simulation parameters are tabulated in Table 2.1 similar to those used in [9]. The characteristics of the vehicular-A channel model are discussed in Appendix A.

2.9.2 CCDF Performance

First, the CCDFs of the PAPR for the DFT-SC-FDMA, and the OFDMA systems have been evaluated and compared for different pulse-shaping filters and modulation formats. Four-time oversampling has been used, and 10^5 data blocks have been generated to calculate the CCDFs of the PAPR. No pulse shaping was applied in the case of the OFDMA system.

The CCDFs of the PAPR for the DFT-IFDMA, the DFT-LFDMA, and the OFDMA systems are shown in Figure 2.7 with different modulation formats. It is clear that without pulse shaping, the DFT-IFDMA system has a lower PAPR than that of the OFDMA system by about 11.5 dB for QPSK and by about 8 dB for 16-QAM, while the PAPR of the DFT-LFDMA system is lower than that of the OFDMA system by about 3.5 dB for QPSK and by about 2.25 dB for

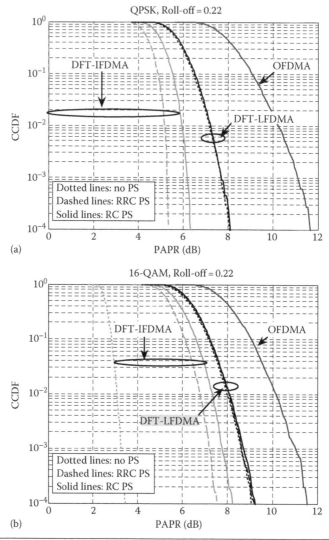

Figure 2.7　Comparison between the CCDFs of the PAPR for the DFT-IFDMA, the DFT-LFDMA, and the OFDMA systems: (a) QPSK and (b) 16-QAM.

16-QAM, but higher than that of the DFT-IFDMA system by about 8 for QPSK and by about 5.75 dB for 16-QAM. With the RC filter with $\gamma = 0.22$, it can be seen that the PAPR increases significantly for the DFT-IFDMA system, whereas the PAPR of the DFT-LFDMA system is nearly independent of the pulse-shaping filter for QPSK and 16-QAM. Figure 2.7 also shows that the PAPR of the DFT-SC-FDMA system depends on the modulation format used.

Table 2.2 PAPR Values at a CCDF $= 10^{-4}$ for the Different Systems

MODULATION FORMAT	PULSE-SHAPING	DFT-IFDMA (dB)	DFT-LFDMA (dB)	OFDMA
QPSK	None	0	8	11.5 dB
	RC	6.5	8	N/A
	RRC	5.3	8	N/A
16-QAM	None	3.5	9.25	11.5 dB
	RC	8	9.25	N/A
	RRC	7.5	9.25	N/A

A full comparison of the PAPR for the different modulation formats at a CCDF $= 10^{-4}$ is given in Table 2.2. It is clear that the DFT-IFDMA and the DFT-LFDMA systems have lower PAPR values than the OFDMA system.

The impact of the pulse-shaping filter on the PAPR performance of the DFT-SC-FDMA signals with different Resource Unit (RU#) allocations is investigated in the following experiments as in [36]. Figures 2.8 and 2.9 show the impact of the radio resources allocation on the PAPR of the DFT-SC-FDMA signals for different mapping

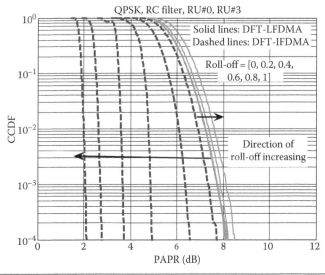

Figure 2.8 Comparison between the CCDFs of PAPR for the DFT-IFDMA and the DFT-LFDMA systems using RC pulse shaping with different roll-off factors for RU#0 and RU#3.

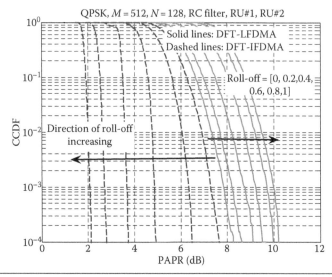

Figure 2.9   Comparison between the CCDFs of PAPR for the DFT-IFDMA and the DFT-LFDMA systems using RC pulse shaping with different roll-off factors for RU#1 and RU#2.

schemes. QPSK is considered. In the DFT-LFDMA system, the PAPRs of RU#0 and RU#1 are the same as those of RU#3 and RU#2, respectively. Thus, Figure 2.8 gives plots for the CCDFs of the PAPR of RU#0 and RU#3, and Figure 2.9 gives plots for the CCDFs of the PAPR of RU#1 and RU#2. In the DFT-IFDMA system, the PAPR decreases as the roll-off factor increases, because the convolution of random samples with the RC filter gives a smaller output peak power with the increase in the roll-off factor. It is also clear that the CCDF of the PAPR of the DFT-IFDMA signal does not change with the choice of the RU# as shown in Figures 2.8 and 2.9, because the DFT-IFDMA time-domain samples remain random with a phase rotation that is related to the choice of the RU#.

2.9.3 Impact of the Input Block Size

The impact of the input block size on the PAPR of the DFT-SC-FDMA system has been studied and shown in Figure 2.10 for different modulation formats. $M = 512$, RU#0, and an RC filter with $\gamma = 0.22$ have been considered. It is observed that the PAPR of the

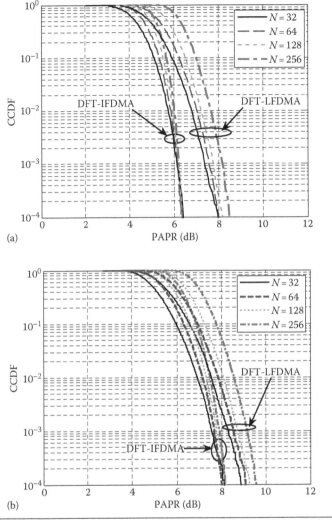

Figure 2.10 Comparison between the CCDFs of the PAPR for the DFT-IFDMA and the DFT-LFDMA systems with different input block sizes: (a) QPSK and (b) 16-QAM.

DFT-LFDMA system increases as the input block size N increases, whereas the PAPR of the DFT-IFDMA system is insensitive to the input block size. On the other hand, the higher the input block size, the higher is the PAPR value. As a result, improvements in the PAPR performance can be achieved by decreasing the input block size.

Table 2.3 PAPR Values at a CCDF = 10^{-4} for the DFT-IFDMA and the DFT-LFDMA Systems with Different Input Block Sizes

INPUT BLOCK SIZE	DFT-IFDMA		DFT-LFDMA	
	QPSK (dB)	16-QAM (dB)	QPSK (dB)	16-QAM (dB)
$N = 32$	6.5	8	8	8.75
$N = 64$	6.5	8	8.25	9
$N = 128$	6.5	8	8.5	9.2
$N = 256$	6.5	8	8.5	9.5

To illustrate the impact of the input block size on the PAPR of the DFT-SC-FDMA system, the measured PAPR values at a CCDF = 10^{-4} are tabulated in Table 2.3. It is clear that the PAPR of the DFT-LFDMA system increases as the input block size N increases. However, it is found that the PAPR of the DFT-IFDMA system is insensitive to the input block size.

2.9.4 Impact of the Output Block Size

The impact of the output block size on the PAPR of the DFT-IFDMA and the DFT-LFDMA systems has been studied and shown in Figure 2.11 for different modulation formats. $N = 128$, and an RC filter with $\gamma = 0.22$ have been used to obtain these results. It is clear that the PAPR of the DFT-IFDMA system is insensitive to the output block size, whereas the PAPR of the DFT-LFDMA system depends on the output block size. On the other hand, the higher the output block size, the smaller is the PAPR value. As a result, improvements in the PAPR performance can be achieved by increasing the output block size. The PAPR in the DFT-LFDMA system for $M = 512$ is nearly the same as that for $M = 1024$. Thus, the output block size $M = 512$ is more suitable for the DFT-LFDMA system to provide a good PAPR performance. This indicates that the output block size must be chosen carefully in order to provide a good PAPR performance for the DFT-LFDMA system.

To illustrate the impact of the output block size on the PAPR of the DFT-LFDMA and the DFT-IFDMA systems, the measured PAPR values are tabulated in Table 2.4. It is clear that the PAPR of the DFT-LFDMA system decreases as the output block size M increases.

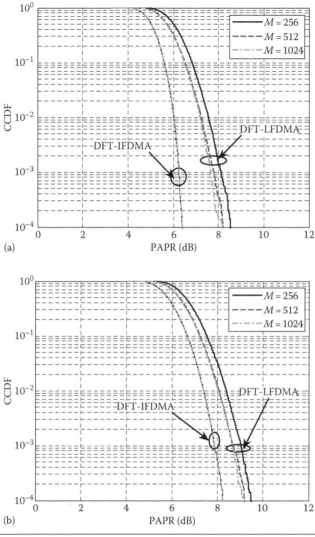

Figure 2.11   Comparison between the CCDFs of the PAPR for the DFT-IFDMA and the DFT-LFDMA systems with different output block sizes: (a) QPSK and (b) 16-QAM.

Table 2.4   PAPR Values at a CCDF $= 10^{-4}$ for the DFT-IFDMA and the DFT-LFDMA Systems with Different Output Block Sizes

OUTPUT	DFT-IFDMA		DFT-LFDMA	
BLOCK SIZE	QPSK (dB)	16-QAM (dB)	QPSK (dB)	16-QAM (dB)
$M=256$	6.5	8.25	8.5	9.5
$M=512$	6.5	8.25	8	9.25
$M=1024$	6.5	8.25	8	9.25

2.9.5 Impact of the Power Amplifier

Figures 2.12 and 2.13 show the impact of the PA on the BER performance of the DFT-IFDMA, and the DFT-LFDMA systems with
different modulation formats. Rapp's solid-state PA model with $p = 2$
has been used. A convolutional code with a memory length of 7 and an
octal generator polynomial of (133,171) has been chosen as the channel code. At an IBO = 10 dB, it is clear that there is no effect of the
nonlinearity, because the PA is operating in the linear region. When
we close the gap and move toward the 1 dB IBO, the effect becomes
more pronounced, especially for 16-QAM. It is also clear that as the
IBO decreases from 5 to 1 dB, the BER performance degrades, significantly, especially for 16-QAM. As a result, the BER performance
is relatively robust to the nonlinear PA for an IBO equal to 10 dB.

The impact of the PA on the performance of the DFT-SC-FDMA
system over a fixed wireless channel has also been investigated in [44].
From the simulation results obtained in [44], it is clear that the saturation level of the PA has a significant effect on the BER performance,
especially with high-order modulation formats. Simulation results
have demonstrated that the saturation level threshold of the PA in
the DFT-SC-FDMA system depends on the subcarriers mapping

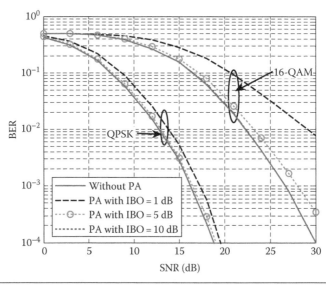

Figure 2.12   BER vs. SNR for the DFT-IFDMA system with different values of the IBO of the
power PA.

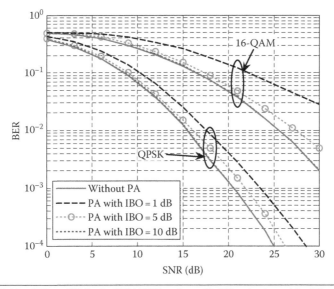

Figure 2.13 BER vs. SNR for the DFT-LFDMA system with different values of the IBO of the PA.

method as well as the modulation format. For the DFT-LFDMA system, it was found that the best choice of the saturation level threshold is –3 dB for both QPSK and 16-QAM. For the DFT-IFDMA system, it was found that the best choice of the saturation level threshold is –4 dB for the QPSK scheme and –3 dB for 16-QAM. This indicates that a PA should be designed carefully for the DFT-SC-FDMA system in order to provide a good performance.

3

DCT-SC-FDMA System

3.1 Introduction

As discussed in Chapter 2, up to now, only the DFT is used to implement the SC-FDMA system. However, it is possible to use other sinusoidal transforms such as the DCT, which is the main objective of this chapter. The DCT-SC-FDMA system is not studied in the literature. Recently, the DCT-OFDM system has received much attention, because it offers certain advantages over the DFT-OFDM system [45–47]. The main advantage of the DCT lies in its excellent spectral energy compaction property, which makes most of the samples transmitted close to zero leading to a reduction in the effect of ISI. In addition, it uses only real arithmetic rather than the complex arithmetic used in the DFT. This reduces the signal processing complexity, especially for real modulation signaling, where the DFT processing still uses complex arithmetic and suffers from I/Q imbalance problem [46,47].

FDE is used in the DCT-SC-FDMA scheme to mitigate the effect of the frequency-selective channel. The length of the CP in the DCT-SC-FDMA system is the same as that in the DFT-SC-FDMA system. In addition, no redundancy exists in the proposed DCT-SC-FDMA system due to the use of the DCT. Thus, the spectral efficiency in the DCT-SC-FDMA system is equal to that in the DFT-SC-FDMA system.

The main contributions of this chapter are the use of the DCT in the SC-FDMA system, the derivation of expressions for the time-domain signals in the DCT-SC-FDMA system, and the study of the PAPR and the BER performance of the DCT-SC-FDMA system.

3.2 DCT

3.2.1 Definition of the DCT

The DCT is a Fourier-related transform similar to the DFT, but it uses only real arithmetic [47,48]. We will consider the type-II DCT because of its large energy compaction property.

Unlike the conventional DFT, a single set of cosinusoidal functions $\cos(2\pi n f_\Delta t)$, where $n = 0$, 1, $N - 1$, and $0 < t < T$ is used in the DCT. f_Δ is the subcarrier spacing. T is the symbol period. The minimum f_Δ required to satisfy the orthogonality condition

$$\int_0^T \sqrt{\frac{2}{T}}\cos(2\pi k f_\Delta t)\sqrt{\frac{2}{T}}\cos(2\pi n f_\Delta t)\,dt = \begin{cases} 1, & k=n \\ 0, & k \neq n \end{cases} \quad (3.1)$$

is $1/2T$. In the DFT, the minimum f_Δ, required to satisfy the orthogonality condition

$$\int_0^T \sqrt{\frac{1}{T}}e^{-j2\pi k f_\Delta t}\sqrt{\frac{1}{T}}e^{j2\pi n f_\Delta t}\,dt = \begin{cases} 1, & k=n \\ 0, & k \neq n \end{cases} \quad (3.2)$$

is $1/T$.

The bandwidth required for the DCT-based system with complex-valued modulation formats is the same as that of the DFT-based system with the same number of subcarriers. However, for a DCT-based system with real-valued modulation formats, one can use a single-sideband transmission technology to improve the bandwidth efficiency. In this case, the bandwidth of a DCT-based system can be only half that of the DFT-based system with the same number of subcarriers [47]. After the DCT, sampling the continuous-time signal at time instants $T(2n+1)/2N$ gives a discrete-time sequence [47] as follows:

$$X(k) = \sqrt{\frac{2}{N}}\beta(k)\sum_{n=0}^{N-1} x(n)\cos\left(\frac{\pi k(2n+1)}{2N}\right), \quad k=0,\dots,N-1 \quad (3.3)$$

where

$x(n)$ is the nth sample of the input signal

$\beta(k)$ can be expressed as follows [33,47]:

$$\beta(k) = \begin{cases} \dfrac{1}{\sqrt{2}} & k=0 \\[2mm] 1 & k=1,2,\ldots,N-1 \end{cases} \tag{3.4}$$

After the inverse DCT (IDCT), the discrete-time signal can be expressed as follows [33,47]:

$$x(n) = \sqrt{\frac{2}{N}} \sum_{k=0}^{N-1} X(k)\beta(k)\cos\left(\frac{\pi k(2n+1)}{2N}\right), \quad n=0,\ldots,N-1 \tag{3.5}$$

3.2.2 Energy Compaction Property of the DCT

The DCT is used in several data compression applications because of the property that is frequently referred to as "energy compaction." Specifically, the DCT of a finite-length sequence often has its coefficients more highly concentrated at the low-frequency indices. So, the DCT coefficients at high frequencies can be set to zero without a significant impact on the energy of the signal. More details for the energy compaction property of the DCT are found in [33].

3.3 DCT-SC-FDMA System Model

The structure of the proposed DCT-SC-FDMA system is depicted in Figure 3.1. A DCT-SC-FDMA system with U users communicating with a base station through independent multipath fading channels is assumed. A total of M subcarriers is also assumed and each user is assigned N subcarriers. At the transmitter side, the modulated symbols are grouped into blocks, each containing N symbols, and an N-point DCT is performed. Then, the subcarriers are mapped in the frequency domain. After that, the M-point IDCT is performed. The signal after the IDCT is in the time domain. Thus, adding a CP to this signal makes the frequency-domain equalization at the receiver side possible. After adding a CP of length N_C to the resulting signal, the signal is transmitted through the wireless channel.

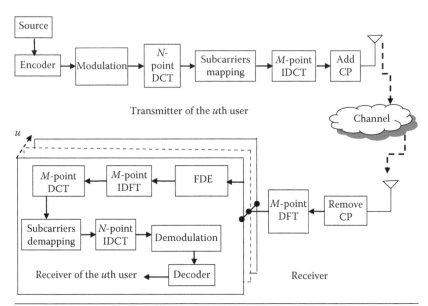

Figure 3.1 Structure of the proposed DCT-SC-FDMA system over a frequency-selective channel.

In matrix notation, the transmitted signal of the uth user $(u = 1, 2, ..., U)$ can be formulated as follows:

$$\tilde{x}^u = P_{\text{add}} D_M^{-1} M_T^u D_N x^u \tag{3.6}$$

where

x^u is an $N \times 1$ vector containing the modulated symbols of the uth user

D_N is an $N \times N$ DCT matrix

D_M^{-1} is an $M \times M$ IDCT matrix

M_T^u is an $M \times N$ matrix describing the subcarriers mapping of the uth user. $M = Q.N$, where Q is the maximum number of users that can transmit simultaneously

P_{add} is an $(M + N_C) \times M$ matrix, which adds a CP of length N_C

The entries of M_T^u for both the DCT-LFDMA and the DCT-IFDMA systems are given in (3.7) and (3.8), respectively:

$$M_T^u = [0_{(u-1)N \times N} ; I_N ; 0_{(M-uN) \times N}] \tag{3.7}$$

$$M_T^u = [0_{(u-1) \times N} ; u_1^T ; 0_{(Q-u) \times N} ; \cdots ; 0_{(u-1) \times N} ; u_N^T ; 0_{(Q-u) \times N}] \tag{3.8}$$

where

I_N and $\mathbf{0}_{Q' \times N}$ matrices denote the $N \times N$ identity matrix, and the $Q' \times N$ all-zero matrix, respectively

$\mathbf{u}_l$ ($l = 1, 2, \ldots, N$) denotes the unit column vector, of length N, with all-zero entries except at l

P_{add} can be represented as follows:

$$P_{add} = [C, I_M]^T \qquad (3.9)$$

where

$$C = [\mathbf{0}_{N_C \times (M - N_C)}, I_{N_C}]^T \qquad (3.10)$$

At the receiver side, the CP is removed from the received signal and the received signal can be written as follows:

$$r = \sum_{u=1}^{U} H_C^u \bar{x}^u + n \qquad (3.11)$$

where

$\bar{x}^u$ is an $M \times 1$ vector representing the block of the transmitted symbols of the uth user

H_C^u is an $M \times M$ circulant matrix describing the multipath channel between the uth user and the base station

n is an $M \times 1$ vector describing the additive noise

It contains independent identically distributed (i.i.d) zero-mean AWGN. Now, substituting Equation 3.5 in Equation 3.11 and applying the DFT, we get

$$R = \sum_{u=1}^{U} \Lambda^u F_M \bar{x}^u + N \qquad (3.12)$$

where

Λ^u is an $M \times M$ diagonal matrix containing the DFT of the circulant sequence of H_C^u

N is the DFT of n

F_M is an $M \times M$ DFT matrix

The generic M-point DFT matrix has entries $\left[F_M\right]_{p,q} = e^{-j2\pi \frac{pq}{M}}$, and its inverse is $F_M^{-1} = \dfrac{1}{M} F_M^H$.

Figure 3.1 depicts the DCT-SC-FDMA receiver for the U users. One can see that for each user, a separate detection is performed. So, the detection process for the uth user only will be discussed in the rest of this section.

After that, the FDE, the M-point IDFT, and the DCT-SC-FDMA demodulation operations are performed to provide the estimate of the modulated symbols as follows:

$$\hat{x}^u = D_N^{-1} M_R^u D_M F_M^{-1} W^u R \tag{3.13}$$

where

W^u is the $M \times M$ FDE matrix of the uth user

M_R^u is the $N \times M$ subcarriers demapping matrix of the uth user

The entries of M_R^u for both the DCT-LFDMA and the DCT-IFDMA systems are given by taking the transpose of (3.7) and (3.8), respectively, that is, $M_R^u = M_T^{u^T}$. In terms of the interference, Equation 3.13 can be rewritten as follows:

$$\hat{x}'' = A^u x^u + \bar{A}^u x^u + \hat{n} \tag{3.14}$$

where

$$A^u = \mathrm{diag}(D_N^{-1} M_R^u D_M F_M^{-1} W^u \Lambda^u F_M D_M^{-1} M_T^u D_N) \tag{3.15}$$

$$\bar{A}^u = D_N^{-1} M_R^u D_M F_M^{-1} W^u \Lambda^u F_M D_M^{-1} M_T^u D_N - A^u \tag{3.16}$$

$$\hat{n} = D_N^{-1} M_R^u D_M F_M^{-1} W^u N \tag{3.17}$$

From Equation 3.14, it is found that only the first term contains the desired data, the second term is due to the residual ISI, and the third term is a noise. Based on Equation 3.14, the signal-to-interference and noise ratio (SINR) of the nth symbol ($n = 1, \ldots, N$) in the uth user's received signal can be expressed as follows:

$$\text{SINR}^u(n) = \frac{\left|A^u(n,n)\right|^2}{\sum_{\substack{k=1 \\ k\neq n}}^{N}\left|\overline{A}^u(n,k)\right|^2 + \sigma_{\hat{n}}^2} \tag{3.18}$$

where $\sigma_{\hat{n}}^2$ is the variance of $\hat{n}$. Finally, the demodulation and the decoding processes are performed.

3.4 Complexity Evaluation

The fast DCT algorithm in [49] provides fewer computational steps than the fast DFT algorithm. Hence, the complexities of the two transmitter schemes are comparable. At the receiver, which is the base station, the complexity is slightly higher than that of the DFT-SC-FDMA system, because the receiver still uses the DFT and the IDFT for the one-tap frequency-domain equalizer. However, the increase in the receiver complexity in the uplink can be tolerated considering the advantages of the DCT-SC-FDMA system.

It is possible to reduce the complexity of the DCT-SC-FDMA receiver by performing the equalization process after the IDCT as shown in Figure 3.2 by feeding a symmetrically extended sequence

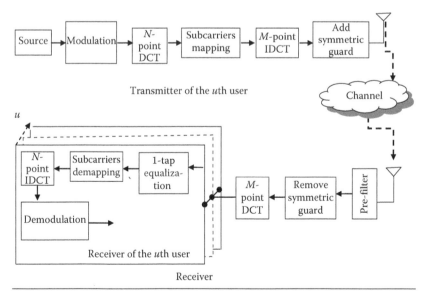

Figure 3.2 Structure of the simplified DCT-SC-FDMA system over a frequency-selective channel.

into the DCT-SC-FDMA system, which means avoiding the use of the IDFT and DFT at the receiver [50]. However, the data transmission efficiency of the resulting scheme decreases to half its value [47]. The proposed scheme in Figure 3.1 eliminates the ISI and avoids decreasing the transmission efficiency, while the complexity of the receiver slightly increases by adding a DFT–IDFT pair. Thus, the structure of a DCT-SC-FDMA system in Figure 3.1 will be adopted.

3.5 Time-Domain Symbols of the DCT-SC-FDMA System

This section gives a derivation of the time-domain symbols in the DCT-SC-FDMA system for each subcarriers mapping scheme.

3.5.1 Time-Domain Symbols of the DCT-IFDMA System

For the DCT-IFDMA system, the frequency-domain samples after the subcarriers mapping $\overline{X}(l)$ can be described as in Equation 3.10. The time-domain symbols $\overline{x}(m)$ are obtained by taking the IDCT of $\overline{X}(l)$.

Let $m = N \cdot q + n$, where $0 \le q \le Q - 1$ and $0 \le n \le N-1$. Then

$$\overline{x}(m) = \overline{x}(N \cdot q + n) = \sqrt{\frac{2}{M}} \sum_{l=0}^{M-1} \overline{X}(l)\beta(l)\cos\left(\frac{\pi l(2m+1)}{2M}\right)$$

$$= \frac{1}{\sqrt{Q}}\sqrt{\frac{2}{N}} \sum_{k=0}^{N-1} X(k)\beta(k)\cos\left(\pi kq\right)\cos\left(\frac{\pi k(2n+1)}{2N}\right) \quad (3.19)$$

$\overline{x}(m)$ in Equation 3.19 can be written as follows:

$$\overline{x}(m) = \begin{cases} \dfrac{1}{\sqrt{Q}}x(n), & q \text{ is even} \\[2em] \dfrac{1}{\sqrt{Q}}\sqrt{\dfrac{2}{N}}\displaystyle\sum_{k=0}^{N-1}(-1)^k X(k)\beta(k)\cos\left(\dfrac{\pi k(2n+1)}{2N}\right), & q \text{ is odd} \end{cases}$$

$$(3.20)$$

Equation 3.20 indicates that the resulting symbols after the IDCT are simply a repetition of the original input symbols in the time domain.

Therefore, the PAPR in the DCT-IFDMA system is the same as that in the conventional single-carrier system and in the conventional DFT-IFDMA system. When the subcarriers allocation starts from the vth subcarrier $(0 < v \leq Q - 1)$, then

$$\bar{X}(l) = \begin{cases} X(l/Q - v) & l = Q.k + v \, (0 \leq k \leq N - 1) \\ 0 & \text{otherwise} \end{cases} \qquad (3.21)$$

Thus, Equation 3.20 can be modified as follows:

$$\bar{x}(m) = \sqrt{\frac{2}{M}} \sum_{l=0}^{M-1} \bar{X}(l)\beta(l)\cos\left(\frac{\pi l(2m+1)}{2M}\right) \qquad (3.22)$$

Equation 3.22 can be written as follows:

$$\bar{x}(m) = \sqrt{\frac{1}{2M}} \sum_{k=0}^{N-1} X(k)\beta(k)\left[\cos\left(\frac{\pi k(2m+1)}{2N}\right)\cos\left(\frac{\pi v(2m+1)}{2M}\right)\right.$$
$$\left. - \sin\left(\frac{\pi k(2m+1)}{2N}\right)\sin\left(\frac{\pi v(2m+1)}{2M}\right)\right] \qquad (3.23)$$

If $v = 0$, Equation 3.23 reduces to Equation 3.20. From Equation 3.23, we can see that the resulting symbols after the IDCT are simply a repetition of the modulated symbols $x(n)$ in the time domain, when $v = 0$. Therefore, the PAPR characteristics of the DCT-IFDMA system are expected to be the same as those of the conventional DFT-IFDMA system.

3.5.2 Time-Domain Symbols of the DCT-LFDMA System

For the DCT-LFDMA system, the frequency-domain samples after the subcarriers mapping $\bar{X}(l)$ can be described as in Equation 3.13. The time-domain symbols $\bar{x}(m)$ are obtained by taking the IDCT of $\bar{X}(l)$.

Let $m = Qn + q$, where $0 \le n \le N - 1$ and $0 \le q \le Q - 1$. Then

$$\bar{x}(m) = \bar{x}(Qn + q) = \sqrt{\frac{2}{M}} \sum_{l=0}^{M-1} \bar{X}(l)\beta(l)\cos\left(\frac{\pi l(2m+1)}{2M}\right)$$

$$= \frac{1}{\sqrt{Q}}\sqrt{\frac{2}{N}} \sum_{k=0}^{N-1} X(k)\beta(k)\cos\left(\frac{\pi k(2Qn+2q+1)}{2QN}\right) \quad (3.24)$$

Substituting $X(k) = \sqrt{\frac{2}{N}}\beta(k)\sum_{p=0}^{N-1} x(p)\cos\left(\frac{\pi k(2p+1)}{2N}\right)$ in Equation 3.24, we get

$$\bar{x}(m) = \bar{x}(Qn + q) = \frac{1}{\sqrt{Q}}\frac{2}{N} \sum_{k=0}^{N-1}\left[\beta^2(k)\sum_{p=0}^{N-1} x(p)\cos\left(\frac{\pi k(2p+1)}{2N}\right)\right]$$

$$\times \cos\left(\frac{\pi k(2Qn+2q+1)}{2M}\right) \quad (3.25)$$

It can be seen from Equation 3.25 in the time domain that the resulting symbols after the IDCT are the sum of the all input symbols in the input block with different real weightings, which would increase the PAPR.

Figure 3.3 shows the amplitude of the signal for each subcarriers mapping scheme for $M = 512$, $N = 128$, QPSK, and $Q = 4$. It is clear that there are more fluctuations and higher peaks in the DCT-LFDMA system.

3.6 Simulation Examples

The performance of the proposed DCT-SC-FDMA system has been evaluated by simulations assuming perfect channel knowledge. For comparison purposes, the DFT-SC-FDMA and the OFDMA systems have also been simulated.

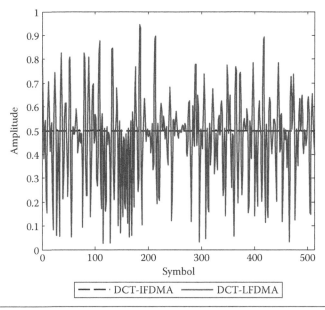

Figure 3.3 Amplitude of the signals in the DCT-SC-FDMA system.

3.6.1 Simulation Parameters

A DCT-SC-FDMA system with 4 users has been considered. A convolutional code with a memory length of 7 and an octal generator polynomial of (133,171) has been chosen as the channel code. The channel model used for simulations is the vehicular-A outdoor channel [43]. The fading was modeled as being quasi-static. Simulation parameters are tabulated in Table 3.1.

3.6.2 BER Performance

Figure 3.4 illustrates the BER performance of the DCT-SC-FDMA system compared to the DFT-SC-FDMA, and the OFDMA systems with different subcarriers mapping schemes and modulation formats. It can be observed that the DCT-SC-FDMA system provides a significant BER performance improvement over the DFT-SC-FDMA and the OFDMA systems for QPSK. At a BER = 10^{-3} with QPSK, the performance gain is about 2 dB for the DCT-LFDMA system, and about 1 dB for the DCT-IFDMA system, when compared to those of

Table 3.1 Simulation Parameters

Simulation method	Monte Carlo
System bandwidth	5 MHz
Modulation type	QPSK and 16-QAM
CP length	20 samples
M	512
N	128
Subcarriers spacing	9.765625 kHz
Channel coding	Convolutional code with rate = 1/2
Subcarriers mapping technique	Localized and interleaved
Channel model	Vehicular-A outdoor channel
Channel estimation	Perfect and MMSE channel estimator
Equalization	MMSE

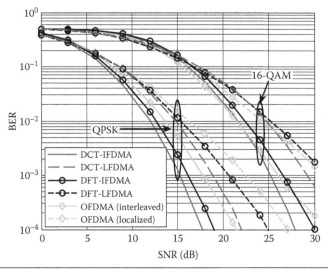

Figure 3.4   BER vs. SNR for the DFT-SC-FDMA, the DCT-SC-FDMA, and the OFDMA systems.

the DFT-LFDMA system and the DFT-IFDMA system, respectively. This is attributed to the energy concentration property of the DCT. The DCT operation distributes more energy to the desired symbol and less energy to the interference. Therefore, the desired symbol suffers less interference coming from neighboring symbols in the DCT-SC-FDMA system than that in the DFT-SC-FDMA system. Thus, the ISI is reduced in the DCT-SC-FDMA system, which reduces the BER.

For 16-QAM, the performance of the DCT-SC-FDMA system is also better than that of the DFT-SC-FDMA and the OFDMA systems. The performance of the DFT-SC-FDMA system is worse than that of the OFDMA and the DCT-SC-FDMA systems for 16-QAM.

This is due to the noise enhancement problem in single-carrier systems. At a BER = 10^{-3}, the SNR improvements for the DCT-SC-FDMA system as compared to the DFT-SC-FDMA, and the OFDMA systems are tabulated in Table 3.2.

Figure 3.5 gives a plot for the BER performance of the DFT-SC-FDMA system, the DCT-SC-FDMA system, and the discrete sine transform–based SC-FDMA (DST-SC-FDMA) system. It is clear that the DCT-SC-FDMA system provides a better BER performance

Table 3.2 SNR Improvement of the DCT-SC-FDMA System at a BER = 10^{-3}

PROPOSED SYSTEM	SNR IMPROVEMENT AS COMPARED TO THE DFT-SC-FDMA SYSTEM		SNR IMPROVEMENT AS COMPARED TO THE OFDMA SYSTEM	
	QPSK (dB)	16-QAM (dB)	QPSK (dB)	16-QAM (dB)
DCT-IFDMA	1	1.5	2.75	1
DCT-LFDMA	2	1.5	3	1

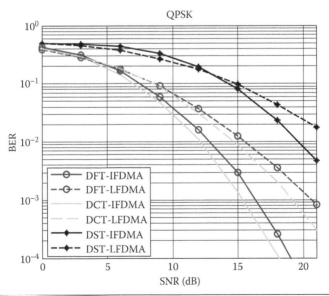

Figure 3.5 BER vs. SNR for the DFT-SC-FDMA, the DCT-SC-FDMA, and the DST-SC-FDMA systems.

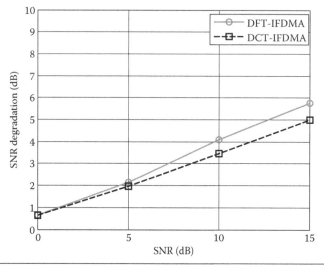

Figure 3.6   SNR degradation vs. SNR for the DFT-SC-FDMA and the DCT-SC-FDMA systems.

than the other systems. The DST-SC-FDMA system provides the worst performance as compared to the other systems.

To justify the results obtained in Figures 3.4 and 3.5, Figure 3.6 gives a plot for the SNR degradation of the DCT-SC-FDMA system and the DFT-SC-FDMA system over 200 channel realizations. The SNR degradation is defined as the ratio between the SNR and the SINR, i.e., SNR degradation (dB) = $10 \log_{10}(\text{SNR/SINR})$. Interleaved subcarriers mapping has been considered. It can be observed that the degradation in the SNR in the DCT-SC-FDMA system is lower than that in the DFT-SC-FDMA system, especially at high SNR values. This is attributed to the energy concentration property of the DCT.

3.6.3 CCDF Performance

The CCDFs of the PAPR for the DCT-SC-FDMA, the DFT-SC-FDMA, and the OFDMA systems have been evaluated and compared for different modulation formats. 10^5 data blocks have been generated to calculate the CCDFs of the PAPR. First, a comparison study of the PAPR is introduced. In Figure 3.7, plots of the CCDFs of the PAPR for the DCT-SC-FDMA, the DFT-SC-FDMA, and the OFDMA systems are shown with different subcarriers mapping and modulation formats. No pulse shaping

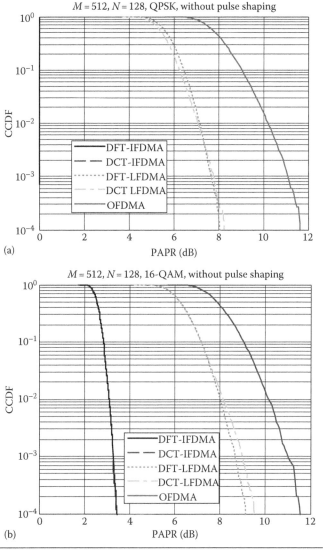

Figure 3.7   Comparison between the CCDFs of the PAPR for the DFT-SC-FDMA, the DCT-SC-FDMA, and the OFDMA systems with different subcarriers mapping techniques: (a) QPSK and (b) 16-QAM.

was applied in Figure 3.7. It is clear that the DCT-IFDMA system has a lower PAPR than that of the OFDMA system by about 11.5 dB for QPSK and about 8 dB for 16-QAM, while the PAPR of the DCT-LFDMA system is lower than that of the OFDMA system by about 3.3 dB for QPSK and 2 dB for 16-QAM, but higher than that of the DCT-IFDMA system by 8.2 dB for QPSK and 6 dB for 16-QAM. It is also found that the DFT-IFDMA and the DCT-IFDMA systems have nearly similar PAPR performances, regardless of the modulation format. Moreover, it is noted that the PAPR of the DCT-SC-FDMA system depends on the modulation format.

The effect of the pulse-shaping filter on the PAPR performance in the DCT-SC-FDMA system has been studied for the different subcarriers mapping techniques and modulation formats, and the results are shown in Figure 3.8 for $M = 512$, $N = 128$, and an RC filter with roll-off factor of 0.22. In this figure, four-time oversampling has been used when calculating the PAPR. It can be seen that the PAPR increases significantly for the DCT-IFDMA and the DFT-IFDMA systems, especially with QPSK. It is also clear that the PAPR of the DCT-LFDMA system increases by about 0.5 dB for QPSK and 16-QAM, whereas the PAPR of the DFT-LFDMA system is nearly independent of the pulse shaping for QPSK and 16-QAM.

A full comparison of PAPR at a CCDF = 10^{-4} for different modulation formats is shown in Table 3.3. From this table, it is clear that the PAPR of the DCT-LFDMA system is slightly higher than that of the DFT-LFDMA system, whereas the PAPR of the DCT-IFDMA system is nearly the same as that of the DFT-IFDMA system.

The impact of the radio resources allocation has also been investigated. Figures 3.9 through 3.12 demonstrate the impact of the radio resources allocation on the PAPR of the DCT-SC-FDMA system with different subcarriers mapping techniques. It is clear that the impact of the radio resources allocation is not the same as that in the DFT-SC-FDMA system. In the DCT-SC-FDMA system, the PAPR depends on the radio resources allocation for both the DCT-IFDMA and the DCT-LFDMA systems.

In the DCT-IFDMA system, the PAPR of RU#0 decreases with the increase in the roll-off value similar to the case of the DFT-IFDMA system. This can be explained with the aid of Equation 3.20, which indicates that the resulting symbols after the IDCT are simply a repetition of the original input symbols in the time domain. This is similar to the case of the DFT-IFDMA system. On the other hand, the convolution of random samples with the RC filter gives a smaller output peak power with the increase in the roll-off factor,

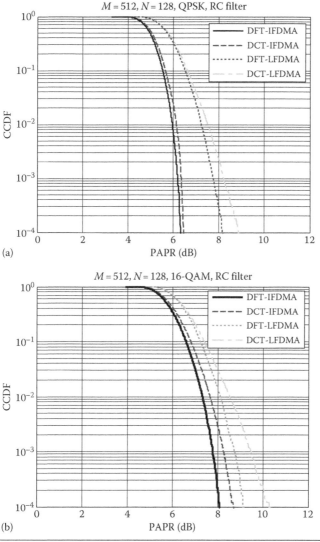

Figure 3.8 Comparison between the CCDFs of the PAPR for the DFT-SC-FDMA and the DCT-SC-FDMA systems with different subcarriers mapping techniques: (a) QPSK and (b) 16-QAM.

and hence the PAPR decreases as the roll-off factor increases. Therefore, the PAPR characteristics of RU#0 for the DCT-IFDMA system are the same as the PAPR characteristics of the DFT-IFDMA system. However, the PAPR of RU#1, RU#2, and RU#3 for the DCT-IFDMA system increases with the increase in the roll-off factor and the PAPR characteristics seem to be similar to those of the DCT-LFDMA system. This can also be seen from Equation 3.23 when

Table 3.3   PAPR Values at a CCDF = 10^{-4} for the Different Systems

MODULATION FORMAT	PULSE SHAPING	DFT-IFDMA (dB)	DFT-LFDMA (dB)	DCT-IFDMA (dB)	DCT-LFDMA (dB)	OFDMA
QPSK	None	0	8	0	8.2	11.5 dB
	RC	6.5	8.2	6.5	8.8	N/A
16-QAM	None	3.5	9.25	3.5	9.5	11.5 dB
	RC	8	9.25	8.75	10	N/A

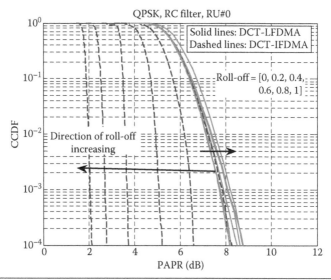

Figure 3.9   Comparison between the CCDFs of the PAPR for the DCT-IFDMA and the DCT-LFDMA systems with an RC filter and different roll-off factors for RU#0.

$v \neq 0$. In other words, the amplitudes of RU#1, RU#2, and RU#3 do not have constant envelopes as in RU#1. Thus, after the pulse-shaping filter, the output peak power of the resulting signal increases with the increase in the roll-off factor.

A full comparison of PAPR for different modulation formats at a CCDF = 10^{-4} and $\gamma = 0.22$ is given in Table 3.4.

Table 3.4 illustrates that the PAPR of the DCT-LFDMA system is nearly the same as that of the DFT-LFDMA system, whereas the PAPR of the DCT-IFDMA system is the same as that of the DFT-IFDMA system for the first user and higher by about 2 dB for the other users. This table indicates that the PAPR of all systems is also high, and it is possible to reduce the PAPR using the PAPR reduction techniques.

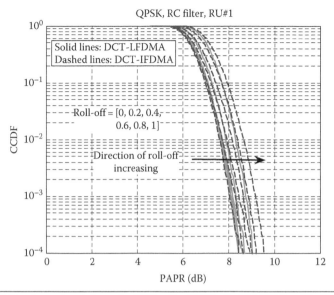

Figure 3.10   Comparison between the CCDFs of the PAPR for the DCT-IFDMA and the DCT-LFDMA systems with an RC filter and different roll-off factors for RU#1.

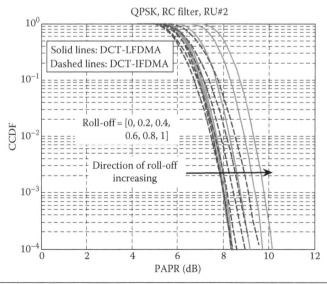

Figure 3.11   Comparison between the CCDFs of the PAPR for the DCT-IFDMA and the DCT-LFDMA systems with an RC filter and different roll-off factors for RU#2.

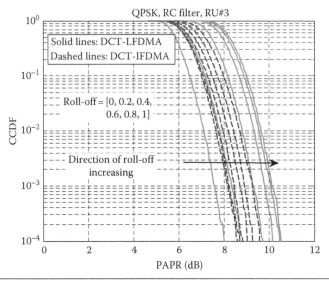

Figure 3.12   Comparison between the CCDFs of the PAPR for the DCT-IFDMA and the DCT-LFDMA systems with an RC filter and different roll-off factors for RU#3.

Table 3.4   PAPR Values at a CCDF $= 10^{-4}$ and $\gamma = 0.22$ for the Different Systems

MAPPING SCHEME	USER 1 (dB)	USER 2 (dB)	USER 3 (dB)	USER 4 (dB)
DFT-LFDMA	8.2	8.5	8.25	8.5
DFT-IFDMA	6.5	6.5	6.5	6.5
DCT-LFDMA	8.5	8.5	8.5	8.75
DCT-IFDMA	6.5	8.5	8.5	8.5

3.6.4 Impact of the Input Block Size

The impact of the input block size on the PAPR performance of the DCT-IFDMA and the DCT-LFDMA systems has been studied and shown in Figure 3.13 for different modulation formats. $M = 512$, RU#0, and an RC filter with $\gamma = 0.22$ have been considered. We can observe that the PAPR of the DCT-LFDMA system decreases as the block size N increases, whereas the PAPR of the DCT-IFDMA system is insensitive to the input block size. As a result, improvements in the PAPR performance can be achieved by increasing the input block size.

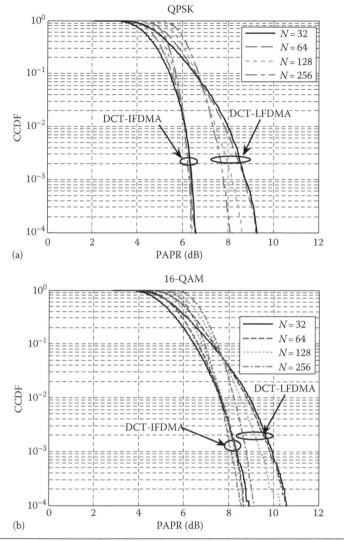

Figure 3.13 Comparison between the CCDFs of the PAPR for the DCT-IFDMA and the DCT-LFDMA systems with different input block sizes: (a) QPSK and (b) 16-QAM.

To illustrate the impact of the input block size on the PAPR of the DCT-SC-FDMA system, the measured PAPR values at a CCDF = 10^{-4} are tabulated in Table 3.5. It is clear that the PAPR of the DCT-LFDMA system decreases as the input block size N increases. However, we can see that the PAPR of the DCT-IFDMA system is insensitive to the input block size.

Table 3.5 PAPR Values at a CCDF $= 10^{-4}$ for the DCT-IFDMA and the DCT-LFDMA Systems with Different Input Block Sizes

INPUT BLOCK SIZE	DCT-IFDMA		DCT-LFDMA	
	QPSK (dB)	16-QAM (dB)	QPSK (dB)	16-QAM (dB)
$N=32$	6.5	8.75	9.25	10.5
$N=64$	6.5	8.5	9.25	10.5
$N=128$	6.5	8.5	8.8	10
$N=256$	6.5	8.5	8	9

3.6.5 Impact of the Output Block Size

The impact of the output block size on the PAPR of the DCT-IFDMA and the DCT-LFDMA systems has also been studied and shown in Figure 3.14 for different modulation formats. $N=128$, RU#0, and an RC filter with $\gamma=0.22$ have been used to obtain these results. It is clear that the PAPR of the DCT-IFDMA is insensitive to the output block size, whereas the PAPR of the DCT-LFDMA signals depends on the output block size. On the other hand, the higher the output block size, the higher is the PAPR value. As a result, improvements in the PAPR performance can be achieved by decreasing the output block size.

To illustrate the impact of the output block size on the PAPR of the DCT-IFDMA and DCT-LFDMA systems, the measured PAPR values at a CCDF $= 10^{-4}$ are tabulated in Table 3.6. It is clear that the PAPR of the DCT-LFDMA system increases as the output block size M increases, whereas the DCT-IFDMA system is insensitive to the output block size.

3.6.6 Impact of the Power Amplifier

Figures 3.15 and 3.16 show the impact of the PA on the BER performance of the DCT-IFDMA and the DCT-LFDMA systems for different modulation formats. Rapp's solid-state PA model with $p=2$ has been used. At an IBO $= 10$ dB, it can be seen that there is no effect of the nonlinearity, because the PA is operating in the linear region. It is also clear that as the IBO decreases from 10 to 5 dB, the BER performance degrades significantly, especially for 16-QAM. Thus, the BER performance for QPSK is relatively robust to the nonlinear PA for an IBO of 10 dB [51].

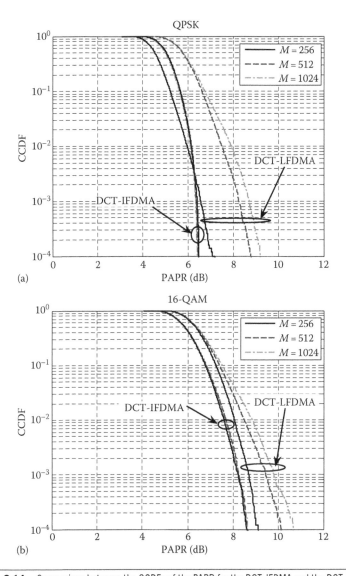

Figure 3.14   Comparison between the CCDFs of the PAPR for the DCT-IFDMA and the DCT-LFDMA systems with different output block sizes: (a) QPSK and (b) 16-QAM.

Table 3.6   PAPR Values at a CCDF $= 10^{-4}$ for the DCT-IFDMA and the DCT-LFDMA Systems with Different Output Block Sizes

OUTPUT BLOCK SIZE	DCT-IFDMA		DCT-LFDMA	
	QPSK (dB)	16-QAM (dB)	QPSK (dB)	16-QAM (dB)
$M=256$	6.5	8.5	7	9
$M=512$	6.5	8.5	8.8	10
$M=1024$	6.5	8.5	9.25	10.5

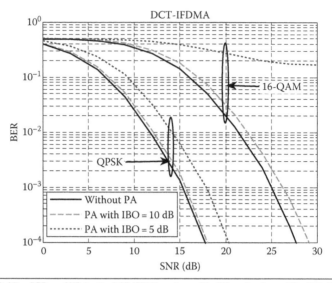

Figure 3.15   BER vs. SNR for the DCT-IFDMA system with different values of the IBO of the PA.

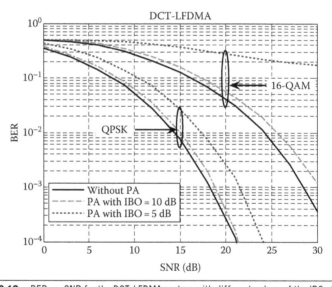

Figure 3.16   BER vs. SNR for the DCT-LFDMA system with different values of the IBO of the PA.

4

TRANSCEIVER SCHEMES FOR SC-FDMA SYSTEMS

4.1 Introduction

As shown in Chapters 2 and 3, even though the DFT-SC-FDMA and the DCT-SC-FDMA transmitted signals are characterized by low signal envelope fluctuations, the performance degradation due to the nonlinear amplification may substantially affect the link performance of these systems. It has been found that the PAPR of these systems is high, and it is possible to enhance the immunity of these systems to the nonlinear amplification effects by reducing the PAPR using the PAPR reduction methods. In the literature, several methods to reduce the PAPR in OFDM systems have been proposed [52–57]. However, the DFT-SC-FDMA and the DCT-SC-FDMA systems are relatively new systems and reducing their PAPR is still an open issue. It is known that applying the PAPR reduction methods directly reduces the PAPR, but it also degrades the BER performance. Therefore, it is urgent to look for an efficient method that aims to reduce the PAPR and simultaneously enhance the BER performance of the DFT-SC-FDMA and the DCT-SC-FDMA systems.

In the literature, improving both the PAPR performance and BER performance of the DFT-SC-FDMA and the DCT-SC-FDMA systems has not been studied. This motivated us to introduce a transceiver scheme in this chapter that reduces the PAPR and enhances the BER performance of the DFT-SC-FDMA and the DCT-SC-FDMA systems. This transceiver system uses the wavelet transform to decompose the transmitted signal into approximation and detail components. The approximation component can be clipped or companded, while the detail component is left unchanged because of its sensitivity to noise. In this transceiver, wavelet filter banks at the transmitter and the receiver demonstrate the ability to reduce the distortion in the reconstructed signal, while retaining all the significant features present in the signal. The performance of the presented system has been investigated with different PAPR reduction methods. Simulation results show that the wavelet-based system with a hybrid clipping and companding method provides a significant performance enhancement when compared with the conventional systems, while the complexity of the system is slightly increased. It is also noted that the performance of the wavelet-based system is better than that of the OFDMA system, even with high-order modulation schemes such as 16-QAM. This indicates that the immunity of the wavelet-based system to the noise enhancement is higher than that of the conventional systems.

4.2 PAPR Reduction Methods

There are several methods to reduce the PAPR in OFDM systems [52–57]. These methods are applicable in the DFT-SC-FDMA and DCT-SC-FDMA systems. The clipping method is the simplest one, but it causes additional clipping noise, which degrades the system performance. An improved version of the clipping method is the clipping and filtering version, which can remove the out-of-band radiation, but it causes some peak regrowth, letting the signal samples to exceed the clipping level at some points [52]. The second method is the companding, which was originally designed for speech processing [53]. It is the most attractive PAPR reduction method for multicarrier transmission due to its good performance and low complexity. The authors in [58,59] have introduced a hybrid method comprising companding and clipping. They have shown that merging both methods can provide a better PAPR performance as well as a better BER performance for OFDM systems. In [58], the μ-law companding method was used, whereas in [59] the uniform companding method was used. The compression function of the uniform companding method transforms the distribution of the transmitted OFDM signals, which is assumed to be Gaussian, into a uniform distribution [59]. However, the compression function of the μ-law enlarges small amplitude signals [53].

Recently, researchers have tried to develop the methods implemented in OFDM systems to further reduce the PAPR in the DFT-SC-FDMA system. In [60], the clipping method has been suggested to reduce the PAPR in the DFT-SC-FDMA system. The results in [60] indicted that the PAPR in the DFT-SC-FDMA system can be further reduced by applying the PAPR reduction methods implemented in OFDM systems. In the literature, reducing the PAPR in the DFT-SC-FDMA system with a companding method has not been studied. We try to optimize the performance of the wavelet-based transceiver schemes with companding or clipping using the BER as a cost function to be minimized. Since the distribution of DFT-SC-FDMA signal samples is not Gaussian [9], the μ-law companding is used to reduce the PAPR. The best values of the companding coefficient and the clipping ratio (CR) can be determined through simulations.

4.2.1 Clipping Method

Clipping is the simplest method that can limit the amplitude of the signal to a desired threshold value. It is a nonlinear process and may cause in-band distortion and out-of-band radiation. The in-band distortion causes degradation in the BER [52]. In the clipping method, when the amplitude exceeds a certain threshold, it is hard-clipped, while the phase is saved. Namely, when we assume the phase of the baseband DFT-SC-FDMA signal $\bar{x}(m)$ to be $\phi(m)$, and the threshold to be A, the output signal after clipping will be given as follows:

$$\bar{x}(m) = \begin{cases} Ae^{j\phi(m)} & |\bar{x}(m)| > A \\ \bar{x}(m) & |\bar{x}(m)| \leq A \end{cases} \quad (4.1)$$

The CR is a useful means to represent the clipping level. It is the ratio between the maximum power of the clipped signal and the average power of the unclipped signal. If the IDFT output signal is normalized, the unclipped signal power is 1. If we clip the IDFT output

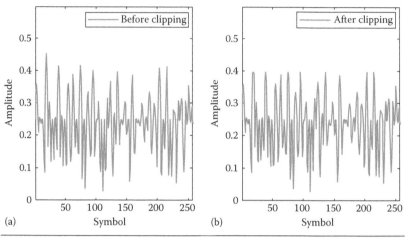

Figure 4.1   Amplitudes of the DFT-LFDMA signals: (a) before clipping and (b) after clipping.

at level A, then the CR is $A^2/1 = A^2$. The amplitudes of the DFT-LFDMA signal before and after clipping are plotted in Figure 4.1 with $N = 128$, $Q = 4$, $M = 512$, QPSK, and CR = 4 dB.

4.2.2 Companding Method

The companding method uses a compander to reduce the signal amplitude. Such approach can effectively reduce the PAPR with a less computational complexity [53]. The corresponding transmitter and receiver need a compander and an expander, respectively, which of course slightly increases the hardware cost. In our simulations, the companding will be performed according to the well-known μ-law. The companding process can be expressed as follows [53]:

$$x_c(m) = V_{max} \frac{\ln\left[1 + \mu \dfrac{|\bar{x}(m)|}{V_{max}}\right]}{\ln[1 + \mu]} \operatorname{sgn}[\bar{x}(m)] \qquad (4.2)$$

where

$$V_{max} = \max(|\bar{x}(m)|), \qquad m = 0, 1, \ldots, M - 1 \qquad (4.3)$$

μ is the companding coefficient
$x_c(m)$ is the companded sample
$\bar{x}(m)$ is the original sample

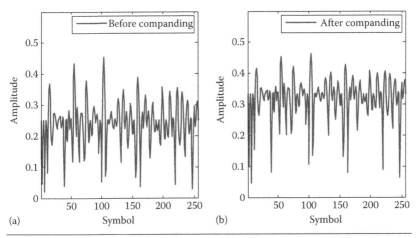

Figure 4.2   Amplitudes of the DFT-LFDMA signals: (a) before companding and (b) after companding.

The expansion process is simply the inverse of Equation 4.2 as follows:

$$\hat{x}(m) = \frac{V_{max}}{\mu}\left[\exp\left(\frac{|x_c(m)|\ln(1+\mu)}{V_{max}} - 1\right)\right]\operatorname{sgn}(x_c(m)) \qquad (4.4)$$

where $\hat{x}(m)$ is the estimated sample after expansion. The amplitudes of the DFT-LFDMA signals before and after companding are plotted in Figure 4.2 with $N = 128$, $Q = 4$, $M = 512$, QPSK, and $\mu = 4$. It is clear that the μ-law companding scheme only expands the amplitudes of small signals.

4.2.3 Hybrid Clipping and Companding

Hybrid clipping and companding comprises clipping followed by μ-law companding. By exploiting the clipping in the first step, the hybrid method can reduce the PAPR of SC-FDMA signals. Moreover, the companding in the second step further reduces the PAPR of the DFT-SC-FDMA signals. Consequently, we expect that the hybrid method effectively reduces the PAPR, while the complexity of the system is slightly increased.

4.3 Discrete Wavelet Transform

Wavelet decomposition is simply the process of decomposing a signal into approximation and detail components using wavelet filters [61–63]. These filters have finite lengths and due to this property,

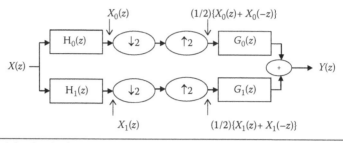

Figure 4.3   One-level wavelet decomposition and reconstruction.

wavelet filter banks can perform local analysis of signals. An important property of the wavelet transform is the perfect reconstruction
(PR), which is the process of rebuilding a decomposed signal into its
original transmitted form without deterioration. The discrete wavelet
transform (DWT) is computed by low-pass and high-pass filtering of
the discrete time signal as shown in Figure 4.3.

4.3.1 Implementation of the DWT

The DWT is normally implemented by a binary tree of filters as
shown in Figure 4.3. The art of finding a good wavelet lies in the
design of the set of filters, H_0, H_1, G_0, and G_1 to achieve various trade-
offs between spatial- and frequency-domain characteristics, while
satisfying the PR condition [61].

Now, we are going to discuss the PR condition. In Figure 4.3, the
process of decimation by 2 at the output of H_0 and H_1 effectively sets
all odd samples of these signals to zero. For the low-pass branch,
this is equivalent to multiplying $x_0(m)$ by $\frac{1}{2}\left(1+(-1)^m\right)$. Hence $X_0(z)$
is converted to $\frac{1}{2}\{X_0(z)+X_0(-z)\}$. Similarly, $X_1(z)$ is converted to
$\frac{1}{2}\{X_1(z)+X_1(-z)\}$. Thus, the expression for $Y(z)$ is given by [61] as
follows:

$$
\begin{aligned}
Y(z) &= \frac{1}{2}\{X_0(z)+X_0(-z)\}G_0(z)+\frac{1}{2}\{X_1(z)+X_1(-z)\}G_1(z) \\
&= \frac{1}{2}X(z)\{H_0(z)G_0(z)+H_1(z)G_1(z)\} \\
&\quad +\frac{1}{2}X(-z)\{H_0(-z)G_0(z)+H_1(-z)G_1(z)\}
\end{aligned}
\tag{4.5}
$$

The first PR condition requires aliasing cancellation and forces the above term in $X(-z)$ to be zero. Hence $\{H_0(-z)G_0(z) + H_1(-z)G_1(z)\} = 0$, which can be achieved if [61]

$$H_1(z) = z^{-k}G_0(-z) \text{ and } G_1(z) = z^k H_0(-z) \quad (4.6)$$

where k must be odd (usually $k = \pm 1$).

The second PR condition is that the transfer function from $X(z)$ to $Y(z)$ should be unity [61]:

$$\{H_0(z)G_0(z) + H_1(z)G_1(z)\} = 2 \quad (4.7)$$

If we define a product filter $P(z) = H_0(z)G_0(z)$ and substitute from Equation 4.6 into Equation 4.7, then the PR condition becomes [61]:

$$H_0(z)G_0(z) + H_1(z)G_1(z) = P(z) + P(-z) = 2 \quad (4.8)$$

This needs to be true for all z, and since the odd powers of z in $P(z)$ cancel with those in $P(-z)$, it requires that $p_0 = 1$ and $p_m = 0$ for all m even and nonzero. The polynomial $P(z)$ should be a zero-phase polynomial to minimize distortion. In general, $P(z)$ is of the following form [61]:

$$P(z) = \cdots + p_5 z^5 + p_3 z^3 + p_1 z + 1 + p_1 z^{-1} + p_3 z^{-3} + p_5 z^{-5} + \cdots \quad (4.9)$$

The design method for the PR filters can be summarized in the following steps [61]:

1. Choose p_1, p_3, p_5, ... to give a zero-phase polynomial $P(z)$ with good characteristics.
2. Factorize $P(z)$ into $H_0(z)$ and $G_0(z)$ with similar low-pass frequency responses.
3. Calculate $H_1(z)$ and $G_1(z)$ from $H_0(z)$ and $G_0(z)$.

To simplify this procedure, we can use the following relation [61]:

$$P(z) = P_t(Z) = 1 + p_{t,1}Z + p_{t,3}Z^3 + p_{t,5}Z^5 + \cdots \quad (4.10)$$

where

$$Z = \frac{1}{2}\left(z + z^{-1}\right) \quad (4.11)$$

4.3.2 Haar Wavelet Transform

The Haar wavelet is the simplest wavelet type. In the discrete form, the Haar wavelet is related to a mathematical operation called the Haar transform [61]. Like all wavelet transforms, the Haar transform decomposes a discrete signal into two components of half the original signal length. At each decomposition level, the high-pass filter produces the detail component denoted by $d(m)$, while the low-pass filter produces the coarse approximation component denoted by $a(m)$. The filtering and decimation process continues until the desired decomposition level is reached. The maximum number of levels depends on the length of the signal. The DWT of the original signal is then obtained by concatenating all the coefficients, $a(m)$ and $d(m)$, starting from the last level of decomposition.

The Haar wavelet uses the simplest possible $P_t(Z)$ with a single zero at $Z = -1$. It is represented as follows [61]:

$$P_t(Z) = 1 + Z \quad \text{and} \quad Z = \frac{1}{2}\left(z + z^{-1}\right) \tag{4.12}$$

Thus

$$P(z) = \frac{1}{2}\left(z + 2 + z^{-1}\right) = \frac{1}{2}(z+1)\left(1 + z^{-1}\right) = G_0(z)H_0(z) \tag{4.13}$$

We can find $H_0(z)$ and $G_0(z)$ as follows:

$$H_0(z) = \frac{1}{2}\left(1 + z^{-1}\right) \tag{4.14}$$

$$G_0(z) = (z+1) \tag{4.15}$$

Using Equation 4.6 with $k = 1$; we get

$$G_1(z) = zH_0(-z) = \frac{1}{2}z\left(1 - z^{-1}\right) = \frac{1}{2}(z-1) \tag{4.16}$$

$$H_1(z) = z^{-1}G_0(-z) = z^{-1}(-z+1) = \left(z^{-1} - 1\right) \tag{4.17}$$

The approximation and detail coefficients can be expressed as follows:

$$a(m) = \sum_{k=-\infty}^{\infty} x(k)H_0(2m-k) \tag{4.18}$$

$$d(m) = \sum_{k=-\infty}^{\infty} x(k)H_1(2m-k) \tag{4.19}$$

4.4 Wavelet-Based Transceiver Scheme

The wavelet-based transceiver architecture uses the DWT to decompose the transmitted signal into an approximation component and a detail component. The approximation component can be clipped and/or companded, while the detail component is left unchanged because of its sensitivity to noise. The wavelet-based system in this book will be called the hybrid wavelet DFT-SC-FDMA (HW-DFT-SC-FDMA) system if the hybrid method is used, the clipped wavelet DFT-SC-FDMA (clipped W-DFT-SC-FDMA) system if the clipping method is used, and the companded wavelet DFT-SC-FDMA (companded W-DFT-SC-FDMA) system if the companding method is used. The Haar wavelet will be used in simulations.

4.4.1 Mathematical Model

In this subsection, the mathematical model of the wavelet-based transceiver scheme is introduced and investigated for the DFT-SC-FDMA system. The extension to the DCT-SC-FDMA system will also be discussed at the end of this subsection. The wavelet-based scheme uses the DWT to provide better BER and PAPR performances. The main purpose of the DWT is to decompose the transmitted signal into an approximation component and a detail component. The approximation component contains most of the signal energy, and the detail component contains a small portion of this energy. So, the noise effect on the detail component is more severe than its effect on the approximation component. In addition, the wavelet filter bank at the transmitter and the receiver exhibits exceptionally good frequency characteristics and demonstrates the ability to reduce distortion in the reconstructed signals, while retaining all the significant features present in the signals.

The block diagram illustrating the HW-DFT-SC-FDMA system is shown in Figure 4.4. It consists of a DFT-SC-FDMA modulator

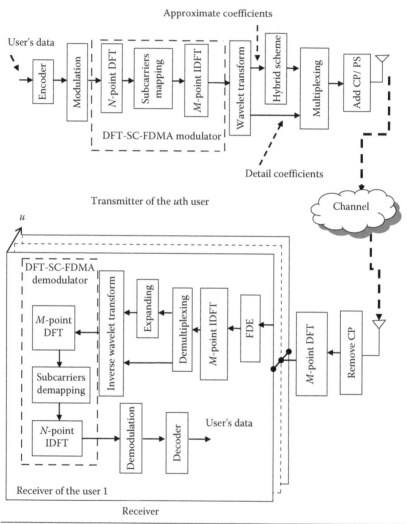

Figure 4.4 The HW-DFT-SC-FDMA system.

followed by the DWT and a PAPR reduction step. The steps performed in this system can be summarized as follows:

1. The encoding and the modulation processes are applied on the source data.

2. The modulated signal passes through the DFT-SC-FDMA modulator explained in Chapter 3.

3. The resulting signal from the DFT-SC-FDMA modulator is transformed by the DWT before transmission. The output of the DWT consists of two components: the approximation component and the detail component given by Equations 4.18 and 4.19.

4. Only the approximation component passes through the PAPR reduction step as shown in Figure 4.4, because the approximation component contains most of the signal energy and hence large amplitudes. As a result, if companding or clipping is to be made on both components, this will affect the detail component more seriously than the approximation component. The amplitudes of the approximation and detail components are plotted in Figure 4.5. This step can be summarized as follows:

   a. *Clipping*: Signal peaks exceeding a certain level are clipped with a proper CR as given by Equation 4.1.

   b. *Companding*: The μ-law companding is used to effectively reduce the PAPR as given by Equation 4.2. The best values of μ and CR are determined through simulations.

   Figure 4.6 shows the amplitudes of the approximation coefficients for the proposed HW-DFT-SC-FDMA system with different PAPR reduction methods and a one-level Haar wavelet decomposition. $N = 128$, $M = 512$, $CR = 4$ dB, and $\mu = 4$ have been considered. It is clear that the hybrid system clips out the peak signals and enhances the small signals.

5. The resulting signal from the PAPR reduction scheme is multiplexed with the detail component to produce the signal $\check{x}^u$, and the CP is added in order to prevent the IBI.

6. Finally, the resulting signal is transmitted through the wireless channel.

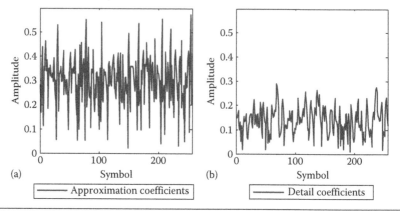

Figure 4.5 Amplitudes of wavelet components: (a) approximation component and (b) detail component.

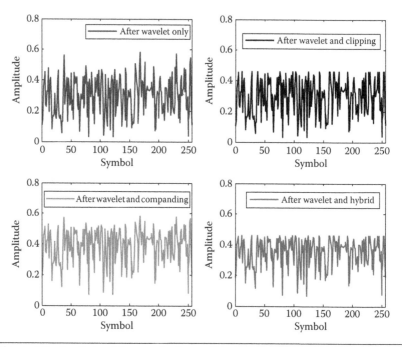

Figure 4.6 Amplitudes of the approximation coefficients for the different proposed schemes.

7. At the receiver, the CP is removed from the received signal and it can be represented by

$$r = \sum_{u=1}^{U} H_C^u \breve{x}^u + n \qquad (4.20)$$

where

> r is an $M \times 1$ vector representing the received signal at the base station
>
> $\check{x}^u$ is an $M \times 1$ vector representing the transmitted signal of the uth user
>
> n contains i.i.d AWGN samples having zero mean and variance, σ_n^2
>
> H_C^u is an $M \times M$ circulant matrix describing the multipath channel

8. The received signal is then transformed into the frequency domain via an M-point DFT as follows:

$$R = \sum_{u=1}^{U} \Lambda^u \check{X}^u + N \tag{4.21}$$

where R, $\check{X}^u$, and N are the DFTs of r, $\check{x}^u$, and n, respectively.

9. The FDE is performed followed by an M-point IDFT, demultiplexing, expanding, and an inverse DWT. The output of the FDE can be expressed as follows:

$$\hat{\check{X}}^u = W^u R \tag{4.22}$$

where

> $\hat{\check{X}}^u$ is an $M \times 1$ vector representing the frequency-domain estimate of the transmitted signal
>
> W^u is the $M \times M$ FDE matrix

The expansion process can be made using Equation 4.4.

10. The resulting signal after the inverse DWT (IDWT) is transformed into frequency domain via an M-point DFT.

11. Finally, the subcarriers demapping, the N-point IDFT, the demodulation, and the decoding processes are performed.

For DCT-SC-FDMA, the system is called the HW-DCT-SC-FDMA system. The structure of the HW-DCT-SC-FDMA system is similar to that in Figure 4.4, but with the DFT and the IDFT modules in the DFT-SC-FDMA modulator and demodulator replaced by the DCT and IDCT modules, respectively.

The main advantage of the wavelet-based system is that the wavelet filter banks at the transmitter and the receiver exhibit exceptionally good frequency characteristics and demonstrate the ability to reduce distortion in the reconstructed signals, while retaining all the significant features present in the signals. Moreover, with the unique time–frequency localization feature of wavelets, the BER performance of communication systems can be improved. Hence, the wavelet-based system can provide more PAPR reduction and a better BER performance as compared to the conventional systems.

4.4.2 Two-Level Decomposition

For the two-level decomposition, the approximation component of the first level is the input to a second wavelet decomposition step. The output of the two-level decomposition consists of three signals, which represent the approximation coefficients, the detail coefficients from the second decomposition, and the detail coefficients from the first decomposition. Only, the approximation coefficients are passed through the PAPR reduction step.

4.4.3 Complexity Evaluation

The complexities of the wavelet-based transceiver schemes are higher than those of the conventional schemes. However, the increase in complexity is tolerable considering the benefits of power reduction at the mobile station, the BER gain, and the robustness to channel estimation errors, as we will see in the simulation results. The complexities of the proposed HW-DFT-SC-FDMA and HW-DCT-SC-FDMA systems are nearly the same.

4.5 Simulation Examples

Computer simulations have been carried out to evaluate the performance of the wavelet-based system with different wireless channels and different PAPR reduction methods.

4.5.1 Simulation Parameters

Uplink DFT-SC-FDMA and DCT-SC-FDMA systems with 4 users have been considered. The vehicular-A [43] and the SUI-3 channels [64] have been adopted in the simulations. The BER has been evaluated using the Monte Carlo simulation method. The SUI-3 channel is one of the six channel models adopted by the IEEE 802.16a standard for evaluating the performance of broadband wireless systems in the 2–11 GHz band [64]. It is discussed in Appendix A. The fading was modeled as being quasi-static over a DFT block. A convolutional code with a memory length of 7 and an octal generator polynomial of (133,171) has been chosen as the channel code. The simulation parameters are listed in Table 4.1. As shown in Chapter 3, the localized systems incur higher PAPRs compared to the interleaved systems. Thus, we will be concerned in this chapter with the localized systems only.

4.5.2 Results of the DFT-SC-FDMA System

First, the best values of the companding coefficient and the CR in the wavelet-based system have been determined through simulations. The objective is to find the best values that reduce the PAPR and provide a good BER performance when compared with that of the W-DFT-SC-FDMA system without clipping or companding. The variation of the BER of the companded W-DFT-SC-FDMA system with the companding coefficient for different SNR values is shown in Figure 4.7.

Table 4.1 Simulation Parameters

Simulation method	Monte Carlo
System bandwidth	5 MHz
Modulation type	QPSK, and 16-QAM
CP length	20 samples
M	512
Subcarriers spacing	9.765625 kHz
N	128
Wavelet transform	Haar wavelet
Companding scheme	μ-law companding
Channel coding	Convolutional code with code rate $= 1/2$
Subcarriers mapping technique	Localized
Channel model	SUI-3 and vehicular-A outdoor channel
Channel estimation	Perfect and MMSE channel estimator
Equalization	MMSE

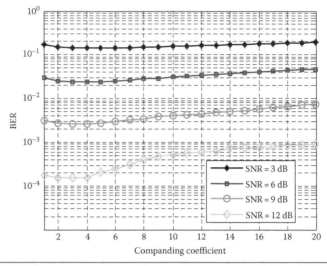

Figure 4.7   BER vs. the companding coefficient for the companded W-DFT-SC-FDMA system.

The BER performance is nearly independent of the companding coefficient for low SNR values. The best value of the companding coefficient lies in the interval [1, 4] for different SNR values. The appropriate choice of the companding coefficient is very important for high SNR values.

The variation of the BER of the clipped W-DFT-SC-FDMA system with the CR is shown in Figure 4.8 for different SNR values. The

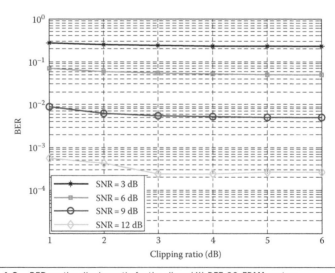

Figure 4.8   BER vs. the clipping ratio for the clipped W-DFT-SC-FDMA system.

value of the CR is better to be chosen greater than or equal to 3 dB for different SNR values. The appropriate choice of the CR is very important for high SNR values. Thus, for the rest of experiments we will use $\mu = 4$ for the companded W-DFT-SC-FDMA system and CR = 3 dB for the clipped W-DFT-SC-FDMA system.

In Figures 4.9 and 4.10, the two parameters, μ and CR, have been taken into account simultaneously for the SUI-3 and the vehicular-A channel models. The figures show the variation of the BER with both the CR and the companding coefficient. SNR = 12 dB and SNR = 15 dB have been considered in Figures 4.9 and 4.10, respectively. From the two figures, it is clear that the region of the minimum BER is nearly the same for the two models. This indicates that the HW-DFT-SC-FDMA system is not too sensitive to the channel model chosen in the simulation. The region of the minimum BER lies in the area with μ varying from 1 to 4 and CR varying from 3 to 6 dB as indicated by the solid lines in the figures. The difference between the BER values in this region is nearly negligible.

However, it is known that the PAPR depends on the values of CR and μ. The larger the value of μ, the lower is the PAPR. On the other hand, the smaller the value of CR, the lower is the PAPR. Thus, in

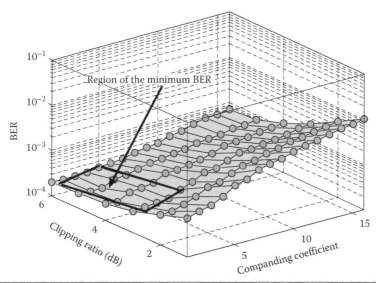

Figure 4.9 Variation of the BER with both the clipping ratio and the companding coefficient for the HW-DFT-SC-FDMA system, when the SUI-3 channel model is used.

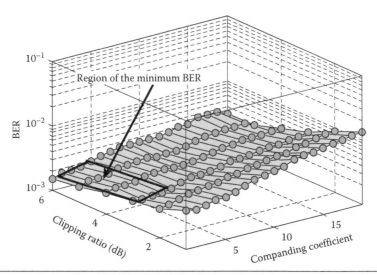

Figure 4.10 Variation of the BER with both the clipping ratio and the companding coefficient for the HW-DFT-SC-FDMA system, when the vehicular-A channel model is used.

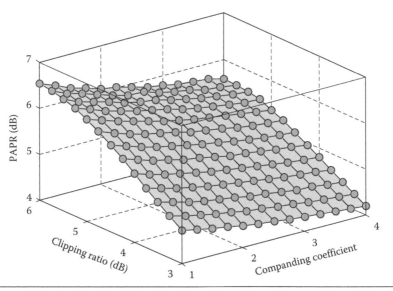

Figure 4.11 Variation of the PAPR with both the clipping ratio and the companding coefficient in the area of interest for the HW-DFT-SC-FDMA system.

Figure 4.11, we show the variation of the PAPR inside this area of interest in order to determine the best choice of μ and CR.

Figure 4.11 shows the variation of the PAPR with the CR and the companding coefficient for the HW-DFT-SC-FDMA system, when μ and CR lie inside the region of minimum BER indicated

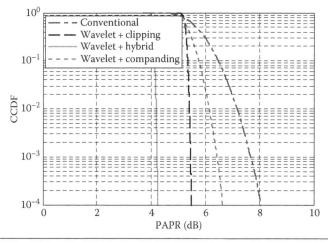

Figure 4.12   CCDFs of the PAPR for the HW-DFT-SC-FDMA system with different PAPR reduction methods and QPSK modulation.

in Figures 4.9 and 4.10. It is clear that with $\mu = 4$ and CR = 3 dB, the value of the PAPR is the lowest as compared with that of the other values of μ and CR in this region. From Figures 4.9 through 4.11, we conclude that $\mu = 4$ and CR = 3 dB is the best choice for the HW-DFT-SC-FDMA system. Thus, we will use $\mu = 4$ and CR = 3 dB for the rest of experiments.

Figures 4.12 and 4.13 illustrate the CCDF of the PAPR of the wavelet-based system with different PAPR reduction methods and compare it with that of the conventional DFT-SC-FDMA system for QPSK and 16-QAM, respectively. It is observed that the HW-DFT-SC-FDMA system provides a considerable reduction in the PAPR when compared to the other systems. These figures also show that the wavelet-based system with companding or clipping also provides a more reduction in PAPR than that of the conventional DFT-SC-FDMA system.

Figure 4.14 shows the variation of the BER with the SNR for the HW-DFT-SC-FDMA system, the companded W-DFT-SC-FDMA system, the clipped W-DFT-SC-FDMA system, and the conventional DFT-SC-FDMA system over the SUI-3 channel. For the coded case, a rate 1/2 convolutional code with constraint length 7 and octal generator polynomial (133,171) has been used. From these results, we can clearly see that the wavelet-based system

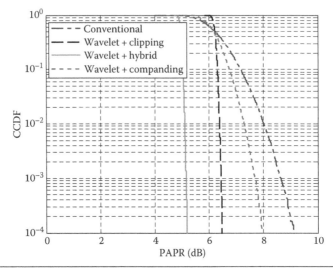

Figure 4.13 CCDFs of the PAPR for the HW-DFT-SC-FDMA system with different PAPR reduction methods and 16-QAM modulation.

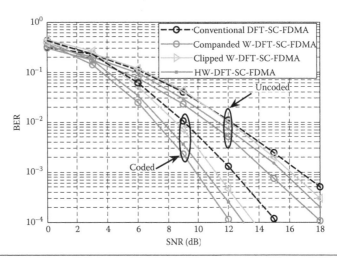

Figure 4.14 BER vs. SNR for the HW-DFT-SC-FDMA system with different PAPR reduction methods.

with the different PAPR reduction methods performs better than the conventional DFT-SC-FDMA system, especially with coding. It is also clear that the performance of the wavelet-based system can be improved by using more elaborate channel coding algorithms. At a BER = 10^{-4} in the coding case, the companded W-DFT-SC-FDMA system provides about a 3 dB gain over the conventional

Table 4.2   SNR Improvement at a BER $= 10^{-3}$ and PAPR Improvement at a CCDF $= 10^{-4}$ over the SUI-3 Channel

SYSTEM	SNR IMPROVEMENT (dB)	PAPR IMPROVEMENT (dB)
Companded W-DFT-SC-FDMA	2.5	1.25
Clipped W-DFT-SC-FDMA	1.5	2.5
HW-DFT-SC-FDMA	2	3.75

DFT-SC-FDMA system. The corresponding gain of HW-DFT-SC-FDMA system is about 2 dB.

From the previous results obtained from Figures 4.12 and 4.14, at a BER $= 10^{-3}$ and a CCDF $= 10^{-4}$, the SNR and the PAPR improvements for all wavelet-based schemes are given in Table 4.2. It is clear that the HW-DFT-SC-FDMA system provides about a 2.5 dB PAPR reduction as compared with the companded W-DFT-SC-FDMA system and about 1.25 dB PAPR reduction as compared with the clipped W-DFT-SC-FDMA system.

It has also been found that the HW-DFT-SC-FDMA system provides a better BER performance than the clipped W-DFT-SC-FDMA system, but compared with the companded W-DFT-SC-FDMA system, it is slightly worse by about 0.5 dB, which is acceptable. Thus, our recommendation is to use the HW-DFT-SC-FDMA system for wireless channels because of its low PAPR. The low PAPR leads to an increase in the range of coverage of wireless devices.

The BER performance of the HW-DFT-SC-FDMA system has been studied and compared with that of the conventional DFT-SC-FDMA and the OFDMA systems over the vehicular-A channel and the results are shown in Figure 4.15. It is clear that the wavelet-based system outperforms the conventional DFT-SC-FDMA and the OFDMA systems, even with 16-QAM. On the other hand, the wavelet-based system has a better performance with sufficient robustness to the noise enhancement problem with the higher-order modulation techniques, because its frequency diversity is higher than that of the conventional system due to the wavelet transform. At a BER $= 10^{-3}$ and QPSK, the wavelet-based system provides about 7.5 dB performance gain when compared with the conventional DFT-SC-FDMA system.

At a BER $= 10^{-2}$, the SNR improvement for the HW-DFT-SC-FDMA system as compared to the DFT-SC-FDMA and the

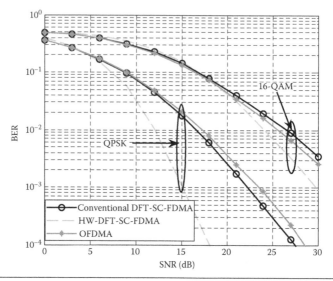

Figure 4.15   BER vs. SNR for the HW-DFT-SC-FDMA system over the vehicular-A channel.

Table 4.3   SNR Improvement at a BER $= 10^{-2}$ over a Vehicular-A Channel

MODULATION	SNR IMPROVEMENT AS COMPARED TO THE DFT-SC-FDMA SYSTEM (dB)	SNR IMPROVEMENT AS COMPARED TO THE OFDMA SYSTEM (dB)
QPSK	5	5.5
16-QAM	2.5	1.5

OFDMA systems is given in Table 4.3. Tables 4.2 and 4.3 illustrate that the HW-DFT-SC-FDMA system provides higher SNR improvements for the vehicular-A channel than those for the SUI-3 channel for QPSK. This indicates that the wavelet-based system is more suitable for the mobile case.

Figure 4.16 shows a comparison between the Haar and Daubechies (db1) wavelets for the HW-DFT-SC-FDMA system over a vehicular-A channel with different wavelet decomposition levels. MMSE equalization and a convolutional code with rate 1/2 have been used. It is clear that increasing the number of wavelet decomposition levels leads to an enhancement in the BER performance of the HW-DFT-SC-FDMA system. It can also be seen that the one-level Daubechies

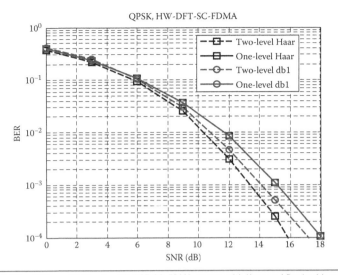

Figure 4.16   BER vs. SNR for the HW-DFT-SC-FDMA system with Haar and Daubechies wavelets and different decomposition levels.

wavelet decomposition provides the same BER as the one-level Haar wavelet decomposition. However, the two-level Haar wavelet decomposition provides a better BER performance than that of the Daubechies wavelet decomposition. This is attributed to the fact that the length of the Haar wavelet filters is shorter than that of the Daubechies wavelet filters. In fact, the process of inverse wavelet transform is a deconvolution process, which inverts the effect of the wavelet transform filters. From the basics of signal processing, deconvolution of a signal affected by a long filter in the presence of noise leads to noise enhancement. This is the reason why the two-level Haar wavelet decomposition gives better results than the two-level Daubechies wavelet decomposition. At a BER = 10^{-4}, the HW-DFT-SC-FDMA system with the two-level Haar wavelet provides a performance gain of about 2 dB as compared to that of the one-level Haar wavelet case. Thus, the wavelet-based system with a Haar wavelet transform is the best choice.

Figure 4.17 shows the CCDF of the PAPR for the HW-DFT-SC-FDMA system with the Haar wavelet with one and two decomposition levels. It is clear that the one-level decomposition achieves the lowest possible PAPR. As the decomposition level increases, most of the signal energy will be concentrated in a smaller number of samples

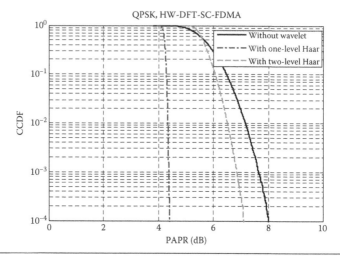

Figure 4.17 CCDFs of the PAPR for the HW-DFT-SC-FDMA system with the Haar wavelet and different decomposition levels.

representing the approximation coefficients of the signal, which increases the probability of existence of large amplitude samples.

From Figures 4.16 and 4.17, it is clear that there is a need for a trade-off between the PAPR performance and the BER performance. Increasing the decomposition level leads to an enhancement in the BER performance and a degradation in the PAPR performance. We can say that the wavelet-based system with one-level wavelet decomposition is better to use because of its ability to provide good BER and PAPR performances with a low complexity.

4.5.3 Results of the DCT-SC-FDMA System

Figure 4.18 shows the variation of the BER with both the CR and the companding coefficient for the HW-DCT-SC-FDMA system. SNR = 15 dB and QPSK have been used. It is clear that the region of the minimum BER for the HW-DCT-SC-FDMA system lies in the area with μ varying from 1 to 6 and CR varying from 4 to 7 dB as indicated in Figure 4.18.

From Figure 4.19, it is clear that the value of the PAPR with μ = 6 and CR = 4 dB is the lowest as compared with that of the other values of μ and CR in this region for the HW-DCT-SC-FDMA system. From Figure 4.19, we can conclude that μ = 6 and CR = 4 dB are the

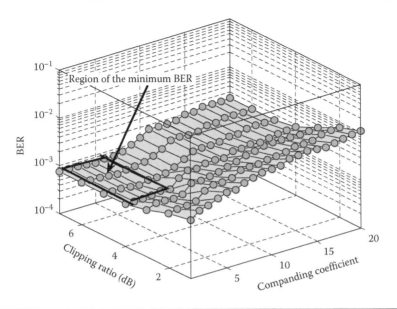

Figure 4.18   Variation of the BER with both the clipping ratio and the companding coefficient for the HW-DCT-SC-FDMA system.

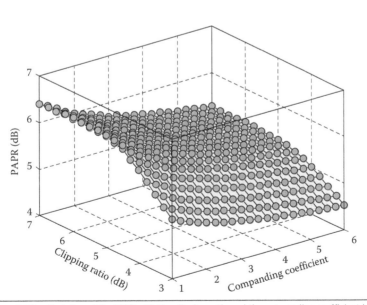

Figure 4.19   Variation of the PAPR with the clipping ratio and the companding coefficient in the area of interest for the HW-DCT-SC-FDMA system.

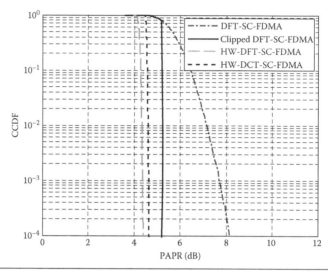

Figure 4.20   CCDFs of the PAPR for different systems.

best choice for the proposed HW-DCT-SC-FDMA system. Thus, these values will be used for the rest of experiments.

Figure 4.20 shows the CCDFs of the PAPR for different systems. This figure illustrates the CCDFs of the PAPR of the HW-DFT-SC-FDMA and HW-DCT-SC-FDMA systems and compares it with the DFT-SC-FDMA system with clipping and the conventional DFT-SC-FDMA system. QPSK has been utilized. It is observed that the HW-DFT-SC-FDMA and HW-DCT-SC-FDMA systems provide considerable reductions in the PAPR when compared to the other systems. This figure also shows that the DFT-SC-FDMA system with clipping also provides a better PAPR performance than that of the conventional DFT-SC-FDMA system. However, the in-band distortion due to the clipping method degrades the BER performance of the DFT-SC-FDMA system. On the other hand, the HW-DFT-SC-FDMA and HW-DCT-SC-FDMA systems are able to provide better BER and PAPR performances.

In Figure 4.21, the variation of the BER with the SNR is shown for all systems. MMSE channel estimation and equalization have been utilized. It is clear from Figure 4.21 that the DCT-SC-FDMA system provides a better BER performance than that of the DFT-SC-FDMA system. This is attributed to the energy compaction property of the DCT, which reduces the ISI. From Figure 4.21, it is also clear

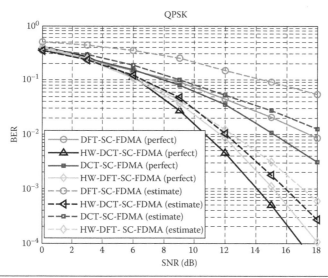

Figure 4.21 BER vs. SNR for different systems.

that the HW-DCT-SC-FDMA and the HW-DFT-SC-FDMA systems provide better BER performance than the conventional DCT-SC-FDMA and the conventional DFT-SC-FDMA systems.

It can be also noted that the HW-DCT-SC-FDMA system provides the best BER performance when compared to the other systems. At a $BER = 10^{-2}$, the HW-DFT-SC-FDMA and HW-DCT-SC-FDMA systems provide about 6 and 7 dB improvements over the conventional DFT-SC-FDMA system, respectively.

The effect of the channel estimation on the performance of different systems is also shown in Figure 4.21, which reveals that the wavelet-based architectures are more robust to channel estimation errors than the conventional DFT-SC-FDMA system. This indicates that these architectures are more suitable for uplink transmission in wireless mobile communications.

Figure 4.22 shows a comparison between the HW-DCT-SC-FDMA and the HW-DFT-SC-FDMA systems with different wavelet decomposition levels. It is clear that increasing the number of wavelet decomposition levels leads to an enhancement in the BER performance of the two systems. At a $BER = 10^{-4}$, the HW-DFT-SC-FDMA system with two-level Haar wavelet provides a performance gain of about 2 dB as compared to that of the one-level Haar wavelet case. At the same value of the BER, the HW-DCT-SC-FDMA

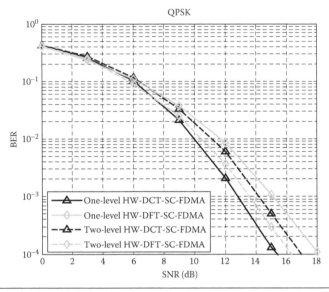

Figure 4.22 BER vs. SNR for the HW-DCT-SC-FDMA and the HW-DFT-SC-FDMA systems with different decomposition levels.

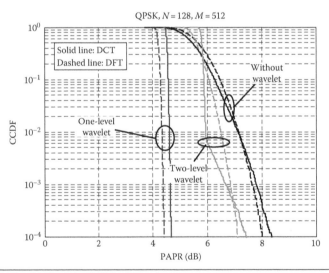

Figure 4.23 CCDFs of the PAPR for different systems.

system with two-level Haar wavelet provides a performance gain of about 2 dB as compared to that of the one-level Haar wavelet case.

Figure 4.23 shows the CCDFs of the PAPR for different systems. It is clear that a one-level wavelet decomposition achieves the lowest possible PAPR for both the HW-DCT-SC-FDMA and the HW-DFT-SC-FDMA systems. As the number of decomposition levels increases, most of the signal energy is concentrated in a small number of samples representing the approximation coefficients of the signal, which increases the probability of existence of large amplitude samples.

From Figures 4.22 and 4.23, it is clear that there is a need for a trade-off between the PAPR performance and the BER performance. Increasing the number of decomposition levels leads to an enhancement in the BER performance and a degradation in the PAPR performance. We can say that the HW-DCT-SC-FDMA and HW-DFT-SC-FDMA systems with one-level wavelet decomposition are better to use because of their ability to provide good BER and PAPR performances with a low complexity. The results in this chapter are published in [65,66].

5

CARRIER FREQUENCY OFFSETS IN SC-FDMA SYSTEMS

5.1 Introduction

As explained in Chapters 3 and 4, in the DFT-SC-FDMA and DCT-SC-FDMA systems, each user employs a different set of orthogonal subcarriers as in the OFDMA system. For perfect time and frequency synchronization, the different orthogonal subcarrier sets for the different users make it possible to avoid the MAI [9,14]. However, similar to the OFDMA system, the DFT-SC-FDMA and the DCT-SC-FDMA systems also suffer from frequency mismatches between the transmitter and the receiver. As a result, in these systems, the CFOs disrupt the orthogonality between subcarriers and give rise to ICI and MAI among users because of the loss of subcarriers orthogonality [18,67]. Moreover, carrier frequency synchronization for the uplink systems is more difficult, since the frequency recovery for a certain user may result in the misalignment of the other synchronized users [67]. Thus, research on the suppression of the interference caused by several different CFOs is of great importance.

The problem of CFOs in multicarrier systems was extensively investigated in the literature [68–73]. There are two synchronization methods to reduce CFOs effect, namely feedback and compensation. In the first method, the estimated frequency offset values are fed back to the subscriber stations on a control channel so that they can adjust their transmission parameters [69]. However, the main disadvantage of this method is the bandwidth loss due to the need for a control channel. This method is not applicable in an environment involving a fast-changing Doppler shift. The second method compensates for the effect of the CFOs on the received signal to recover the ideal waveform, which can be received when the transmitter and the receiver are frequency synchronized. The single-user detector and the circular-convolution detector are two compensation-based synchronization schemes for the uplink OFDMA system [70]. In these detectors, error floors, which are mainly induced by the large MAI, appear. Thus, interference cancellation methods are suggested to mitigate the MAI [73]. In [73], a circular-convolution detector is used to estimate the MAI in the frequency domain. Then, PIC is used to regenerate and remove the MAI from the original signal.

For the uplink DFT-SC-FDMA system, there have been a few papers in the literature that address the issue of the CFOs problem [18,67,74–77]. In [18], the impact of the CFOs compensation on the performance of the MIMO DFT-SC-FDMA system was investigated. Frequency offset estimation for high-speed users in E-UTRA uplink was proposed and investigated in [67]. In [74], the authors presented an equalizer to mitigate the impact of the residual MAI after the CFOs compensation process in the DFT-SC-FDMA system. A joint suppression method for the phase noise and CFOs by block-type pilots for the MIMO DFT-SC-FDMA system was discussed in [75]. In [76], the derivation of the FDE model was performed taking into account the residual MAI after the CFOs compensation. An interference self-cancellation method for the DFT-SC-FDMA system was proposed and discussed in [77]. Up to now, to the best of our knowledge, no work has been reported on the joint equalization and CFOs compensation for the DFT-SC-FDMA and the DCT-SC-FDMA systems. This motivated us to do the research work in this chapter and to develop it for the MIMO DFT-SC-FDMA system in Chapter 6.

The main objective of this chapter is to deal with the CFOs problem and propose an efficient compensation scheme for the DFT-SC-FDMA and the DCT-SC-FDMA systems. First, we describe the baseband signal model for the DFT-SC-FDMA and the DCT-SC-FDMA systems with the general, interleaved, and localized subcarriers mapping techniques in the presence of CFOs, and also derive an expression for the user-by-user received SIR after the separation of users' data. Then, we present an MMSE scheme, which jointly performs equalization and CFOs compensation.

A low-complexity implementation of the MMSE scheme using a banded-matrix factorization is presented. Finally, a hybrid scheme comprising the MMSE scheme and PIC is also investigated for further enhancement of the performance of the systems with interleaved subcarriers mapping. As will be shown, these schemes are able to maintain a good performance even in the presence of estimation errors.

The banded-matrix approximation is also considered in [13,78]. The difference between the MMSE scheme and the scheme in [13] is that the scheme in [13] is performed for all users simultaneously, whereas the MMSE scheme is performed for each user separately. Moreover, the MMSE scheme is performed after the demapping process, whereas the scheme in [13] is performed before the demapping process. This makes the overall complexity of the MMSE scheme depend on the number of users and the type of the subcarriers mapping technique, whether localized or interleaved. As a result, a low complexity as compared to the scheme in [13] can be achieved, especially when the number of users is small, as will be seen in Section 5.4. On the other hand, the complexity of the scheme in [13] depends on neither the number of users nor the subcarriers mapping technique, since it is performed before the demapping process. Furthermore, the scheme in [13] requires a separate equalization process to compensate for the effect of the wireless channel, whereas the MMSE scheme jointly performs the equalization and the CFOs compensation.

In [78], a PIC is used to avoid the performance degradation in [13] for small banded-matrix bandwidths. The authors in [78] have shown that the computational complexity of the scheme in [13] can be minimized without any degradation in system performance for small banded-matrix bandwidths using PIC. For high banded-matrix bandwidths, the computational complexity of the scheme in [78] is higher than that of the scheme in [13], and thus applying the PIC is not an advantage. However, the scheme in [78] requires robust channel and CFOs estimation algorithms in order to avoid the error propagation problem due to the PIC algorithm [4]. For the comparison purpose, the circular-convolution detector [70,73], the single-user detector [70,73], and the scheme in [13] are simulated for the DFT-SC-FDMA and the DCT-SC-FDMA systems.

5.2 System Models in the Presence of CFOs

5.2.1 DFT-SC-FDMA System Model

We consider a DFT-SC-FDMA system with U users (terminals) communicating at the same time with a fixed base station through independent multipath Rayleigh-fading channels as shown in Figure 5.1. At the transmitter side, the system model is similar to that in Chapter 2.

At the receiver side, assuming perfect time synchronization, the received signal at the base station can be written as follows:

$$\tilde{r} = \sum_{u=1}^{U} \tilde{E}^u H^u \tilde{x}^u + \tilde{n} \qquad (5.1)$$

where

$\tilde{E}^u$ is an $(M+N_C) \times (M+N_C)$ diagonal matrix with elements $[\tilde{E}^u]_{m,m} = e^{j2\pi\varepsilon_u m/M}$, $m = 0,\ldots,$ $M+N_C-1$, which describe the CFO of the uth user. $\varepsilon_u = \Delta f T$ is the normalized CFO of the uth user. Δf is the carrier frequency offset

H^u is an $(M+N_C) \times (M+N_C)$ matrix describing the channel of the uth user

$\tilde{x}^u$ is an $(M+N_C) \times 1$ vector containing the transmitted samples of the uth user

$\tilde{n}$ is an $(M+N_C) \times 1$ vector containing the noise

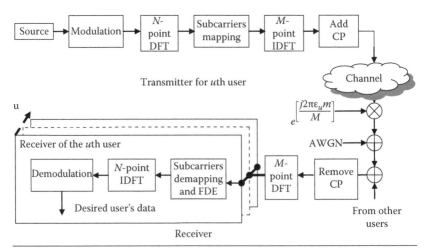

Figure 5.1   Structure of the uplink DFT-SC-FDMA system in the presence of CFOs.

After the removal of the CP, the received signal becomes

$$r = P_{\text{rem}}\tilde{r} = \sum_{u=1}^{U} E^u H_C^u \bar{x}^u + n \tag{5.2}$$

where

P_{rem} is an $M \times (M + N_C)$ matrix, which removes the CP

E^u is an $M \times M$ diagonal matrix, which describes the CFO of the uth user after the CP removal

$n = P_{\text{rem}}\tilde{n}$ and $\bar{x}^u = P_{\text{rem}}\tilde{x}^u$ are the noise and the transmitted signal after the CP removal, respectively

H_C^u is an $M \times M$ circulant matrix describing the channel of the uth user

After that, the received signal is transformed into the frequency domain via an M-point DFT as follows:

$$R = \sum_{u=1}^{U} \Pi_{\text{cir}}^u \Lambda^u \bar{X}^u + N \tag{5.3}$$

where

$\Pi_{\text{cir}}^u = F_M E^u F_M^{-1}$ is a circulant matrix representing the interference from the uth user

$\bar{x}^u = F_M \bar{x}^u$ is an $M \times 1$ vector representing the transmitted samples from the uth user after the mapping process

N is the DFT of n

After the demapping process, the equalization and the N-point DFT are applied. The estimated time-domain symbols of the kth user can be written as follows:

$$\hat{x}^k = A^k x^k + \bar{A}^k x^k + \sum_{\substack{u=1 \\ u \neq k}}^{U} B^u x^u + \hat{n} \tag{5.4}$$

The structures of all components of Equation 5.4 are given as follows:

$$A^k = \text{diag}(F_N^{-1} W^k \Pi_d^k \Lambda_d^k F_N) \tag{5.5}$$

$$\bar{A}^k = F_N^{-1} W^k \Pi_d^k \Lambda_d^k F_N - A^k \tag{5.6}$$

$$\hat{n} = F_N^{-1} W^k M_R^k N \qquad (5.7)$$

$$B^u = F_N^{-1} W^k \, \Pi_r^u \Lambda_d^u F_N \qquad (5.8)$$

where

$\Pi_d^k = M_R^k \, \Pi_{cir}^k M_T^k$ is the $N \times N$ interference matrix of the kth user

$\Pi_r^u = M_R^k \, \Pi_{cir}^u M_T^u$ is the $N \times N$ interference matrix from the uth user

$\Lambda_d^u = M_R^u \Lambda^u M_T^u$ is an $N \times N$ diagonal matrix representing the channel of the uth user

W^k is the $N \times N$ diagonal equalization matrix of the kth user

Finally, the detection and the decoding processes take place in the time domain. From Equation 5.4, it is clear that only the first term contains the desired data, the second term is due to the ISI, the third term is an MAI, and the fourth term is a noise. Based on Equation 5.4, the SIR in the nth symbol ($n = 1, \ldots, N$) in the kth user's data can be expressed as follows:

$$\mathrm{SIR}^k(n) = \frac{\left| A^k(n,n) \right|^2}{\displaystyle\sum_{\substack{i=1 \\ i \neq n}}^{N} \left| \bar{A}^k(n,i) \right|^2 + \sum_{\substack{u=1 \\ u \neq k}}^{U} \sum_{i=1}^{N} \left| B^u(n,i) \right|^2} \qquad (5.9)$$

We will consider a DFT-SC-FDMA system with $M = 128$, $N = 32$, and $U = 4$ to plot the SIR of the first symbol of the first user vs. the normalized CFO. Each frequency offset is a random variable with uniform distribution in the period $[-\varepsilon_{max}, \varepsilon_{max}]$, where ε_{max} is the maximum normalized CFO. We consider both the interleaved and the localized subcarriers mapping techniques. The SIR performances are shown in Figure 5.2. It is clear that the SIR performance for the DFT-IFDMA and the DFT-LFDMA systems is degraded with the increase in the CFO. Figure 5.3 shows the effect of interference caused by CFOs in the scatter diagram of a QPSK-modulated DFT-SC-FDMA system for different CFOs values, where the AWGN term is ignored. It clearly shows how the signal constellation can be affected by the CFOs causing an unacceptable BER.

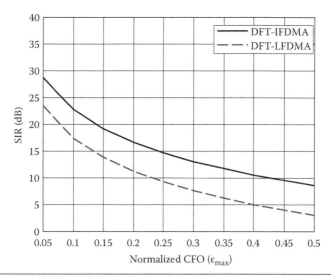

Figure 5.2 SIR vs. ε_{max} for the DFT-SC-FDMA system.

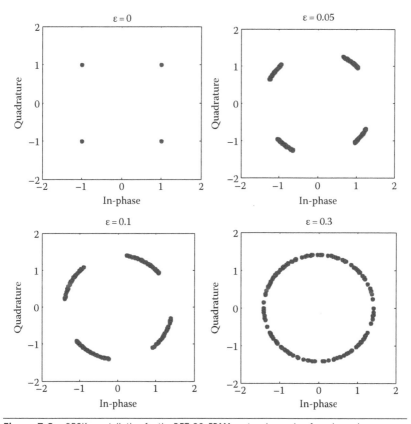

Figure 5.3 QPSK constellation for the DFT-SC-FDMA system in a noise-free channel.

5.2.2 DCT-SC-FDMA System Model

The DCT-SC-FDMA system model is described over a frequency-selective channel. At the transmitter side, the system model is similar to that in Chapter 4. At the receiver side, assuming perfect time synchronization, the received signal at the base station after removing the CP can be expressed as in Equation 5.2, where E^u is an $M \times M$ diagonal matrix with elements $[E^u]_{m,m} = e^{j\pi\zeta_u(2m+1)/2M}$, which describe the CFO matrix of the uth user for the DCT-SC-FDMA system. $\zeta_u = 2T\Delta f$ is the normalized CFO of the uth user with respect to the subcarriers frequency spacing. In a similar manner to Section 5.2.1, the estimated time-domain symbols of the kth user can be expressed as in Equation 5.4. In this case, the structures of all components of Equation 5.4 are modified as follows:

$$A^k = \mathrm{diag}(D_N^{-1} M_R^k D_M F_M^{-1} W^k \Pi_{\mathrm{cir}}^k \Lambda^k F_M D_M^{-1} M_T^k D_N) \quad (5.10)$$

$$\bar{A}^k = D_N^{-1} M_R^k D_M F_M^{-1} W^k \Pi_{\mathrm{cir}}^k \Lambda^k F_M D_M^{-1} M_T^k D_N - A^k \quad (5.11)$$

$$\hat{n} = D_N^{-1} M_R^k D_M F_M^{-1} W^k N \quad (5.12)$$

$$B^u = D_N^{-1} M_R^k D_M F_M^{-1} W^k \Pi_{\mathrm{cir}}^u \Lambda^u F_M D_M^{-1} M_T^u D_N \quad (5.13)$$

where W^k is the $M \times M$ FDE matrix of the kth user.

The following example is used to investigate the interference matrix for the DFT-SC-FDMA and the DCT-SC-FDMA systems. Assuming that the desired subcarrier index is $k = 39$ in 128 subcarriers. Interference coefficients of the DFT-SC-FDMA and the DCT-SC-FDMA systems in the presence of the CFOs are presented in Figures 5.4 and 5.5, respectively. To highlight the difference between these two sequences of the interference coefficients, we show the coefficients for the subcarrier index from 20 to 60. The remaining coefficients are sufficiently small to be ignored, and are not plotted in these figures.

Each CFO is a random variable with uniform distribution in $[-\varepsilon_{\max}, \varepsilon_{\max}]$ for the DFT-SC-FDMA system and in $[-\zeta_{\max}, \zeta_{\max}]$ for the DCT-SC-FDMA system, where $\zeta_{\max}$ and $\varepsilon_{\max}$ are the maximum allowed value of users' CFO.

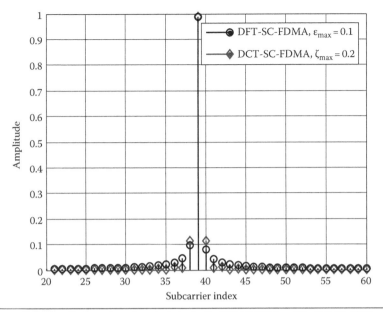

Figure 5.4   Amplitude vs. the subcarrier index for the DFT-SC-FDMA and the DCT-SC-FDMA systems when $\varepsilon_{max} = 0.1$ and $\zeta_{max} = 0.2$.

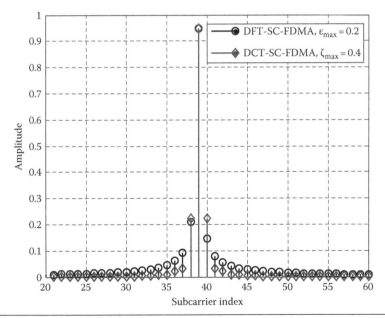

Figure 5.5   Amplitude vs. the subcarrier index for the DFT-SC-FDMA and the DCT-SC-FDMA systems when $\varepsilon_{max} = 0.2$ and $\zeta_{max} = 0.4$.

It is shown that the interference coefficients of the DCT-SC-FDMA system are more highly concentrated near subcarrier index 39 than the interference coefficients of the DFT-SC-FDMA system for the same value of $\Delta f\, T$, because the DCT operation distributes more energy to the desired subcarrier and less energy to the interference than the DFT operation. Therefore, the desired subcarrier suffers less interference coming from neighboring subcarriers in the DCT-SC-FDMA system than that in the DFT-SC-FDMA system. The derivations of the interference coefficients for the DFT-SC-FDMA and the DCT-SC-FDMA systems are in Appendices B and C, respectively.

5.3 Conventional CFOs Compensation Schemes

There are several techniques that were developed to compensate for the CFOs in multicarrier communication systems [68–73]. These techniques may be applicable in the DFT-SC-FDMA and the DCT-SC-FDMA systems. The single-user detector and the circular-convolution detector are the two common compensation-based synchronization schemes for multicarrier systems [70]. In the following subsections, the two detectors will be discussed for the uplink DFT-SC-FDMA system only. The extension to the DCT-SC-FDMA system is straightforward as shown in [79,82].

5.3.1 Single-User Detector

In the single-user detector, the compensation for the CFOs is carried out at the base station for each user, where the sampled sequence is compensated in the time domain, and then DFT-processed for each user. As a result, multiple DFT blocks for each user are used as shown in Figure 5.6. Mathematically, the received sequence in Equation 5.1 is multiplied by a time-domain sequence $\tilde{E}^{k^*}$ before the DFT processing as follows:

$$r' = \tilde{E}^{k^*}\tilde{r} = H^k\tilde{x}^k + \tilde{E}^{k^*}\sum_{\substack{u=1 \\ u \neq k}}^{U} \tilde{E}^u H^u \tilde{x}^u + n' \qquad (5.14)$$

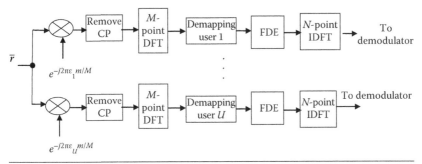

Figure 5.6   The single-user detector for the DFT-SC-FDMA system.

where

$n' = \tilde{E}^{k^*} \tilde{n}$ is the noise after the CFOs correction

$\tilde{E}^{k^*}$ is an $(M+N_C) \times (M+N_C)$ diagonal matrix with elements $[\tilde{E}^{k^*}]_{m,m} = e^{-j2\pi\varepsilon_k m/M}$

Note that in Equation 5.14, the single-user detector completely removes the ICI. However, applying the single-user detector before the DFT tends to cause some performance degradations. This is explained by the fact that the frequency compensation for a certain user before the DFT can increase the frequency offsets in the data of the other users within the DFT block [70,73]. In addition, in the single-user detector, a DFT block is required for each user to detect the information symbols, which leads to an increase in the system complexity.

5.3.2 Circular-Convolution Detector

To reduce the required number of DFT blocks, the compensation for the CFOs can be carried out in the frequency domain as shown in Figure 5.7 [70,73]. Based on the system model in Section 5.2.1,

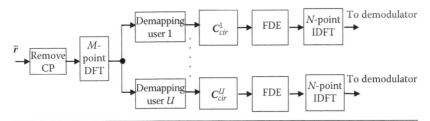

Figure 5.7   The circular-convolution detector for the DFT-SC-FDMA system.

the received signal after the demapping process can be written as follows:

$$R_d^k = \Pi_d^k \Lambda_d^k X^k + \sum_{\substack{u=1 \\ u \neq k}}^{U} \Pi_r^u \Lambda_d^u X^u + N_d \tag{5.15}$$

where $N_d = M_R^k N$ is an $N \times 1$ vector representing the noise after the demapping process. After the circular-convolution detector, the frequency-domain samples of the kth user can be written as follows:

$$\bar{R}^k = W^k C_{cir}^k R_d^k \tag{5.16}$$

where $C_{cir}^k = M_R^k F_M E^{k^*} F_M^{-1} M_T^k$ is an $N \times N$ circular matrix. Finally, an N-point IDFT, demodulation, and decoding processes are performed. In [70,73], it was shown that the circular-convolution detector provides a better performance than the single-user detector with lower complexity. However, error floors, which are mainly induced by the large MAI, also appear. Thus, an interference cancellation scheme was suggested in [73] to mitigate the MAI.

5.4 MMSE Scheme

To avoid the problems associated with the single-user and the circular-convolution detectors such as the residual MAI and the complexity, we present an MMSE equalization scheme for the DFT-SC-FDMA and the DCT-SC-FDMA systems, which can achieve equalization and CFOs compensation simultaneously. It is performed in the frequency domain, and it requires only a single DFT stage for all users. Moreover, the complexity of the MMSE scheme is further reduced using a banded-matrix implementation. The MMSE scheme is derived taking into account both the noise and the MAI. In this section, the MMSE scheme is derived only for the DFT-SC-FDMA system. The extension to the DCT-SC-FDMA system is straightforward as shown in [79].

5.4.1 Mathematical Model

A block diagram of the MMSE scheme is depicted in Figure 5.8. To derive the MMSE matrix of this scheme, Equation 5.15 must be rearranged as follows:

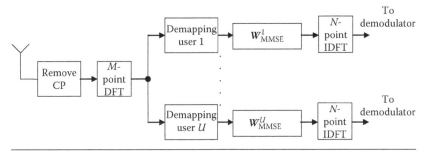

Figure 5.8 The MMSE equalization scheme for the DFT-SC-FDMA system.

$$R_d^k = \Pi_p^k X^k + N_P \tag{5.17}$$

where

$\Pi_p^k = \Pi_d^k \Lambda_d^k$ is the $N \times N$ interference matrix of the kth user

$N_P = \displaystyle\sum_{u=1,u\neq k}^{U} \Pi_R^u X^u + N_d$ is the MAI-plus-noise matrix, where

$\Pi_R^u = \Pi_r^u \Lambda_d^u$ is the $N \times N$ interference matrix from the uth user

Then, we define the error e between the estimated symbols $\hat{X}^k = W_{\text{MMSE}}^k R_d^k$ and the transmitted symbols X^k as follows:

$$e^k = W_{\text{MMSE}}^k R_d^k - X^k \tag{5.18}$$

The equalization matrix of the kth user is determined by the minimization of the following MSE cost function:

$$J^k = E\left\{ \left\| e^k \right\|^2 \right\} = E\left\{ \left\| W_{\text{MMSE}}^k R_d^k - X^k \right\|^2 \right\} \tag{5.19}$$

where $E\{\}$ is the expectation. Solving $\partial J^k / \partial W_{\text{MMSE}}^k = 0$, we obtain that

$$W_{\text{MMSE}}^k = R_X^k \Pi_p^{k^H} (\Pi_p^k R_X^k \Pi_p^{k^H} + R_{N_P}^k)^{-1} \tag{5.20}$$

where R_X^k and $R_{N_P}^k$ are the data and the overall noise (MAI plus noise) covariance matrices of the kth user. We assume that the noise is AWGN with zero mean and covariance σ_n^2, such that $E[N_d N_d^H] = \sigma_n^2 I_N$. Then $R_{N_P}^k$ can be obtained as follows:

$$
R^k_{Np} = E\left\{\left(\sum_{\substack{u=1 \\ u \ne k}}^{U} \Pi^u_R X^u + N_d\right)\left(\sum_{\substack{u=1 \\ u \ne k}}^{U} \Pi^u_R X^u + N_d\right)^H\right\}
$$

$$
= \sum_{\substack{u=1 \\ u \ne k}}^{U} \Pi^u_R R^u_X \Pi^{uH}_R + \sigma^2_n I_N \tag{5.21}
$$

If all users have the same power and the average powers of the received signals on all subcarriers are the same and denoted as σ^2_X, we can have $R^k_X = \sigma^2_X I_N$. Thus, Equation 5.20 can be simplified as follows:

$$
W^k_{\text{MMSE}} = \Pi^{kH}_P\left(\Pi^k_P \Pi^{kH}_P + \sum_{\substack{u=1 \\ u \ne k}}^{U} \Pi^u_R \Pi^{uH}_R + (1/\text{SNR})I_N\right)^{-1}
$$

$$
= \left(\Pi^{kH}_P \Pi^k_P + \sum_{\substack{u=1 \\ u \ne k}}^{U} \Pi^{uH}_R \Pi^{uH}_R + (1/\text{SNR})I_N\right)^{-1} \Pi^{kH}_P \tag{5.22}
$$

where $\text{SNR} = (\sigma^2_X / \sigma^2_n)$. The estimated time-domain symbols of the kth user can be written as follows:

$$
\hat{x}^k = F^{-1}_N W^k_{\text{MMSE}} R^k_d \tag{5.23}
$$

The main advantage of the MMSE scheme is that it minimizes the MAI and the noise. Thus, the residual MAI is lower than that in the single-user and the circular-convolution detectors as we will see in the simulation results. In addition, it requires a single DFT stage for all users, similar to the circular-convolution detector, since it is performed in the frequency domain.

5.4.2 Banded-System Implementation

The MMSE scheme is able to remove the interference for each user. However, calculating and inverting the $N \times N$ matrix is practically difficult for a large DFT size. In the MMSE scheme, the complexity can be reduced, since most of the elements in $\boldsymbol{\Pi}_p^k$ are zeros. Figure 5.9 gives plots for the amplitude of the first row of $\boldsymbol{\Pi}_p^k$ with the DFT-LFDMA and the DFT-IFDMA systems. The subcarrier with index 0 is the desired subcarrier, whereas the other subcarriers represent the interference. We consider a system with $N=32$, $M=128$, and $U=4$. The frequency offset is a random variable with uniform distribution in $[-0.15, 0.15]$.

As shown in Figure 5.9, the amplitude of the interference caused by any subcarrier on the subcarrier 0 decreases as the distance between these subcarriers increases. Thus, a threshold r for the number of subcarriers, beyond which the interference is neglected, can be introduced as a design parameter. From Figure 5.9, we can see that $r=10$ for the DFT-LFDMA system and $r=5$ for the DFT-IFDMA system are the best choices to give a good performance in the MMSE scheme with an acceptable complexity. Thus, these values will be used to study the complexity of the MMSE scheme in Figures 5.10 and 5.11. After this approximation, $\boldsymbol{\Pi}_p^k$ can be written as follows:

$$\boldsymbol{\Pi}_p^k = \boldsymbol{\Pi}_{\text{app}}^k + \boldsymbol{\Pi}_{\text{rem}}^k \tag{5.24}$$

where

$\boldsymbol{\Pi}_{\text{rem}}^k$ is the remaining interference matrix
$\boldsymbol{\Pi}_{\text{app}}^k$ is the approximate interference matrix and can be written as follows:

$$\boldsymbol{\Pi}_{\text{app}}^k(n,n') = \begin{cases} \boldsymbol{\Pi}_p^k(n,n') & \text{if} \quad |n'-n| \leq r \\ \\ 0 & \text{if} \quad |n'-n| > r \end{cases} \tag{5.25}$$

We can rewrite Equation 5.17 as follows:

$$R_d^k = \boldsymbol{\Pi}_{\text{app}}^k X^k + \breve{N} \tag{5.26}$$

where

$$\breve{N} = \boldsymbol{\Pi}_{\text{rem}}^k \Lambda_d^k X^k + N_p \tag{5.27}$$

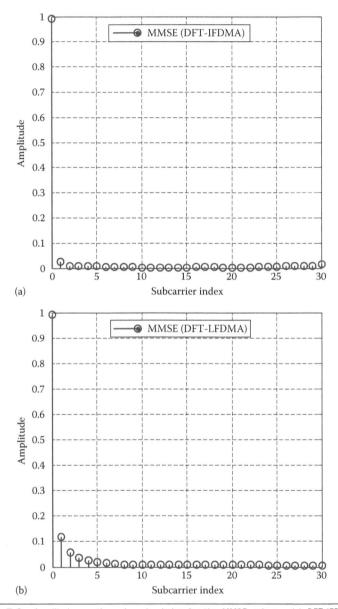

Figure 5.9 Amplitude vs. the subcarrier index for the MMSE scheme: (a) DFT-IFDMA and (b) DFT-LFDMA.

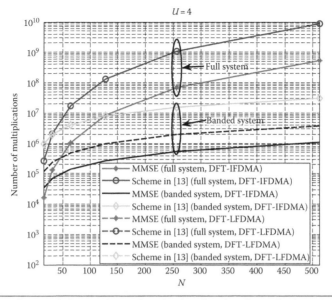

Figure 5.10 Complexity vs. the number of subcarriers for all equalization schemes and implementations.

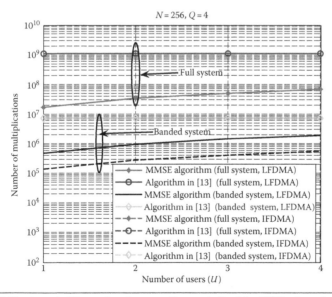

Figure 5.11 Impact of the number of users on the complexity of the MMSE scheme for the DFT-SC-FDMA system with different bandwidths of the banded matrix.

The MMSE solution of Equation 5.26 can be written by replacing $\boldsymbol{\Pi}_p^k$ in Equation 5.22 by $\boldsymbol{\Pi}_{app}^k$.

Based on Equation 5.25, $\boldsymbol{\Pi}_{app}^k$ has also nonzero entries near its diagonal. This kind of matrix is called a banded matrix with bandwidth r [80]. Since $\boldsymbol{\Pi}_{app}^k$ is a banded matrix, $\boldsymbol{\Pi}_{app}^{k^H}\boldsymbol{\Pi}_{app}^k$ is also a banded matrix and its bandwidth is $2r$. In this case, the MMSE scheme can be implemented using a banded-matrix factorization [80]. More details about the banded-matrix algorithm are found in [13,80]. The steps of the banded-matrix implementation of the equalization will be stated in the next section.

5.4.3 Complexity Evaluation

For the MMSE scheme, the inversion of an $N \times N$ matrix for each user is required, which is practically difficult for a large N. The full-system implementation requires a complexity of $O(N^3)$, which is large for a system with a large number of active subcarriers. However, the structure of the approximated interference matrix $\boldsymbol{\Pi}_{app}^k$ is a banded structure. Thus, the banded-matrix implementation is considered to reduce the complexity of the MMSE scheme. The banded-matrix implementation and complexity evaluation are summarized as follows [13,80]:

1. Multiplication of the two banded matrices, $\boldsymbol{\Pi}_{app}^k$ and $\boldsymbol{\Pi}_{app}^{k^H}$. This step requires approximately $2N(2r^2 + 1)$ operations.
2. Lower and upper factorization and forward and feedback substitution. Since $\boldsymbol{\Pi}_{app}^k\ \boldsymbol{\Pi}_{app}^{k^H}$ has a bandwidth of $2r$, this step requires $N(8r^2 + 10r + 1)$ operations.
3. Banded matrix and vector production. This step requires about $2N(4r + 1)$.

The total number of operations required in our banded-matrix implementation for all users is approximately $NU[16r^2 + 26r + 5]$. Hence, for large values of N, the overall complexity is lower than that of the full-system implementation. On the other hand, the total number of operations required in the banded-matrix implementation in [13] for the uplink DFT-SC-FDMA system is $M[16r^2 + 26r + 5]$, where $M = QN$ and $U \leq Q$. The full-system implementation in [13] requires a complexity of $O(M^3)$, which is too large for a system with a large M. The authors in [13] showed that $r = 30$ is the best choice for their scheme. Thus, $r = 30$ will be used for the scheme in [13] in Figures 5.10 and 5.11.

Figure 5.10 gives a comparison between the MMSE scheme and the scheme in [13] for different values of N, when both schemes are applied for the uplink DFT-SC-FDMA system. $U = 4$ and $Q = 4$ are considered. It is clear that the complexity of the MMSE scheme is lower than that of the scheme in [13], especially for the DFT-IFDMA system. It is also clear that the complexity of the banded-matrix implementation of the MMSE scheme is lower than that of the full-system implementation, especially when the value of N is large. At low values of N, the complexity of the full-system implementation of the MMSE scheme is the lowest of all schemes.

The impact of the number of users on the complexity of the MMSE scheme is also illustrated in Figure 5.11. $Q = 4$ and $N = 256$ are considered. It is clear that the complexity of the scheme in [13] is independent of the number of users, whereas the complexity of the MMSE scheme depends on the number of users that are communicating at the same time. The complexity of the MMSE scheme decreases, when the number of users is low. Even, for a full-load case when $U = Q$, the complexity of the MMSE scheme is still lower than that of the scheme in [13].

Figures 5.10 and 5.11 indicate that the complexity of the MMSE scheme is lower than that of the scheme in [13], especially for the DFT-IFDMA system, when the number of users is low. Moreover, for a real implementation, the MMSE scheme is faster than the scheme in [13], since the compensation process is performed for each user separately, and hence the compensation processes of all users can be performed in parallel.

In the DCT-SC-FDMA system, the equalization process is performed after the DFT immediately [79], i.e., before the demapping process. Thus, the total number of operations required in the banded-matrix implementation for the DCT-SC-FDMA system for all users is approximately $MU[16r^2 + 26r + 5]$.

5.5 MMSE+PIC Scheme

It is known that the interleaved subcarriers mapping systems are more sensitive to CFOs than the localized subcarriers mapping systems.

So, at high CFOs and SNR values, the residual MAI after the MMSE scheme may degrade the BER performance of the interleaved subcarriers mapping systems. This is a motivation to look for a more sophisticated interference cancellation scheme so that the MAI can be cancelled out.

5.5.1 Mathematical Model

We can implement a linear combination of the MMSE scheme and a PIC scheme to further reduce the effect of the residual MAI for the interleaved system, as shown in Figure 5.12. In the MMSE+PIC scheme, the MMSE scheme is used to estimate the MAI, which is then regenerated and removed from the original received signal using the PIC in the frequency domain.

The steps of the MMSE+PIC scheme can be summarized as follows:

1. The CP is removed from the received signal.
2. Then, the DFT is applied to the received signal.
3. After that, the MMSE scheme is applied after the subcarriers demapping process to estimate the samples for each user, as we discussed in Section 5.4.

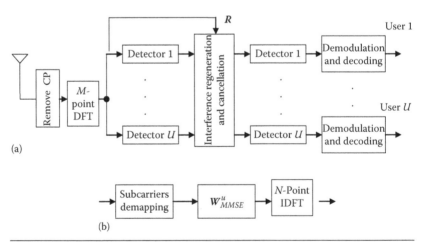

Figure 5.12 Block diagram of the MMSE+PIC scheme. (a) Structure of the MMSE receiver scheme and (b) structure of the detector.

4. The frequency-domain estimates of the interfering users' samples are then transformed into time-domain symbols, and the decision function is applied as follows:

$$\hat{x}^k = f_{\text{dec}}(F_N^{-1}\hat{X}^k) \tag{5.28}$$

where f_{dec} is the decision function. We adopt a hard-decision function.

5. The MAI is regenerated in the frequency domain as follows:

$$R_{\text{MAI}}^k = \sum_{\substack{u=1 \\ u \neq k}}^{U} \Pi_{\text{cir}}^u \Lambda^u M_T^u F_N \hat{x}^u \tag{5.29}$$

6. The MAI is subtracted from the received signal R to get the frequency-domain interference-free signal as follows:

$$R_{\text{free}}^k = R - R_{\text{MAI}}^k \tag{5.30}$$

7. A better estimate of the frequency-domain samples can be obtained by applying the demapping process and the MMSE scheme on the interference-free signal R_{free}^k as follows:

$$\hat{X}^u = W_{\text{MMSE}}^k M_R^k R_{\text{free}}^k \tag{5.31}$$

8. Finally, the DFT, demodulation, and decoding processes are applied to provide a better estimate of the desired data.

The main advantage of the MMSE+PIC scheme is its better BER performance, even at high CFOs. However, the complexity of the receiver, which is the base station, will be increased.

5.6 Simulation Examples

Experiments have been carried out to study the effectiveness of the different schemes in the uplink DFT-SC-FDMA and DCT-SC-FDMA systems. The BER has been evaluated using the Monte Carlo simulation method.

5.6.1 Simulation Parameters

Uplink DFT-SC-FDMA and DCT-SC-FDMA systems with 512 subcarriers have been considered. In these systems, there are four users with 128 subcarriers allocated to each user. The users employ QPSK mapping for their data symbols. The channel model used for simulations is the vehicular-A model [43]. A convolutional code with memory length 7 and octal generator polynomial (133,171) has been chosen as the channel code. The simulation parameters are tabulated in Table 5.1.

5.6.2 Impact of the CFOs

Figures 5.13 and 5.14 present the variation of the BER with the maximum normalized CFO for the DFT-SC-FDMA, the DCT-SC-FDMA, and the OFDMA systems at different values of the SNR. To further demonstrate the feasibility of the simulation results, independent CFOs have been assumed for all uplink users. These figures show that the performance of all systems with different subcarriers mapping techniques deteriorates when the CFOs increase, especially at high SNR values, because at high SNR values, the interference is the dominant. From these figures, it is clear that for small to moderate CFOs, the performance of the DCT-SC-FDMA system is better than that of the DFT-SC-FDMA and the OFDMA systems. It is also noted that the sensitivities of the DFT-SC-FDMA and the DCT-SC-FDMA

Table 5.1   Simulation Parameters

Simulation Method	Monte Carlo
System bandwidth	5 MHz
Modulation type	QPSK
CP length	20 samples
M	512
N	128
Channel coding	Convolutional code with rate $= 1/2$
Subcarriers mapping technique	Interleaved and localized
CFOs	Constant and random
U	4
Channel model	Vehicular-A channel
CFOs estimation	Perfect
Channel estimation	Perfect

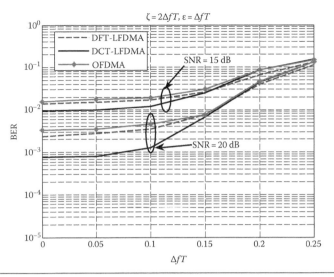

Figure 5.13 BER vs. ΔfT for the localized systems.

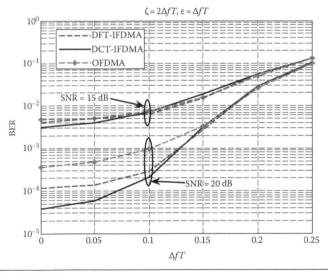

Figure 5.14 BER vs. ΔfT for the interleaved systems.

systems to the CFO are lower than that of the OFDMA system. At high CFOs, the performance of all systems begins to deteriorate.

Figure 5.15 shows a comparison between the SIR of the DFT-IFDMA, the DCT-IFDMA, and the OFDMA systems. One can see that the superiority of the DCT-IFDMA system appears when $\Delta fT < 0.2$. The reason is that the energy-compaction property of

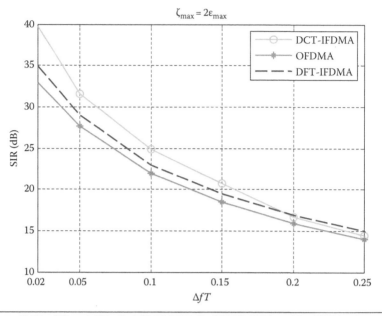

Figure 5.15   SIR vs. ΔfT for the DFT-SC-FDMA, the DCT-SC-FDMA, and the OFDMA systems.

the DCT is feasible for small CFOs values. On the other hand, the desired subcarrier suffers less interference coming from the neighboring subcarriers in the DCT-IFDMA system than in the DFT-IFDMA and the OFDMA systems. The reduced interference leads to a better BER performance, which can be observed from the rest of experiments. This figure also shows that the SIR of the DFT-IFDMA system is higher than that of the OFDMA system.

5.6.3 Results of the MMSE Scheme

5.6.3.1 DFT-SC-FDMA System

The MMSE scheme has been tested for the DFT-SC-FDMA system with constant CFOs. Simulation results in Figure 5.16 are introduced to highlight the effect of the matrix bandwidth on the performance of the MMSE scheme with different subcarriers mapping techniques. In Figure 5.16, we show the performance of the full-system implementation as well as the banded-matrix implementation with $r = 1$, 5, and 10. The values of the CFOs are $\varepsilon_1 = 0.3$, $\varepsilon_2 = -0.1$, $\varepsilon_3 = -0.05$, and $\varepsilon_4 = 0.05$. Comparing the performance of the full-system implementation and the banded-matrix implementation with different bandwidths, we find that

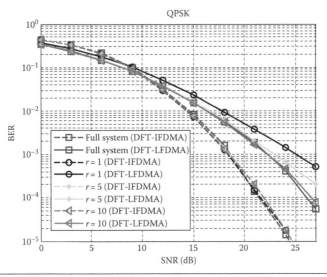

Figure 5.16   BER vs. SNR for the DFT-SC-FDMA system with the MMSE scheme and different banded-matrix bandwidths.

even with $r=1$, the banded-matrix implementation can provide a performance close to that of the full-system implementation, especially for the DFT-IFDMA system. For the DFT-IFDMA system, the banded matrix with $r=5$ is suitable, since it maintains a performance comparable to that of the full-system implementation and can save a significant amount of computational power. For the DFT-LFDMA system, the banded matrix with $r=10$ achieves a trade-off between the required low computational complexity and high performance.

The BER vs. SNR for all schemes is shown in Figure 5.17. For comparison purposes, the results with the circular-convolution detector, with the single-user detector, and with the scheme in [13] are presented. The full-system implementation is used for both the MMSE scheme and the scheme in [13]. It is clear that the MMSE scheme improves the BER performance of the DFT-SC-FDMA system in the presence of CFOs significantly. It outperforms the circular-convolution detector, and the single-user detector, and its performance is nearly the same as that of the scheme in [13]. It can be also seen that the MMSE scheme for the DFT-LFDMA system provides the same BER performance as that without CFOs, while for the DFT-IFDMA system it suffers 1.5 dB loss in the BER performance as compared

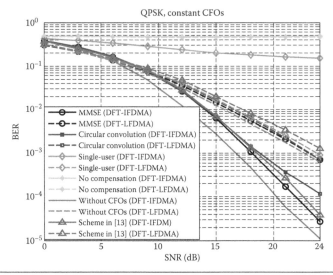

Figure 5.17   BER vs. SNR for the DFT-SC-FDMA system for all schemes with constant CFO.

to that without CFOs. Figure 5.17 also shows that the performance of the DFT-IFDMA system with a single-user detector is the worst of all schemes. This performance degradation is caused by the fact that a CFOs compensation for a user before the DFT increases the frequency offsets in the data of the other users within the DFT block.

To further demonstrate the feasibility of the MMSE scheme, independent CFOs are assumed for all uplink users. Each CFO is a random variable with uniform distribution in [−0.15, 0.15]. The maximum normalized CFO is chosen similar to that in [81]. The CFOs are chosen randomly to simulate a more practical scenario.

The results with the circular-convolution detector, with the single-user detector, with the scheme in [13], and with the MMSE scheme are presented in Figure 5.18. It is observed that all schemes eliminate the effect of the CFOs for the DFT-LFDMA system. For the DFT-IFDMA system, it can be seen that the MMSE scheme is effective to mitigate the CFOs and its performance is the best of all schemes, especially at high SNR values. For the single-user detector and the circular-convolution detector, error floors, which are mainly induced by the large MAI, appear at high SNR values.

5.6.3.2 DCT-SC-FDMA System   The MMSE scheme has been tested for the DCT-SC-FDMA system. Figure 5.19 shows the

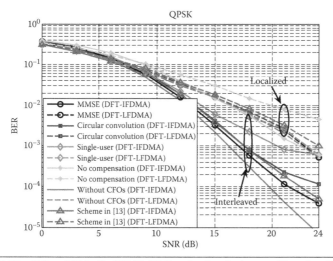

Figure 5.18   BER vs. SNR for the DFT-SC-FDMA system for all schemes with random CFOs.

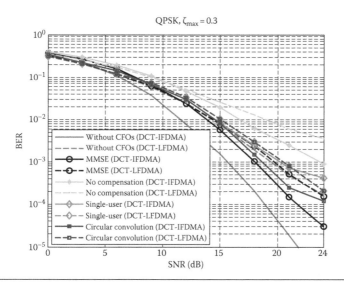

Figure 5.19   BER vs. SNR for the DCT-SC-FDMA system for all schemes with random CFOs.

BER performance comparison between the conventional CFOs compensation schemes and the MMSE scheme for the DCT-SC-FDMA system. The DCT-SC-FDMA system without CFOs and the DCT-SC-FDMA system without CFOs compensation have also been studied for comparison. It is clear that the MMSE scheme significantly outperforms the conventional schemes, especially at high SNR values. Although the performance of the MMSE scheme for the DCT-IFDMA system is superior to all conventional schemes, the performance loss is about 2.5 dB at a BER = 10^{-3}. Adding a PIC stage to the MMSE scheme can avoid this loss and provide a better BER performance, especially at high CFOs values.

5.6.4 Results of the MMSE+PIC Scheme

Based on the results obtained in Figures 5.18 and 5.19, since the MMSE scheme is able to eliminate the effects of the CFOs for the localized systems, the interleaved systems are only considered in Figures 5.20 through 5.23.

5.6.4.1 DFT-SC-FDMA System   Figure 5.20 shows the BER performance of the MMSE and MMSE+PIC schemes for the DFT-IFDMA system. The results without CFOs, and without CFOs compensation, are also presented. $\varepsilon_{max} = 0.15$ has been assumed. It is clear that the MMSE+PIC scheme mitigates the MAI and provides a better performance, even at high SNR values.

In Figure 5.21, the impact of the CFOs on the performance of the DFT-IFDMA system with different receiver schemes is studied at

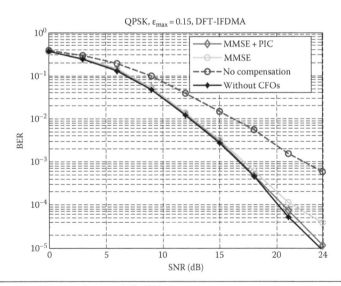

Figure 5.20   BER vs. SNR for the DFT-IFDMA system.

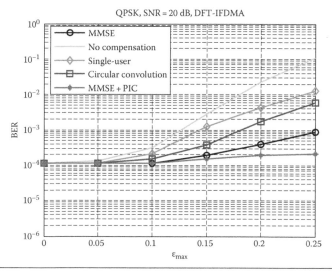

Figure 5.21 BER vs. ε_{max} for the DFT-IFDMA system.

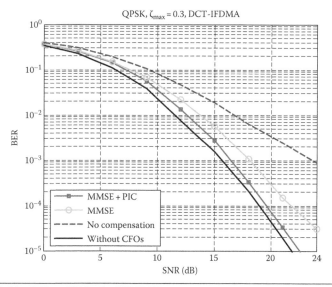

Figure 5.22 BER vs. SNR for the DCT-IFDMA system.

SNR = 20 dB. It is clear that as ε_{max} increases, the effectiveness of all receiver schemes except the MMSE+PIC scheme decreases because of the increased amount of MAI.

It is also clear that the MMSE and MMSE+PIC schemes provide the best performance as compared to the other schemes, especially at

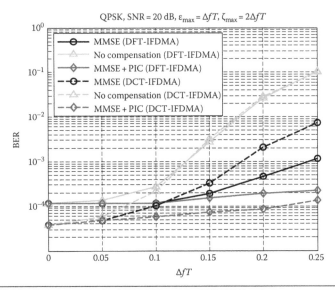

Figure 5.23 BER vs. ΔfT for the DFT-IFDMA and the DCT-IFDMA systems.

high CFOs values, because the residual MAI is low, since they take into account both the noise and the MAI. It is also noted that the MMSE+PIC scheme is able to mitigate the residual MAI for the DFT-IFDMA system, even at high ε_{max}, because the MMSE+PIC scheme removes the MAI using the PIC.

5.6.4.2 DCT-SC-FDMA System Figure 5.22 shows the BER performance of the MMSE and MMSE+PIC schemes for the DCT-IFDMA system. From this figure, it is clear that the PIC can avoid the MAI and provide a better BER performance than the MMSE scheme and its performance is close to the system without CFOs. This is attributed to the ability of the MMSE+PIC scheme to effectively eliminate the MAI. It is observed from Figure 5.22 that the performance loss due to the MMSE+PIC scheme is 0.5 dB at a BER = 10^{-3} when compared to the system without CFOs, which is acceptable.

Figure 5.23 illustrates the BER performance vs. the ΔfT for the DFT-IFDMA and the DCT-IFDMA systems. It can be seen that the performance of the MMSE+PIC scheme is always better than the MMSE scheme, especially at high CFOs values. On the other hand, as ΔfT increases, the effectiveness of the MMSE scheme decreases because of the increased amount of MAI. It is also noted that the

MMSE+PIC scheme is able to mitigate the residual MAI, even at high CFOs values, because it removes the MAI using the PIC stage.

5.6.5 Impact of Estimation Errors

5.6.5.1 DFT-SC-FDMA System The impacts of the CFOs estimation errors and channel estimation errors on the performance of the discussed schemes for the DFT-SC-FDMA system have been investigated and shown in Figures 5.24 and 5.25. The estimated CFOs are obtained by adding a zero-mean independent Gaussian random variable to the true values of the CFOs. The estimated channels are obtained in the same manner. The CFOs of all users have been chosen randomly in the interval [−0.15, 0.15].

Figure 5.24 shows that the BER performance of the MMSE scheme for the DFT-LFDMA system starts to degrade as the standard deviation of the CFOs estimation error δ_{CFO} becomes larger than 0.05 and the standard deviation of the channel coefficients estimation error δ_{Ch} becomes larger than 0.01.

Figure 5.25 shows the performance of the MMSE and MMSE+PIC schemes with the estimation errors for the DFT-IFDMA system.

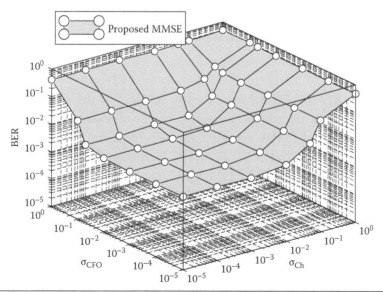

Figure 5.24 BER performance of the DFT-LFDMA system with the MMSE scheme in the presence of estimation errors at SNR = 20 dB.

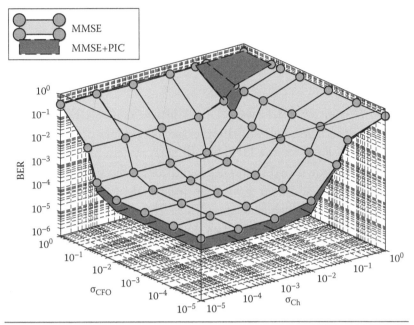

Figure 5.25   BER performance of the DFT-IFDMA system with the MMSE and MMSE+PIC schemes in the presence of estimation errors at SNR = 24 dB.

It can be seen that the performance of both schemes starts to degrade as δ_{CFO} becomes larger than 0.01 and δ_{Ch} becomes larger than 0.01. These limits can be satisfied with the estimation algorithms, and this assures that the discussed compensation schemes are robust to the estimation errors.

5.6.5.2 DCT-SC-FDMA System   The impacts of the channel estimation errors on the performance of the discussed schemes for the DCT-SC-FDMA system have also been studied and shown in Figures 5.26 and 5.27. Figure 5.26 shows that the BER performance of the DCT-LFDMA system with the MMSE scheme starts to degrade as δ_{CFO} becomes larger than 0.05 and δ_{Ch} becomes larger than 0.01. Figure 5.27 shows that the performance of the DCT-IFDMA system with the MMSE and MMSE+PIC schemes starts to degrade as δ_{CFO} becomes larger than 0.05 and δ_{Ch} becomes larger than 0.01. This shows that the discussed compensation schemes are also robust to the estimation errors for the DCT-SC-FDMA system.

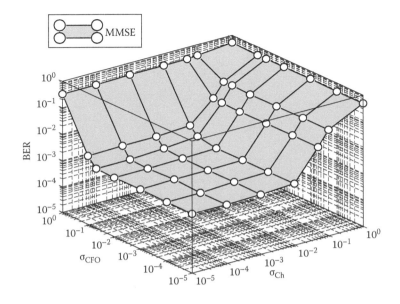

Figure 5.26 BER performance of the DCT-LFDMA system with the MMSE scheme in the presence of estimation errors at SNR = 20 dB.

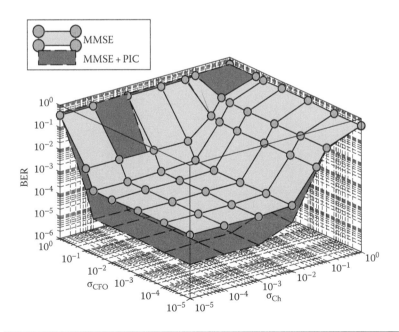

Figure 5.27 BER performance of the DCT-IFDMA system with the MMSE and MMSE+PIC schemes in the presence of estimation errors at SNR = 20 dB.

EQUALIZATION AND CFOS COMPENSATION FOR MIMO SC-FDMA SYSTEMS

6.1 Introduction

In the previous chapter, the problem of CFOs for SISO DFT-SC-FDMA and SISO DCT-SC-FDMA systems was investigated and treated. In this chapter, more complicated systems are considered. We investigate the equalization and CFOs compensation for MIMO DFT-SC-FDMA and MIMO DCT-SC-FDMA systems. MIMO systems are very attractive in wireless communications because the MIMO technique increases spectral efficiency and capacity of wireless systems without a need for a large spectrum [83]. MIMO systems can be categorized into spatial multiplexing MIMO systems and diversity-based MIMO systems [84,85]. Spatial multiplexing MIMO systems transmit different data streams from multiple transmit antennas to increase the data rate. On the other hand, the basic idea of diversity-based MIMO systems is to receive or transmit the same information-bearing signals redundantly to reduce the probability of deep fading.

There are several well-known equalizers that can be used to recover the transmitted signals in MIMO systems such as the maximum likelihood (ML) equalizer, the linear ZF equalizer, and the MMSE equalizer. Even though the ML equalizer is optimal from the minimum symbol error rate perspective, it has a prohibitive computational cost [86]. The computational complexity of the ML equalizer increases exponentially with the number of transmit antennas and the modulation order [86], and hence it is impractical for real applications. On the other hand, the complexities of the ZF and the MMSE equalizers are far lower than that of the ML equalizer, but they lead to substantial performance degradations [87,88]. The main advantage of the ZF equalizer over the MMSE equalizer is that the statistics of the additive noise and the transmitted data are not required in its implementation but it causes noise enhancement. However, the complexities of the ZF and the MMSE equalizers are still high due to the superposition of all of the transmitted streams at each receive antenna.

Recently, the MIMO systems have been already implemented in evolving standards such as IEEE802.16 and LTE [1,17]. Furthermore, the 3GPP standard includes MIMO systems in the standardization procedure of the LTE-advanced [17]. Two main research trends have been adopted to study the DFT-SC-FDMA system with MIMO techniques. The first trend focuses on solving the PAPR reduction problem [60,89,90], and the second one focuses on enhancing the BER performance [91,92]. Among the various MIMO techniques, Space-Frequency Block Coding (SFBC) is a promising technique for implementation in the DFT-SC-FDMA system [89]. However, the use of the SFBC increases the PAPR because of the spatial processing [90]. The complexity of the equalization in these systems increases due to the superposition of all of the transmitted streams at each receive antenna [3,93–95], especially in the presence of CFOs. Moreover, the performance of these systems substantially deteriorates due to the CFOs. So, there is a need to enhance the performance of MIMO systems and to reduce the complexity of the equalization process required in MIMO systems, which are the main objectives of this chapter.

In this chapter, flexible mathematical multiuser MIMO DFT-SC-FDMA and MIMO DCT-SC-FDMA baseband system models in the absence of CFOs are defined. This gives us a tool to study the two different subcarrier assignment strategies and different MIMO system setups. Also, an efficient LRZF for the uplink MIMO DFT-SC-FDMA and MIMO DCT-SC-FDMA systems in the absence of CFOs is discussed. The presented equalization scheme simplifies the matrix inversion process by performing it in two steps. In the first step, the IAI is cancelled. Then, the ISI is mitigated in the second step. A regularization term is added in the second step of the matrix inversion to avoid the noise enhancement. It is known that the ZF equalizer is in fact an inverse filter for the channel. For low-pass channel characteristics, the ZF equalizer is equivalent to a high-pass filter, which amplifies the noise at high frequencies. The signal components at high frequencies are much weaker than noise. So, the amplification of high-frequency components means amplification of noise much more than the high-frequency signal components in the so-called noise enhancement phenomenon. In the discussed scheme, regularization changes the nature of the inverse filter from a high-pass nature to a band-pass nature, rejecting the high-frequency noise. So, the noise enhancement is avoided leading to a BER reduction.

Then, the chapter extends the baseband system model toward user-specific CFOs and also derives a user-by-user received SIR. Two joint equalization and CFOs compensation schemes are presented for the uplink MIMO DFT-SC-FDMA system in the presence of CFOs. The LRZF scheme is developed for the MIMO DFT-SC-FDMA system in the presence of CFOs. It is derived to jointly perform the equalization and the CFOs compensation processes. It is now called joint LRZF (JLRZF) equalization. The MMSE equalization in Chapter 5 is also developed for the MIMO DFT-SC-FDMA system in the presence of CFOs. It is now called joint MMSE (JMMSE) equalization and its MMSE weights are derived taking into account the MAI and the noise.

6.2 MIMO System Models in the Absence of CFOs

In this section, the uplink block-based system models for the spatial multiplexing DFT-SC-FDMA (SM DFT-SC-FDMA), the SFBC DFT-SC-FDMA, the SFBC DCT-SC-FDMA, and the spatial multiplexing DCT-SC-FDMA (SM DCT-SC-FDMA) systems in the absence of CFOs are described. In these MIMO systems, two types of interference should be suppressed at the base station: the IAI, which exists between the different mobile unit antennas and the ISI resulting from the multipath distortion. The MIMO techniques are implemented in a similar way to those in [89].

6.2.1 SM DFT-SC-FDMA System Model

An SM DFT-SC-FDMA system with U users is considered. Each user is equipped with N_t transmit antennas and the base station has N_r receive antennas. $N_r = 2$ and $N_t = 2$ are assumed. The structure of the 2×2 SM DFT-SC-FDMA system is depicted in Figure 6.1.

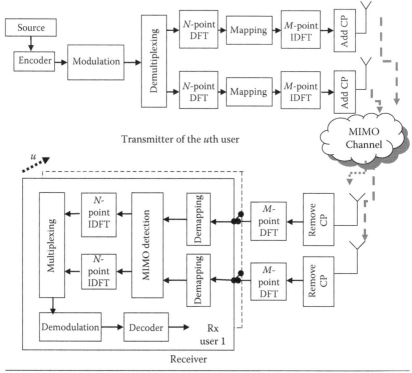

Figure 6.1   2×2 SM DFT-SC-FDMA transmitter and receiver.

We assume perfect time and frequency synchronization. As illustrated in Figure 6.1, the modulated signal is demultiplexed into two output vectors as follows:

$$\boldsymbol{x}_1^u = \left[s_1^u, s_3^u, \ldots, s_{2N-1}^u \right]^T \tag{6.1}$$

$$\boldsymbol{x}_2^u = \left[s_2^u, s_4^u, \ldots, s_{2N}^u \right]^T \tag{6.2}$$

where

$s_1^u, s_2^u, \ldots, s_{2N-1}^u$ and s_{2N}^u are the modulated symbols of the uth user

$\boldsymbol{x}_j^u$ is the signal vector transmitted by the jth ($j = 1, 2, \ldots, N_t$) transmit antenna of the uth user

On each transmit antenna, the output of the demultiplexing stage for the uth user is transformed into frequency domain via an N-point DFT and then mapped according to the subcarriers mapping method used. All the transmit antennas of the uth user map their data to the same subcarriers, leaving the unoccupied subcarriers for other users to use. The mapped frequency-domain sequence at each transmit antenna is transformed back into time domain via an M-point IDFT process.

The transmitted signal from the jth transmit antenna of the uth user can be formulated as follows:

$$\overline{x}_j^u = F_M^{-1} M_T^u F_N x_j^u \tag{6.3}$$

where x_j^u is an $N \times N$ vector representing the demultiplexed output of the uth user at the jth antenna. After the IDFT, a CP of length N_C must be appended at the end of each block.

After the removal of the CP, the received signals at the N_r receive antennas can be expressed as follows:

$$r_M = \sum_{u=1}^{U} H_M^u \overline{x}_M^u + n_M \tag{6.4}$$

The structure of H_M^u in (6.4) is given as follows:

$$H_M^u = \begin{bmatrix} H_{11}^u & H_{12}^u \\ H_{21}^u & H_{22}^u \end{bmatrix} \tag{6.5}$$

where

$\overline{x}_M^u = [\overline{x}_1^u, \overline{x}_2^u]^T$ and $\overline{x}_j^u$ is given by Equation 6.3

$r_M = [r_1, r_2]^T$, where r_i is an $M \times 1$ vector containing the received block at the ith ($i = 1, 2, \ldots, N_r$) receive antenna

$n_M = [n_1, n_2]^T$, where n_i is an $M \times 1$ vector representing the noise at the ith receive antenna

H_{ij}^u is an $M \times M$ circulant matrix describing the multipath channel between the jth transmit antenna and the ith receive antenna of the uth user

Figure 6.1 depicts the MIMO DFT-SC-FDMA receiver for the U users. It is clear that a separate detection is performed for each user. So, after the demapping process, the received signal of the kth user is given by

$$R_M^k = \Lambda_M^k X_M^k + N_M \tag{6.6}$$

The structure of Λ^k_M can be expressed as follows:

$$\Lambda^k_M = \begin{bmatrix} \overline{\Lambda}^k_{11} & \overline{\Lambda}^k_{12} \\ \overline{\Lambda}^k_{21} & \overline{\Lambda}^k_{22} \end{bmatrix} \tag{6.7}$$

where

$\overline{\Lambda}^k_{11}$ is an $N \times N$ diagonal matrix representing the DFT of the channel after the demapping process

$\boldsymbol{R}^k_M = [\boldsymbol{R}^k_1, \boldsymbol{R}^k_2]^T$, where $\boldsymbol{R}^k_i$ is an $N \times 1$ vector representing the received block at the ith receive antenna after the demapping process of the kth user

$\boldsymbol{X}^k_M = [\boldsymbol{X}^k_1, \boldsymbol{X}^k_2]^T$, where $\boldsymbol{X}^k_j$ is the DFT of $\boldsymbol{x}^k_j$

$\boldsymbol{N}_M = [\boldsymbol{N}_1, \boldsymbol{N}_2]^T$, where $\boldsymbol{N}_i$ is an $M \times 1$ vector containing the noise in the frequency domain after the demapping process at the ith receive antenna

After that, the equalization process is performed and the resulting signals are transformed into time domain via an N-point IDFT. Finally, the multiplexing, the demodulation, and the decoding processes are applied.

6.2.2 SFBC DFT-SC-FDMA System Model

STBC and SFBC are also other types of the MIMO techniques, which can be used to increase the diversity and to improve the BER performance [83,90]. In the high mobility case, STBC cannot be applied, since the requirement of a quasi-static channel response between adjacent data blocks may no longer hold. So, in this case, SFBC can be applied. The difference between the system models of the SM DFT-SC-FDMA and the SFBC DFT-SC-FDMA systems is that in the SFBC DFT-SC-FDMA system, the modulated signal is first transformed into frequency domain via an N-point DFT. Then, the SFBC encoder is used to provide two different signals. After that, the resulting signals are mapped. In the 2×2 DFT-SC-FDMA system with classical SFBC, the encoder gives two output vectors as follows:

$$\boldsymbol{X}^u_1 = [s^u_1, s^u_2, \ldots, s^u_{N-1}, s^u_N]^T \tag{6.8}$$

$$X_2^u = [-s_2^{u*}, s_1^{u*}, \ldots, -s_N^{u*}, s_{N-1}^{u*}]^T \qquad (6.9)$$

where $s_1^u, s_2^u, \ldots, s_{N-1}^u$ and s_N^u are now the samples resulting from the DFT operator. At the receiver side, the SFBC decoder process is applied after the MIMO equalization process.

6.2.3 SFBC DCT-SC-FDMA System Model

The structure of the 2×2 SFBC DCT-SC-FDMA system is depicted in Figure 6.2. At the transmitter side, the transmitted signal is constructed similar to that in the previous section. The difference is that the N-point DFT and the M-point IDFT are replaced by the N-point DCT and the M-point IDCT, respectively.

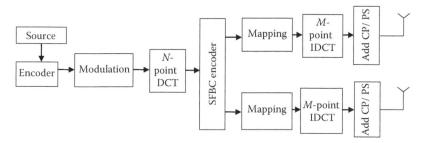

Transmitter for the uth user

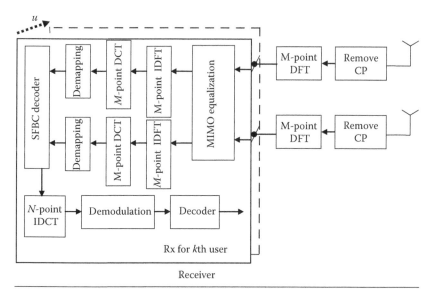

Receiver

Figure 6.2 2×2 SFBC DCT-SC-FDMA transmitter and receiver.

At the receiver side, the CP is removed from the received signals and the resulting signals are then transformed into frequency domain. Figure 6.2 shows that the MIMO equalization process is performed after the M-point DFT immediately. So, the complexity of the equalization process in this case will be higher than that in the DFT-SC-FDMA system, as we will see in Section 6.4.2. Then, the M-point IDFT, the M-point DCT, the subcarriers demapping, and the SFBC decoding operations are performed. Finally, the N-point IDCT, the demodulation, and the decoding processes are applied.

6.2.4 SM DCT-SC-FDMA System Model

The difference between the system models of the SFBC DCT-SC-FDMA and the SM DCT-SC-FDMA systems is that in the SM DCT-SC-FDMA system, the modulated signal is first demultiplexed. Then, the DCT-SC-FDMA modulation is applied at each transmit antenna. At the receiver side, the multiplexing process is applied after the DCT-SC-FDMA demodulation.

6.3 MIMO Equalization Schemes

This section is devoted to the explanation of the MIMO equalization schemes. In MIMO systems, two types of interference should be suppressed at the base station: the IAI, which exists between the different mobile unit antennas, and the ISI due to the multipath distortion. The optimal MIMO receiver is the ML detector, but its complexity increases exponentially with the number of transmit antennas and the modulation order. Time-domain detection techniques entail high complexity, especially in channels with large delay spread. It remains a challenging task to handle the combined effect of these types of interference efficiently [95]. FDE schemes are known to be excellent candidates for severe time-dispersive channels, allowing good performance and implementation complexity that is much lower than those of traditional time-domain equalization techniques [95]. In this section, we will discuss two FDE receivers, namely MIMO ZF equalization and MIMO MMSE equalization. The detection process for the kth user only is derived in the following sections.

6.3.1 MIMO ZF Equalization Scheme

Conventional MIMO ZF equalization is based on applying the Moore–Penrose pseudo-inverse of the effective channel matrix, which requires a large computational complexity and results in noise enhancement, especially for large MIMO systems. The ZF filter matrix of the kth user can be written as follows:

$$\boldsymbol{W}^k = (\Lambda_M^{k^H} \Lambda_M^k)^{-1} \Lambda_M^{k^H} \quad (6.10)$$

The ZF linear equalizer nulls the IAI and the ISI. However, noise enhancement occurs, and the performance of this equalizer is significantly affected by the noise-boosting effect.

6.3.2 MIMO MMSE Equalization Scheme

Given the statistics of the additive noise and the users' data, a better equalizer is the one that can minimize the MSE and partially remove the ISI. This equalizer is called the MMSE equalizer. It is generally preferred to the ZF linear equalizer because of its better treatment to noise. The MMSE solution for the kth user for the DFT-SC-FDMA system is given by

$$\boldsymbol{W}^k = (\Lambda_M^{k^H} \Lambda_M^k + (1/\text{SNR})\boldsymbol{I}_{2N})^{-1} \Lambda_M^{k^H} \quad (6.11)$$

where $\boldsymbol{I}_{2N}$ is a $2N \times 2N$ identity matrix. The main disadvantage of the MMSE equalizer is that the estimation of the SNR is required.

6.4 LRZF Equalization Scheme

In this section, the mathematical model of the LRZF equalizer is derived. The complexity of this equalizer is also discussed.

6.4.1 Mathematical Model

In this section, the problems associated with the MIMO ZF and the MIMO MMSE equalizers are solved, by utilizing the LRZF equalization scheme for the MIMO DFT-SC-FDMA system. It involves a two-step filtering process to efficiently equalize the received signal with a low-complexity matrix inversion process. In the first step, the LRZF scheme cancels the IAI. Then, the ISI is mitigated in the second step. The LRZF scheme also avoids the problem of noise enhancement in the direct-inversion ZF equalization scheme by adding a regularization term in the second step. Moreover, the estimation of the SNR is not required. The LRZF equalization scheme for the kth user can be described as follows:

1. *IAI cancellation.* In this step, the IAI is cancelled by applying the matrix W_{IAI}^{k}. This step can be expressed as follows:

$$\breve{X}_{M}^{k} = W_{\text{IAI}}^{k} R_{M}^{k} \tag{6.12}$$

where R_{M}^{k} is given by Equation 6.6. W_{IAI}^{k} can be expressed as follows:

$$W_{\text{IAI}}^{k} = \begin{bmatrix} I_{N} & -A_{2}^{k} \\ -A_{1}^{k} & I_{N} \end{bmatrix} \tag{6.13}$$

and

$$A_{1}^{k} = \overline{\Lambda}_{21}^{k} \overline{\Lambda}_{11}^{k^{-1}} \tag{6.14}$$

$$A_{2}^{k} = \overline{\Lambda}_{12}^{k} \overline{\Lambda}_{22}^{k^{-1}} \tag{6.15}$$

After the first step, MIMO signals for the kth user are separated.

2. *ISI cancellation.* In this step, the ISI is cancelled by applying the matrix W_{ISI}^{k}. This step can be performed as follows:

$$\hat{X}_{M}^{k} = W_{\text{ISI}}^{k} W_{\text{IAI}}^{k} R_{M}^{k} = W_{\text{ISI}}^{k} (W_{\text{IAI}}^{k} \Lambda_{M}^{k}) X_{M}^{k} + W_{\text{ISI}}^{k} W_{\text{IAI}}^{k} N_{M}$$

$$= W_{\text{ISI}}^{k} \begin{bmatrix} \overline{\Lambda}_{11}^{k} - A_{2}^{k} \overline{\Lambda}_{21}^{k} & 0_{N \times N} \\ 0_{N \times N} & \overline{\Lambda}_{22}^{k} - A_{1}^{k} \overline{\Lambda}_{12}^{k} \end{bmatrix} X_{M}^{k} + W_{\text{ISI}}^{k} W_{\text{IAI}}^{k} N_{M}$$

$$= W_{\text{ISI}}^{k} \begin{bmatrix} Q_{1}^{k} & 0_{N \times N} \\ 0_{N \times N} & Q_{2}^{k} \end{bmatrix} X_{M}^{k} + W_{\text{ISI}}^{k} W_{\text{IAI}}^{k} N_{M} \tag{6.16}$$

where

$$Q_1^k = \overline{\Lambda}_{11}^k - A_2^k \overline{\Lambda}_{21}^k \qquad (6.17)$$

$$Q_2^k = \overline{\Lambda}_{22}^k - A_1^k \overline{\Lambda}_{12}^k \qquad (6.18)$$

To avoid singularities in the direct inversion of Q_1^k and Q_2^k, we arrange W_{ISI}^k as follows:

$$W_{\text{ISI}}^k = \begin{bmatrix} \left(Q_1^{k^H} Q_1^k + \alpha I_N\right)^{-1} Q_1^{k^H} & 0_{N \times N} \\ 0_{N \times N} & \left(Q_2^{k^H} Q_2^k + \alpha I_N\right)^{-1} Q_2^{k^H} \end{bmatrix} \qquad (6.19)$$

where α is a regularization parameter. It is used to avoid the noise enhancement in the conventional ZF equalization. Note that it is possible to construct the W_{ISI}^k as follows:

$$W_{\text{ISI}}^k = \begin{bmatrix} Q_1^{k-1} & 0_{N \times N} \\ 0_{N \times N} & Q_2^{k-1} \end{bmatrix} \qquad (6.20)$$

However, the resulting scheme in this case is equivalent to the conventional ZF equalizer, i.e., $W_{\text{ISI}}^k W_{\text{IAI}}^k = \Lambda_M^{k^{-1}}$, with lower complexity. Moreover, the resulting scheme suffers from the noise enhancement problem. So, in W_{ISI}^k we have added the regularization term to avoid this problem. The best value of α depends on the value of 1/SNR, which is the MMSE equalizer weight [19]. But the problem associated with the MMSE equalizer is the estimation of the SNR, which is not known at the receiver. To avoid this problem, it is better to choose α as a constant [4,19]. The main advantages of the LRZF scheme lie in its lower complexity, as we will see in the following section, and its lower BER, when compared with the conventional ZF equalization method that uses the direct matrix inversion.

6.4.2 Complexity Evaluation

6.4.2.1 DFT-SC-FDMA System   From the computational complexity perspective, the complexity of the LRZF equalization scheme is reduced by breaking the ZF matrix inversion process into two steps. As a result, the algorithm avoids the direct channel matrix inversion. For the conventional ZF equalization scheme described by Equation 6.10, the inverse of a $2N \times 2N$ matrix for each user is required, which is time-consuming for large N. The implementation of Equation 6.10 requires approximately $8N^3$ operations, which is practically impossible for large N [80]. So, the complexity of the conventional ZF equalization is of $O(N^3)$, which is the same as that of the MMSE equalization. The ML decoder compares the received signal vector of size $N \times 1$ with all the 2^{vN} candidates, resulting in a complexity of $O(N \times 2^{vN})$, where v is the modulation order, which is too large as compared to the other schemes [88].

For the LRZF scheme, only the inverse of $N \times N$ diagonal matrices is required, which has a complexity of $O(N)$ [80]. In order to study the complexity of the LRZF scheme in more detail, it is better to describe it as follows:

$$W_{\text{LRZF}} = W_{\text{ISI}} W_{\text{IAI}} = \begin{bmatrix} \bar{Q}_1^{-1} & -\bar{Q}_1^{-1} A_2^k \\ -\bar{Q}_2^{-1} A_1^k & \bar{Q}_2^{-1} \end{bmatrix} \tag{6.21}$$

where

$$\bar{Q}_1^{-1} = \left(Q_1^{kH} Q_1^k + \alpha I \right)^{-1} Q_1^{kH}$$
$$\bar{Q}_2^{-1} = \left(Q_2^{kH} Q_2^k + \alpha I \right)^{-1} Q_2^{kH}$$

Now, we can count the number of operations required for the LRZF scheme as follows.

Since $\bar{\Lambda}_{11}^k$ is a diagonal matrix, we can calculate $\bar{\Lambda}_{11}^{k-1}$ with N operations. Also, since $\bar{\Lambda}_{11}^{k-1}$, $\bar{\Lambda}_{12}^k$, $\bar{\Lambda}_{21}^k$, and $\bar{\Lambda}_{22}^k$ are diagonal matrices, we can calculate Q_2^k and A_1^k with $3N$ and $2N$ operations, respectively. Similarly, we can calculate $\bar{\Lambda}_{22}^{k-1}$, Q_1^k, and A_2^k with N, $3N$, and $2N$ operations, respectively. Since Q_1^k and Q_2^k are also diagonal matrices, we can calculate $\bar{Q}_1^{-1}$ and $\bar{Q}_2^{-1}$ with $6N$ operations. Finally, calculating the products $\bar{Q}_1^{-1} A_1^k$ and $\bar{Q}_2^{-1} A_2^k$, which are diagonal matrices, requires $2N$ operations. As a result, the total number of operations required to implement the LRZF scheme in Equation 6.21 is $14N$. Hence, the complexity of the LRZF scheme is of $O(N)$. This means that the overall complexity of the LRZF scheme for each user is much lower than those of the ML equalizer, the MMSE equalizer, and the conventional ZF equalizer.

6.4.2.2 DCT-SC-FDMA System   In the DCT-SC-FDMA system, the equalization process is performed after the DFT, immediately, i.e., before the demapping process, and the inversion of a $2M \times 2M$ matrix is required. As a result, the total number of operations required to perform the LRZF equalization for the DFT-SC-FDMA system is $14M$.

Figure 6.3 gives a comparison between the ZF and the LRZF schemes for different values of the input block size N for the DFT-SC-FDMA and the DCT-SC-FDMA systems. $Q = 4$ is considered. In Figure 6.3, the complexity of the MMSE equalization scheme is not illustrated, because it has the same complexity as the ZF equalization scheme.

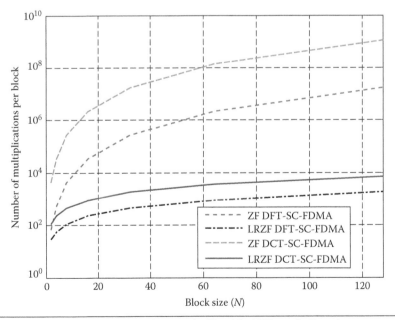

Figure 6.3   Complexity comparison between the ZF and the LRZF equalization schemes.

It is clear that the complexity of the LRZF scheme for both systems is much lower than that of the conventional ZF equalizer, especially when the value of N is large.

6.5 MIMO System Models in the Presence of CFOs

In uplink MIMO systems, the effect of CFOs strongly degrades the process of signal detection if it is not compensated beforehand due to the IAI between the different mobile unit antennas, the ISI, and the MAI. Moreover, in the uplink, since different CFOs between all mobile terminals occur, the CFOs compensation for one user may cause interference to the other users. This makes the problem of CFOs more serious in uplink MIMO systems than in uplink SISO systems. So, there is a need for an accurate equalization scheme in uplink MIMO systems.

In this section, the baseband signal model in Section 6.2 is extended for the MIMO DFT-SC-FDMA system in the presence of CFOs, and the user-by-user received SIR is also derived.

6.5.1 System Model

The transmitted signal from the jth transmit antenna of the uth user can be formulated as in Section 6.2. At the receiver side, the received signal in Equation 6.4 can be modified to include the impact of the CFOs as follows:

$$r_M = \sum_{u=1}^{U} E_M^u H_M^u \bar{x}_M^u + n_M \tag{6.22}$$

where E_M^u is given as follows:

$$E_M^u = I_2 \otimes E^u \tag{6.23}$$

where E^u is an $M \times M$ diagonal matrix with elements $[E^u]_{m,m} = e^{j 2 \pi \varepsilon_u m / M}$, $m = 0, \ldots, M-1$, which describes the CFO matrix of the uth user. ε_u is the CFO of the uth user normalized to the subcarriers spacing. $\otimes$ denotes the Kronecker product. After the demapping process, the received signal of the kth user is given by

$$R_M^k = \Pi_M^k \Lambda_M^k X_M^k + \sum_{\substack{u=1 \\ u \neq k}}^{U} \Pi_M^u \Lambda_M^u X_M^u + N_M \tag{6.24}$$

where Λ_M^u is given by Equation 6.7. The structures of Π_M^k and Π_M^u in Equation 6.24 can be expressed as follows:

$$\Pi_M^k = I_2 \otimes \Pi_d^k \tag{6.25}$$

$$\Pi_M^u = I_2 \otimes \Pi_r^u \tag{6.26}$$

where Π_d^k and Π_r^u are defined in Chapter 5.

After that, the impact of the multipath channel and the CFOs must be mitigated, which is the main objective of this chapter. So, this process will be discussed in detail in the following sections. Finally, the signal is transformed into the time domain via an N-point DFT and the multiplexing process is performed, followed by the demodulation and decoding processes. The signal after the DFT can be expressed as follows:

$$\hat{x}_M^k = (I_2 \otimes F_N^{-1}) W_M^k R_M^k \tag{6.27}$$

where W_M^k is the MIMO detection matrix, which represents the equalization and the CFOs compensation processes.

6.5.2 Signal-to-Interference Ratio

In this section, an analytical expression of the SIR for the uplink MIMO DFT-SC-FDMA system in the presence of CFOs is derived. The estimated time-domain symbols of the kth user in Equation 6.27 can be rewritten as follows:

$$\hat{x}_M^k = A_M^k x_M^k + \bar{A}_M^k x_M^k + \sum_{\substack{u=1 \\ u \neq k}}^{U} B_M^u x_M^u + \hat{n}_M \tag{6.28}$$

The structures of the components of Equation 6.28 are given as follows:

$$A_M^k = \mathrm{diag}((I_2 \otimes F_N^{-1}) W_M^k \Pi_M^k \Lambda_M^k (I_2 \otimes F_N)) \tag{6.29}$$

$$\bar{A}_M^k = (I_2 \otimes F_N^{-1}) W_M^k \Pi_M^k \Lambda_M^k (I_2 \otimes F_N) - A_M^k \tag{6.30}$$

$$\widehat{n}_M = (I_2 \otimes F_N^{-1}) W_M^k N_M \qquad (6.31)$$

$$B_M^u = (I_2 \otimes F_N^{-1}) W_M^k \Pi_M^u \Lambda_M^u (I_2 \otimes F_N) \qquad (6.32)$$

From Equation 6.28, it is found that only the first term contains the desired data, the second term is due to the ISI and the IAI, the third term is an MAI, and the fourth term is a noise. Based on Equation 6.28, the SIR of the nth symbol ($n = 1, \ldots, 2N$) in the kth user received signal can be expressed as follows:

$$\text{SIR}_M^k(n) = \frac{\left| A_M^k(n,n) \right|^2}{\sum_{\substack{l=1 \\ l \neq n}}^{2N} \left| \bar{A}_M^k(n,l) \right|^2 + \sum_{\substack{u=1 \\ u \neq k}}^{U} \sum_{l=1}^{2N} \left| B_M^u(n,l) \right|^2} \qquad (6.33)$$

6.6 Joint Equalization and CFOs Compensation Schemes

In this section, we discuss two joint equalization and CFOs compensation schemes for the uplink MIMO DFT-SC-FDMA system in the presence of CFOs. These schemes are derived taking into account the noise, the ISI, and the MAI. As a result, the equalizers mitigate the effects of both the multipath channel and the CFOs at the same time. Moreover, they are performed in the frequency domain. So, they require only a single DFT at each receiver antenna for all users.

6.6.1 JLRZF Equalization Scheme

In this section, the LRZF scheme in Section 6.4 is developed for the MIMO DFT-SC-FDMA system in the presence of CFOs. It is derived to jointly perform the equalization and CFOs compensation processes. This scheme will be referred to as the JLRZF equalizer. It is performed in two steps in order to reduce the complexity, as in Section 6.4. To derive the JLRZF equalizer steps, Equation 6.24 must be rearranged as follows:

$$R_M^k = \bar{\Pi}_P^k X_M^k + N_P \qquad (6.34)$$

where

$\bar{\Pi}_P^k = \bar{\Pi}_M^k \Lambda_M^k$ is a $2N \times 2N$ matrix representing the interference

$N_P = \displaystyle\sum_{u=1,u\neq k}^{U} \Pi_P^u X_M^u + N_M$ is the MAI plus noise matrix

The structure of $\bar{\Pi}_P^k$ can be expressed as follows:

$$\bar{\Pi}_P^k = \Pi_M^k \Lambda_M^k = \begin{bmatrix} \Pi_d^k \Lambda_{11}^k & \Pi_d^k \Lambda_{12}^k \\ \Pi_d^k \Lambda_{21}^k & \Pi_d^k \Lambda_{22}^k \end{bmatrix} = \begin{bmatrix} \Pi_{11}^k & \Pi_{12}^k \\ \Pi_{21}^k & \Pi_{22}^k \end{bmatrix} \qquad (6.35)$$

The JLRZF equalization scheme for the kth user can be described as follows:

1. *IAI cancellation.* In this step, the IAI is cancelled by applying the matrix W_1^k. This step can be expressed as follows:

$$\breve{X}_M^k = W_1^k R_M^k \qquad (6.36)$$

   where
   R_M^k is given by Equation 6.34
   W_1^k can be expressed as follows:

$$W_1^k = \begin{bmatrix} I_N & -C_2^k \\ -C_1^k & I_N \end{bmatrix} \qquad (6.37)$$

$$C_1^k = \Pi_{21}^k \Pi_{11}^{k\,-1} \qquad (6.38)$$

$$C_2^k = \Pi_{12}^k \Pi_{22}^{k\,-1} \qquad (6.39)$$

   After the first step, the MIMO signals for the kth user are separated.

2. *ISI and MAI cancellation.* In this step, the impacts of the ISI and the MAI are cancelled by applying the matrix W_2^k. This step can be performed as follows:

$$\hat{X}_M^k = W_2^k W_1^k R_M^k = W_2^k (W_1^k \overline{\Pi}_P^k) X_M^k + W_2^k W_1^k N_P$$

$$= W_2^k \begin{bmatrix} \Pi_{11}^k - C_2^k \Pi_{21}^k & 0_{N \times N} \\ 0_N & \Pi_{22}^k - C_1^k \Pi_{12}^k \end{bmatrix} X_M^k + W_2^k W_1^k N_P \quad (6.40)$$

$$= W_2^k \begin{bmatrix} \Phi_1^k & 0_{N \times N} \\ 0_{N \times N} & \Phi_2^k \end{bmatrix} X_M^k + W_2^k W_1^k N_P$$

where

$$\Phi_1^k = \Pi_{11}^k - C_2^k \Pi_{21}^k \quad (6.41)$$

$$\Phi_2^k = \Pi_{22}^k - C_1^k \Pi_{12}^k \quad (6.42)$$

To avoid singularities in the direct inversion of Φ_1^k and Φ_2^k, we propose the arrangement of W_2^k as follows:

$$W_2^k = \begin{bmatrix} \left(\Phi_1^{k^H} \Phi_1^k + \alpha I_N \right)^{-1} \Phi_1^{k^H} & 0_{N \times N} \\ 0_{N \times N} & \left(\Phi_2^{k^H} \Phi_2^k + \alpha I_N \right)^{-1} \Phi_2^{k^H} \end{bmatrix} \quad (6.43)$$

where α is a regularization parameter. The objective of the regularization parameter in the second step is not only to avoid the problem of noise enhancement, but also to reduce the effect of the MAI. It is important to note that most of the elements in Π_{11}^k, Π_{12}^k, Π_{21}^k, and Π_{22}^k are zeros, and these matrices can be approximated as banded matrices as in Chapter 5. The optimum solution of the JLRZF scheme is derived in Appendix D. It is difficult to obtain a closed form for the covariance matrix of the overall noise. So, it is better to choose α as a constant, which is as close as possible to the optimum value.

6.6.2 JMMSE Equalization Scheme

In this section, the MMSE equalization in Chapter 5 is developed for the MIMO DFT-SC-FDMA system in the presence of CFOs. It will be referred to as the JMMSE equalizer and its MMSE weights are derived taking into account the ISI, the MAI, and the noise. As in

Chapter 5, the equalization coefficients of the kth user are determined to minimize the cost function:

$$J_M^k = E\left\{\left\|e_M^k\right\|^2\right\} = E\left\{\left\|W_{\mathrm{JMMSE}}^k\ R_M^k - X_M^k\right\|^2\right\} \tag{6.44}$$

where $E\{\cdot\}$ is the expectation. Solving $\partial J_M^k / \partial W_{\mathrm{JMMSE}}^k = 0$, we obtain

$$W_{\mathrm{JMMSE}}^k = \overline{\boldsymbol{\Pi}}_P^{k^H}\left(\overline{\boldsymbol{\Pi}}_P^k\ \overline{\boldsymbol{\Pi}}_P^{k^H} + \sum_{\substack{u=1 \\ u \neq k}}^{U} \overline{\boldsymbol{\Pi}}_P^u\ \overline{\boldsymbol{\Pi}}_P^{u^H} + (1/\mathrm{SNR})\boldsymbol{I}_{2N}\right)^{-1} \tag{6.45}$$

The main advantage of the JMMSE equalizer is that the residual interference is lower than that in the conventional schemes as will be seen in the simulation results. In addition, it requires a single DFT operation at each receive antenna for all users.

6.6.3 Complexity Evaluation

For the JLRZF equalizer, only the inverse of $N \times N$ matrices is needed, which requires approximately N^3 operations. However, the structures of the approximated interference matrices are banded. As a result, the inverse of the $N \times N$ matrices can be performed using the banded-matrix factorization, which requires $N[16r^2 + 26r + 5]$ operations, where r is the bandwidth of the banded matrix [80]. In this case, the complexity of the JLRZF equalizer is of $O(N)$ as we discussed in Chapter 5.

For the JMMSE equalizer, the inversion of a $2N \times 2N$ matrix for each user is needed, which requires approximately $8N^3$ operations [80]. On the other hand, the JMMSE equalizer requires a complexity of $O(N^3)$.

6.7 Simulation Examples

Experiments have been carried out to study the effectiveness of the different equalization schemes in the uplink MIMO DFT-SC-FDMA and MIMO DFT-SC-FDMA systems [96–98]. The BER has been evaluated by the Monte Carlo simulation method.

Table 6.1   Simulation Parameters

Simulation method	Monte Carlo
System bandwidth	5 MHz
Modulation type	QPSK
CP length	20 samples
M	512
N	128
Channel coding	Convolutional code with rate = 1/2
Subcarriers mapping technique	Interleaved and localized
MIMO technique	SM and SFBC
N_t	2
N_r	2
U	4
Channel model	Vehicular-A channel
CFOs estimation	Perfect
Channel estimation	Perfect

6.7.1 Simulation Parameters

To evaluate the performance of the different equalizers for the uplink MIMO DFT-SC-FDMA and MIMO DCT-SC-FDMA systems, simulations have been performed over Rayleigh fading MIMO channels. The channels corresponding to different transmit and receive antennas have the same statistics. The channel model used for simulations is the vehicular-A model [43]. A convolutional code with memory length 7 and octal generator polynomial (133,171) has been chosen as the channel code. The simulation parameters are tabulated in Table 6.1.

6.7.2 Absence of CFOs

This section is devoted to explain the performance of the LRZF equalization scheme in the absence of CFOs.

6.7.2.1 Results of the LRZF Equalization Scheme   The impact of the regularization parameter on the performance of the LRZF scheme has been studied for the different MIMO techniques. Figures 6.4 through 6.11 illustrate the BER versus the regularization parameter at different SNR values for the SM DFT-SC-FDMA, SM DCT-SC-FDMA, SFBC DFT-SC-FDMA, and SFBC DCT-SC-FDMA

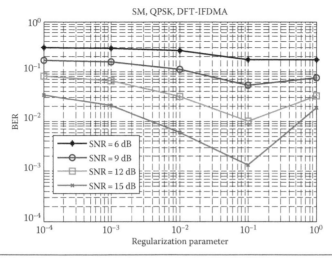

Figure 6.4 BER vs. the regularization parameter for SM DFT-IFDMA system with the LRZF equalizer.

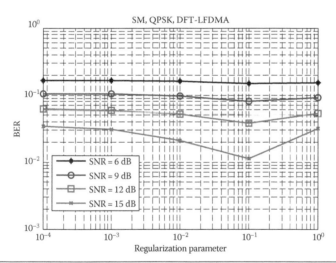

Figure 6.5 BER vs. the regularization parameter for SM DFT-LFDMA system with the LRZF equalizer.

systems with the LRZF scheme. From these figures, it is clear that the best choice of the regularization parameter is $\alpha = 0.1$ regardless of the subcarriers mapping techniques used. On the other hand, the BER performance deteriorates for smaller and larger values of α, because we deal with a nonlinear minimization problem for the MSE or BER, which has a unique minimum. As a result, we have used $\alpha = 0.1$ for the rest of experiments.

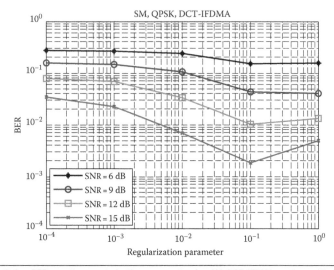

Figure 6.6   BER vs. the regularization parameter for SM DCT-IFDMA system with the LRZF equalizer.

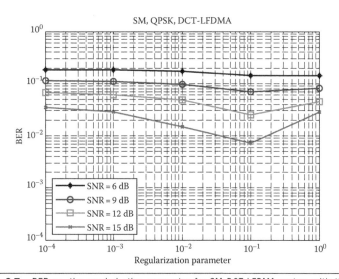

Figure 6.7   BER vs. the regularization parameter for SM DCT-LFDMA system with the LRZF equalizer.

Figures 6.12 through 6.15 show the BER results for MIMO DFT-SC-FDMA and MIMO DCT-SC-FDMA systems with the LRZF scheme for different subcarriers mapping and MIMO techniques. From these results, we can see that the performance of the LRZF scheme significantly outperforms that of the conventional ZF equalization,

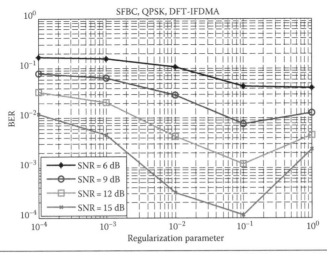

Figure 6.8 BER vs. the regularization parameter for SFBC DFT-IFDMA system with the LRZF equalizer.

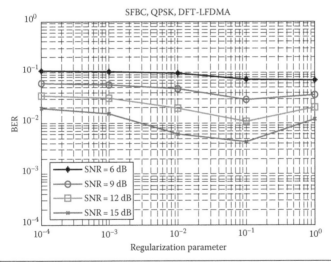

Figure 6.9 BER vs. the regularization parameter for SFBC DFT-LFDMA system with the LRZF equalizer.

especially when the interleaved subcarriers mapping technique is used. At a BER = 10^{-2}, the LRZF scheme for the DFT-IFDMA system provides about 8 dB gain, when compared to the conventional ZF equalization scheme. The same performance gain is also obtained for the DCT-IFDMA system.

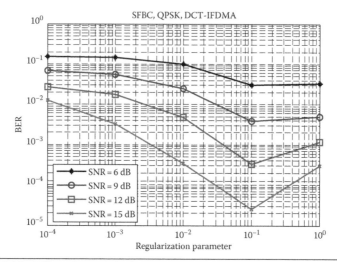

Figure 6.10   BER vs. the regularization parameter for SFBC DCT-IFDMA system with the LRZF equalizer.

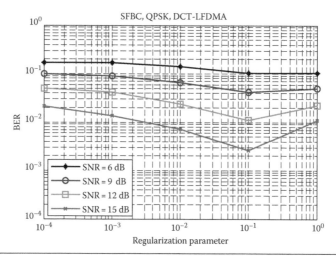

Figure 6.11   BER vs. the regularization parameter for SFBC DCT-LFDMA system with the LRZF equalizer.

As mentioned in Section 6.4, the best theoretical value of α is 1/SNR [4,19]. So, a comparison between $\alpha = 0.1$ and $\alpha = 1/\text{SNR}$ is also shown in Figures 6.12 through 6.15. From these figures, it is clear that there is a slight difference in the performance between $\alpha = 1/\text{SNR}$ (MMSE equalization) and $\alpha = 0.1$ (regularized equalization) at high SNR values. When the SNR is low, the two values nearly

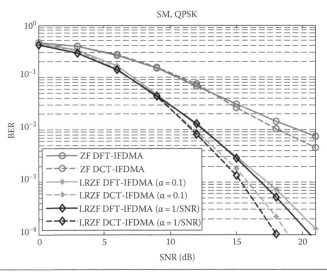

Figure 6.12   BER vs. the SNR for SM DFT-IFDMA and SM DCT-IFDMA systems with different schemes.

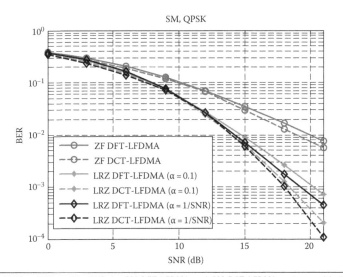

Figure 6.13   BER vs. the SNR for SM DFT-LFDMA and SM DCT-LFDMA systems with different schemes.

give the same performance. As a result, an approximation of $\alpha = 0.1$ is satisfactory.

Figure 6.16 illustrates the CCDF of the PAPR for the SFBC DFT-LFDMA and the SFBC DCT-LFDMA systems. We have generated 10^5 uniform random data blocks to acquire each point in

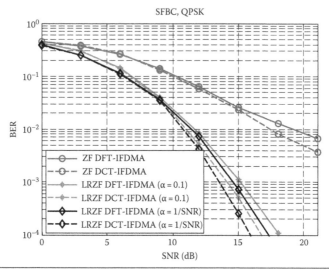

Figure 6.14   BER vs. SNR for SFBC DFT-IFDMA and SFBC DCT-IFDMA systems with different schemes.

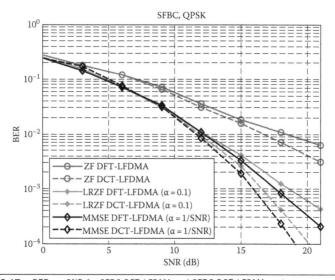

Figure 6.15   BER vs. SNR for SFBC DFT-LFDMA and SFBC DCT-LFDMA systems.

the CCDF of PAPR. From Figure 6.16, it is clear that the SFBC DCT-LFDMA system provides a lower PAPR than the SFBC DFT-LFDMA system.

6.7.2.2 Impact of Estimation Errors   The impact of the channel estimation errors on the BER performance of the DFT-SC-FDMA and

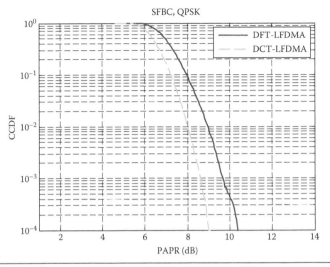

Figure 6.16   CDDF of PAPR for SFBC DFT-IFDMA and SFBC DCT-IFDMA systems.

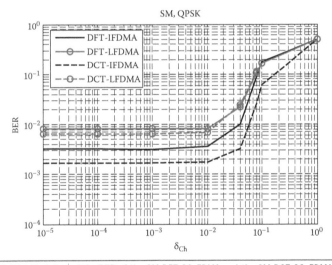

Figure 6.17   Variation of the BER of the SM DFT-SC-FDMA and the SM DCT-SC-FDMA systems with the LRZF scheme in the presence of estimation errors. SNR = 15 dB.

the DCT-SC-FDMA systems with the LRZF equalization scheme has been investigated with different MIMO techniques. An SNR of 15 dB has been used in the simulations. Figures 6.17 and 6.18 show the performance of the LRZF scheme with the channel estimation errors for different MIMO techniques. It is clear that the performance of all

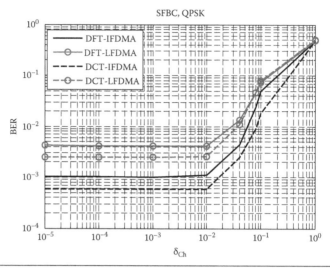

Figure 6.18 Variation of the BER of the SFBC DFT-SC-FDMA and the SFBC DCT-SC-FDMA systems with the LRZF scheme in the presence of estimation errors. SNR = 15 dB.

systems starts to degrade as σ_{Ch} becomes larger than 0.01. This limit can be satisfied with the channel estimation algorithms, which show that the LRZF scheme is robust to estimation errors for different subcarriers mapping and MIMO techniques.

6.7.3 Presence of CFOs

The performances of the JLRZF and JMMSE schemes have been investigated in the presence of CFOs. Independent CFOs have been assumed for all uplink users. Each CFO is a random variable with uniform distribution in [–0.1, 0.1].

6.7.3.1 Results of the JLRZF Equalization Scheme
Based on the results obtained in Section 6.7.2, it has been found that the LRZF equalizer provides a better BER performance than that of the conventional ZF equalizer for different MIMO techniques: SM and SFBC. So, only the SM technique is used for the rest of experiments.

Figure 6.19 gives a plot for the SIR from the first symbol of the first user versus ε_{max} for the SISO DFT-IFDMA and MIMO DFT-IFDMA systems. Equation 6.33 has been used to calculate the SIR. It is clear that the SIR performance of both the SISO and the MIMO

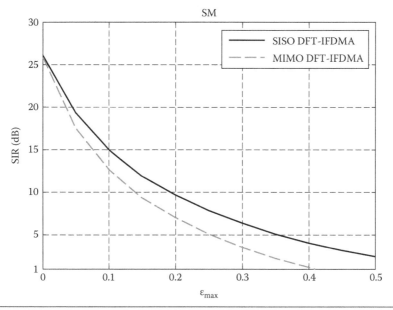

Figure 6.19   SIR vs. ε_{max} for the SISO DFT-IFDMA and the MIMO DFT-IFDMA systems.

scenarios deteriorates with the increase in ε_{max}. Also, it is clear that the impact of the CFOs on the performance of the MIMO DFT-IFDMA system is higher than that on the performance of the SISO DFT-IFDMA system, because the MIMO DFT-IFDMA system also suffers from IAI, which is not found in the SISO scenario.

The effect of the regularization parameter on the performance of the JLRZF scheme has been studied and shown in Figures 6.20 and 6.21 for the DFT-IFDMA and the DFT-LFDMA systems, respectively. Each CFO is a random variable with uniform distribution in [−0.1, 0.1]. It is clear that the best choice for the regularization parameter is $\alpha = 0.1$, regardless of the subcarriers mapping technique used. As a result, we have used $\alpha = 0.1$ for the rest of experiments.

The results without CFOs, without CFOs compensation, with the circular-convolution detector, with the single-user detector, with the joint ZF (JZF) when $\alpha = 0$, and with the JLRZF are presented in Figures 6.22 and 6.23 for different subcarriers mapping techniques. Note that the equalization process after the circular-convolution and the single-user detectors has been performed with the LRZF scheme.

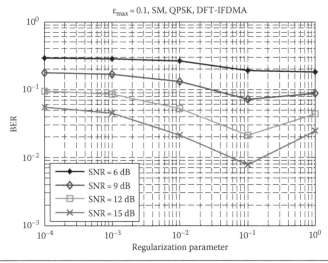

Figure 6.20 BER vs. the regularization parameter for SM DFT-IFDMA system with the JLRZF scheme.

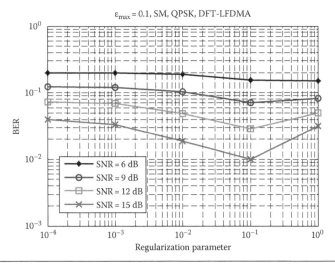

Figure 6.21 BER vs. the regularization parameter for SM DFT-LFDMA system with the JLRZF scheme.

Also, the LRZF scheme has been used to obtain the results of a system without CFOs and without CFOs compensation. It is observed that the JLRZF scheme is effective to mitigate the CFOs and its performance is the best as compared to the other schemes, especially at high SNR values. For the single-user detector, error floors, which are

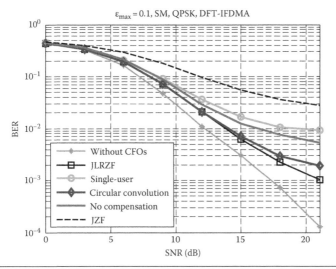

Figure 6.22 BER vs. SNR for SM DFT-IFDMA system with the JLRZF scheme.

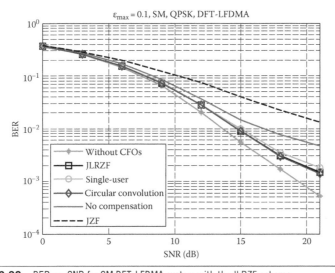

Figure 6.23 BER vs. SNR for SM DFT-LFDMA system with the JLRZF scheme.

mainly induced by the high MAI at moderate and high SNR values, appear. Performance degradation in the single-user detector is caused by the fact that CFOs compensation for a single user before the DFT can increase the frequency offsets in the data of the other users within the DFT block. As a result, for the DFT-IFDMA system, the performance of the single-user detector is bad, even when compared with that of a system without CFOs compensation. This indicates that the single-user detector is not suitable for the uplink MIMO DFT-IFDMA system. From Figure 6.23, it is clear that for the DFT-LFDMA system, all schemes except the JZF provide the same BER performance.

Figures 6.22 and 6.23 show that the JZF equalizer does not operate satisfactorily in the interference-limited environments of the uplink MIMO DFT-SC-FDMA system. The performance of the JZF equalizer is the worst, even when compared with a system without CFOs compensation. On the other hand, the JLRZF scheme avoids the problems associated with the JZF equalizer with lower complexity, because the JLRZF scheme uses a regularization parameter to avoid the noise enhancement and to minimize the MAI.

6.7.3.2 Results of the JMMSE Equalization Scheme The performance of the JMMSE scheme is shown in Figures 6.24 and 6.25 for different subcarriers mapping techniques. $\varepsilon_{max} = 0.1$ has been assumed. Note that the equalization process after the circular-convolution detector and the single-user detector has been performed with the MMSE equalizer. Also, the MMSE equalization has been used to obtain the

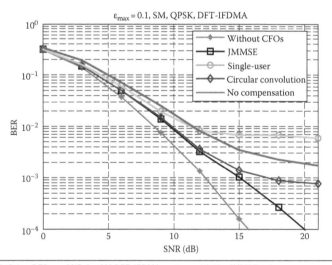

Figure 6.24 BER vs. SNR for SM DFT-IFDMA system with the JMMSE scheme.

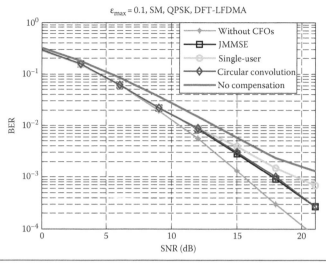

Figure 6.25   BER vs. SNR for SM DFT-LFDMA system with the JMMSE scheme.

results of a system without CFOs and without CFOs compensation. It is clear that the performance of the JMMSE scheme is better than that of the single-user detector, and the circular-convolution detector, because the JMMSE equalizer is derived to minimize the ISI, the MAI, and the noise. It is also clear that an error floor exists in the single-user and circular-convolution detectors due to the MAI.

From Figures 6.22 and 6.24, at a BER = 10^{-3}, the JMMSE scheme provides about 6 dB performance gain as compared with the JLRZF scheme. However, its complexity is higher than that of the JLRZF scheme as discussed in Section 6.6.

6.7.3.3 Impact of Estimation Errors   Figures 6.26 and 6.27 demonstrate the impact of estimation errors on the performance of the JMMSE and the JLRZF schemes for different subcarriers mapping techniques. $\varepsilon_{max} = 0.1$ and SNR = 15 dB have been used. It is clear that the performance of both schemes starts to degrade as σ_{CFO} becomes larger than 0.05 and σ_{Ch} becomes larger than 0.01. These limits can be satisfied with the estimation algorithms, which show that these schemes are robust to estimation errors, especially to the CFOs estimation errors.

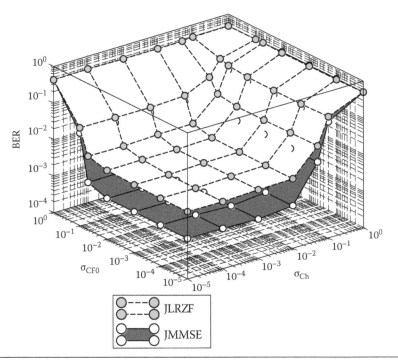

Figure 6.26 Variation of the BER of SM DFT-IFDMA system with the different schemes in the presence of estimation errors.

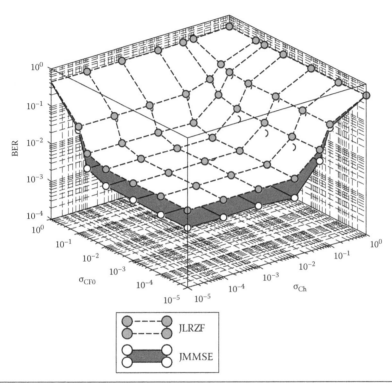

Figure 6.27 Variation of the BER of SM DFT-LFDMA system with the different schemes in the presence of estimation errors.

7

FUNDAMENTALS
OF COOPERATIVE
COMMUNICATIONS

7.1 Introduction

Cooperative communication is a recent paradigm proposed for conveying information in wireless networks, where closely located single-antenna terminals cooperatively transmit and/or receive by forming virtual antenna arrays. It exploits the broadcast nature of the wireless channel and the spatial diversity offered by having spatially separated relay nodes in the network to form what is known as cooperative diversity.

In recent decades, there was an explosive growth of wireless networks motivated by their ability of providing communications anywhere and anytime. Therefore, a high proliferation of wireless services such as mobile communications, wireless local area networks (WLANs), and wireless fidelity (WiFi) networks has emerged. The next-generation systems are expected to provide services with higher transmission rates and quality of service (QoS) compared to the existing ones. The main challenge facing designers in achieving the demands of these future wireless applications is the impairments associated with the wireless channel. One major challenge is the scarcity of the two fundamental resources for communications, namely, energy and bandwidth. Other major challenges for communicating over wireless channels are fading, shadowing, and interference that affect the performance of communications.

Multipath fading is one of the most challenging phenomena in wireless communications. The received signal varies as a result of the destructive and constructive interference of the multipath signals. Destructive interference results in a fading phenomenon, which has a dramatic effect on the overall system performance compared to that caused by additive noise. A common solution to combat multipath fading is through the use of diversity techniques. Diversity in the wireless systems can be achieved through time diversity, frequency diversity, and spatial diversity [23]. Spatial diversity is a particularly attractive technique, because it does not require additional bandwidth or a reduction of the transmission rate. The downside of spatial diversity is that it requires multiple antennas with sufficient separation between the antenna elements, which is difficult to achieve in practice due to size and cost limitations. Cooperative diversity is a kind of spatial diversity that can be obtained without multiple transmit or receive antennas; single-antenna users in the network cooperate with each other to constitute a virtual antenna array. The concept of user cooperation diversity was first introduced in [99,100]. In [101], practical cooperative diversity protocols were proposed and evaluated in terms of outage performance.

An ad-hoc wireless network is a collection of wireless mobile nodes that self-configure to form a network without the aid of any established infrastructure like a base station (BS) in cellular networks. Ad-hoc networks have a wide range of applications such as peer-to-peer wireless data exchange, home networks, and sensor networks. In most of these applications, such as sensor networks, there is an increasing demand for small, low-cost, and battery-powered devices that are deployed over a wide area. Due to the low transmit power, these nodes have limited communication ranges. Therefore, in order to achieve a significant power saving, information should be conveyed to the destination through multiple intermediate nodes. Thus, cooperative communication, in which nodes share their resources to facilitate each other's communication, is essential for these networks.

The conventional cellular systems appear incapable of delivering the high data rates and coverage expected for the future-generation wireless systems. According to the International Telecommunication Union's (ITU) requirements [102], future-generation systems can support peak data rates of 100 Mb/s and 1 Gb/s in high-speed mobility environments (up to 350 km/h), and pedestrian environments (up to 10 km/h). These high data rates require SNRs at the receiver that may be difficult to obtain at the cell edges. Therefore, there must be a trade-off between the data rate and the reliability.

Increasing the data rate reduces reliability, and increasing the reliability requires decreasing the data rate or reducing the coverage area. The most widely used strategy to address these challenges is reducing the cell size by increasing the number of BSs, which in turn increases the deploying costs. Instead, an increasingly attractive solution is to insert fixed relays into the cell, whose purpose is to aid communication from the BS to the mobile station and vice versa. Relay technologies have been actively studied and considered in the standardization process of next-generation mobile communication systems such as the Third Generation Partnership Project (3GPP) Long-Term Evolution (LTE-advanced) and IEEE 802.16j, which is the standard of multihop mobile WiMAX [103,104]. It was shown in [105,106] that relay technologies can effectively improve service coverage and system throughput, especially when multiple relay stations are deployed.

There are some key differences between user cooperation diversity and the traditional antenna array techniques including multiple-input multiple-output (MIMO) systems. First, the cooperative links are not ideal, i.e., their channels may be fading or additive white Gaussian noise (AWGN) channels. Second, each node in cooperative systems has its own resources (e.g., energy) and QoS requirements. Moreover, the number of relay nodes is not known a priori. Therefore, the design of any distributed protocol or coding scheme should take into account these constraints. As will be seen in this book, these differences are not trivial and in order to gain the advantages of MIMO systems, new techniques that are substantially different from classical antenna array schemes must be developed.

Cooperative communication techniques rely on two fundamental phenomena associated with the nature of wireless communications. One is the broadcast nature of the wireless medium and the other is the spatial diversity that can be obtained from the participation of the different nodes involved in the cooperation. Depending on the processing at the relay node, two fundamental methods of relaying at the relay node are considered in the literature, namely amplify-and-forward (AF) and decode-and-forward (DF) protocols [103]. In the AF protocol, the relay simply amplifies the received signal and retransmits it. In the DF protocol, the relay detects the received signal and retransmits a regenerated version of it.

7.2 Diversity Techniques and MIMO Systems

It is known that wireless fading channel characteristics change with time. In communication systems designed around a single signal path between the source and destination, a deep fade on this path is a likely event that needs to be addressed with techniques such as increasing the error correcting capability of the channel coding scheme, reducing the transmission rate, using more elaborate detectors, etc. Nevertheless, these solutions may still fall short for many practical channel realizations. The following sections summarize some relevant work on diversity and multiple antenna systems, which are considered as effective solutions for channel fading.

7.2.1 Diversity Techniques

Viewing the problem of communication through a fading channel with a different perspective, the overall reliability of the link can be significantly improved by providing more than one signal path between the source and the destination, each exhibiting a fading process as much independent from the others as possible. In this way, the chance that there is at least one sufficiently strong path is improved. Those techniques that aim at providing multiple, ideally independent signal paths are collectively known as diversity techniques. The independent channel fading in different wireless links means that significant gains can be achieved using diversity techniques that take advantage of the probabilistic independence of the fading of these interterminal links. Diversity provides robustness against fading by averaging over multiple fading realizations (traditionally in time, frequency, or space) using some diversity combination methods. The application of diversity can result in the probability of error P_e decreasing with the Dth power of the SNR, where D is the diversity order. The probability of error then has the following behavior:

$$P_e \propto (\mathrm{SNR})^{-D} \tag{7.1}$$

The diversity of any scheme is measured through the diversity order D of the system and is defined as the rate of decay of the probability of error with SNR, when SNR tends to infinity using a log–log scale as follows [107]:

$$D = \lim_{\text{SNR}\to\infty} \frac{\log P_e}{\log \text{SNR}} \tag{7.2}$$

The most common diversity methods can be classified as follows:

Time diversity: Time diversity is achieved by transmitting the same information with a time spacing that exceeds the coherence time (defined as the time duration over which two received signals have a strong potential for amplitude correlation) of the channel, so that multiple repetitions of the signal will undergo independent fading conditions. Error correction codes with interleaving can realize the time diversity in the wireless channel. However, the use of error correction codes with large interleavers to achieve the time diversity effect results in bandwidth expansion and large delays.

Frequency diversity: Frequency diversity is achieved by transmitting the same signal over different carrier frequencies, whose separation is larger than the coherence bandwidth (the range of frequencies in which the channel exhibits a flat response) of the channel. This form of diversity also suffers from bandwidth expansion.

Spatial diversity: Spatial diversity utilizes multiple antennas at the transmitter or receiver or at both ends of point-to-point links to achieve diversity. With sufficient spacing, the signals transmitted from multiple antennas experience independent fading and can be coherently combined at the receiver using appropriate signal processing techniques.

In order to benefit from the different transmitted copies of the original information, there are different combination methods at the destination node. In general, the choice of the diversity combination method depends on the desired trade-off between complexity and performance, and the amount of the available channel state information (CSI). Selection combining (SC) just chooses the diversity branch with the highest instantaneous signal amplitude, and is the simplest, but it has the worst performance. Maximal ratio combining (MRC) provides an optimal performance in the presence of AWGN noise, but it is the most complex and requires a perfect CSI. Equal gain combining (EGC) also combines weighted versions of the different diversity branches as in MRC, but it has a reduced complexity since it provides an equal weight to all branches so that it is independent of the knowledge of the instantaneous signal envelope amplitudes. Generalized selection combining (GSC) involves MRC among a selected subset of the overall diversity branches.

In order to investigate the effect of channel fading and also the effect of diversity for improving the performance, a simulation example of two independent flat-fading channels using Jackes isotropic fading channel model [108] has been carried out, and the result is shown in Figure 7.1. It shows the temporal variation of the received signal power. In this example, the carrier frequency is 900 MHz, and the mobile velocity is 100 km/h. For these conditions, the maximum Doppler frequency is 83.3 Hz. It is clear from this figure that reliable communication is difficult to achieve with this type of channel. Note that there are frequent fades of more than 10 dB and about six fades of more than 20 dB for channels 1 and 2. As indicated earlier, diversity techniques are considered as the most effective means for combating the effect of fading. As shown in this figure, the effective fading is greatly reduced using the SC method, which selects the better channel. There are far fewer instances of fading exceeding 10 dB, and therefore the performance is greatly improved.

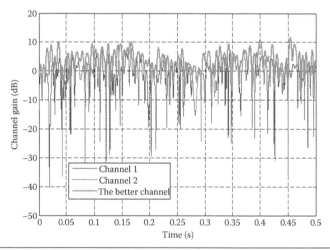

Figure 7.1   Example of two independent flat-fading channels using Jackes model with a carrier frequency of 900 MHz, and a mobile velocity of 100 km/h. The effect of SC is also indicated.

7.2.2 Multiple-Antenna Systems

The spatial diversity utilizing the MIMO systems can significantly improve the capacity and reliability of communication over fading channels using spatial multiplexing and space–time coding [109,110]. When there are multiple antennas at the receiver, the diversity combination techniques and the results of the previous section are directly applicable. However, when there are multiple antennas at the transmitter, more hardware complexity is required because the resulting arrays of signals are superimposed at the receive antennas. There has been significant work on the design of practical coding methods for multiple-antenna systems that can take advantage of the offered spatial diversity of multiple transmit antennas to achieve similar diversity gains to receive antenna arrays, including the space–time codes described in [110–112].

Figure 7.2 shows an example of a multiple-antenna system with M transmit antennas and N receive antennas. It has been proved that in the high SNR regime, the capacity of a MIMO channel with M transmit antennas, N receive antennas, and independent identically distributed (i.i.d) Rayleigh fading between each pair of antennas is given by

$$C(\text{SNR}) \approx \min(M, N) \log(\text{SNR}) \qquad (7.3)$$

where the number of degrees of freedom in the channel is thus the minimum of M and N. It has also been shown that the space–time codes can be designed such that the SER behaves as $c.\text{SNR}^{-MN}$ at high SNR values for some constant c [110,111]. Therefore, the maximum diversity

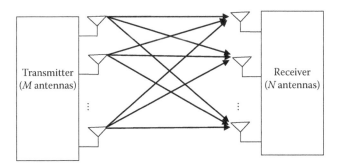

Figure 7.2   MIMO channels.

order is given by $D = MN$, where this is also the number of independent fading signals between the transmitting and receiving terminals.

7.3 Classical Relay Channel

The classical three-terminal relay channel was first introduced in [113] and is shown in Figure 7.3. It assumes that there is a source S that wants to transmit information to a single destination D. Moreover, there is a relay node R that is able to help the destination. The work in [114] analyzed the capacity of this three-node network. Lower and upper bounds for the capacity of this channel with AWGN were developed. It was assumed that the relay works in the full-duplex mode, i.e., the relay receives and transmits, simultaneously, on the same band. Three random coding schemes were introduced, generally known as facilitation, cooperation, and compression. In the facilitation scheme, the relay does not actively help the source, but rather facilitates the source transmission by inducing as little interference as possible. In the cooperation scheme, the relay fully decodes the message and retransmits some information to the destination node. In the compression scheme, the relay encodes a compressed or quantized version of its received signal and transmits it to the destination. The relay channel capacity results of [114] were extended in [115] to the case, where terminals are not able to transmit and receive on the same channel, but operate in the half-duplex mode.

7.4 Cooperative Communication

In this section, we present an overview of a new trend in wireless communications known as cooperative communication, where the users in the wireless network, whether cellular or ad hoc, may increase

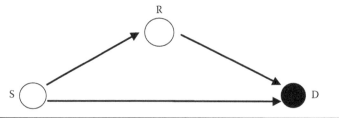

Figure 7.3 Classical relay channel.

their effective QoS (measured at the physical layer by bit error rate (BER), or outage probability) via cooperation. The basic idea behind cooperative communication is the classical relay channel introduced in the previous section. However, the cooperative communication we consider here is different from the relay channel in some respects:

- In the classical relay channel transmission, the relay terminal is an additional terminal, which helps the source (it does not have its own information to transmit), while in the cooperative communication, there are two sources which have to help each other, and both have information to transmit as shown in Figure 7.4.

- Recent developments in cooperative communications are motivated by the concept of diversity in a fading channel, while the authors in [114] analyzed the capacity in an AWGN channel. Therefore, the recent work in cooperation has taken somewhat different emphasis. Some important results for the capacity bounds (achievable rates) in cooperative wireless networks are included in [115,116].

As indicated in Section 7.2, MIMO systems are considered as effective means to achieve spatial diversity without bandwidth expansion as well as to increase the capacity. Unfortunately, the physical size of most mobile hardware devices limits the number of antennas that can be co-deployed due to probabilistic correlation between antennas. This is attributed to the fact that the separation between antennas must be at least on the order of half the wavelength of the carrier frequency to

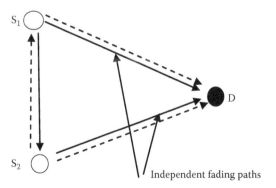

Figure 7.4   Cooperative communication, where each terminal is both a user and a relay.

avoid correlated fading. Cooperative communications refer to the class of techniques, where the benefits of MIMO technology are gained via sharing information between multiple cooperating terminals in a wireless network. Wireless relay networks that employ cooperative diversity have sometimes been referred to as virtual MIMO systems.

The idea of cooperation rises naturally in non-infrastructure networks such as ad-hoc networks and peer-to-peer communication networks. The self-organization feature of ad-hoc networks allows them to adapt to a wide spectrum of applications and network conditions and reduces the cost of configuration and maintenance. Although the success of ad-hoc networks in the commercial domain is limited, some new classes of networks emerged such as mesh networks and sensor networks that share some of the characteristics of ad-hoc networks.

The idea of cooperation has also found support in infrastructure-based broadband networks [117]. Conventionally, infrastructure-based networks follow a single-hop cellular architecture, in which users and the BSs communicate directly. The main challenge in today's wireless broadband networks is to support high rate data communication with continuous coverage at a reduced cost. The scarcity of wireless spectrum encouraged the allocation of high-frequency bands, where power attenuation with distance is more severe. This factor significantly decreases the coverage of a BS. The solution is to deploy more BSs in order to increase the capacity and coverage. However, this trivial solution sometimes called deploying microcells adds to the already high infrastructure and deployment costs. As a result, we face a scenario in which the wireless systems can achieve any two, but not all three, of high capacity, high coverage, and low cost. Integrating cooperative communication to cellular networks and forming hybrid networks emerged as a pragmatic solution to mitigate this problem. Although wireless relays use additional radio resources, they have a lower cost compared to BSs, since they do not require a high-capacity wired connection to the backbone. Multihop relaying is already a part of the standards currently being developed for wireless broadband systems such as 802.16j and 802.16m, and the LTE-advanced, which is an indication of growing consensus of the effectiveness of cooperative communication [103,104].

In general, there are two types of cooperative communication. The conventional and simplest form of cooperation is multihop relaying.

A typical example is the approach described earlier to enhance the coverage of the infrastructure-based networks. This approach can be seen as the simplest form of the cooperative communication. Another type is the cooperation that can be achieved by exploiting two fundamental features of the wireless medium: its broadcast nature, and its ability to achieve diversity through independent channels. It is known as cooperative diversity, and it will be explained in detail in the following section.

7.5 Cooperative Diversity Protocols

As indicated in the previous section, cooperative diversity is a new approach in which several single-antenna nodes cooperate with each other in order to form virtual multiple transmit antennas to achieve the benefits of the MIMO systems. The concept of cooperative diversity was first introduced in [100,101], and it is known as user cooperation. In [102], different protocols that achieve spatial diversity through node cooperation were proposed such as the DF and the AF protocols. In the DF protocol, the relay node decodes the source symbol before retransmission to the destination node. In the AF protocol, the relay amplifies the received signal before retransmission to the destination. We now review, in some detail, the main signaling methods used at the relay node in order to help the source forward its information to the destination node, and compare their performance via simulation.

7.5.1 Direct Transmission

In this case, no intermediate (relay) node between the source and the destination is considered. We introduce this case for the comparison purpose and to show the advantages of the cooperative diversity techniques compared to the direct transmission case. If we assume randomly generated i.i.d zero-mean circularly symmetric complex Gaussian inputs, the maximum mutual information that can be achieved between the source and the destination is given by

$$I_D = \log\left(1 + |h_{s,d}|^2 \, \text{SNR}\right) \tag{7.4}$$

where $h_{s,d}$ is the source-destination fading coefficient and it also captures the effect of path loss and shadowing. Statistically, $h_{s,d}$ is modeled as a zero-mean, independent, circularly symmetric complex Gaussian random variable with variance $\sigma_{s,d}^2$. SNR is the average SNR defined as $\text{SNR} = P/N_0$, where P denotes the average transmitted power per symbol and N_0 is the noise variance per symbol. Note that logarithms are to base two. The outage event for target rate R is given by $I_D < R$. Therefore, the outage probability, in bits per channel use, is then given by

$$P_O^D = \Pr\left[I_D < R\right]$$

$$= \Pr\left[\left|h_{s,d}\right|^2 < \frac{2^R - 1}{\text{SNR}}\right] \qquad (7.5)$$

The probability of outage can be estimated by integrating overall fading channel realizations that result in a rate less than the target rate R.

7.5.2 Amplify and Forward

This is the simplest strategy that can be used at the relay, because the relay node does not do any decoding for the received signal. The relay amplifies the received signal from the source and transmits it to the destination without doing any decision. The main drawback of this strategy is that the relay transmitter amplifies its own receiver noise. When this strategy is applied to the cooperative communication, it is able to achieve a better uncoded BER than that in the case of direct transmission [102]. Additionally, the outage probability of the cooperative communication was also derived, demonstrating that a diversity of order two is obtained for two cooperative users. In this case, the maximum mutual information between the input and the output is given by [102]

$$I_{\text{AF}} = \frac{1}{2}\log\left(1 + \text{SNR}\left|h_{s,d}\right|^2 + f\left(\text{SNR}\left|h_{s,r}\right|^2, \text{SNR}\left|h_{r,d}\right|^2\right)\right) \qquad (7.6)$$

as a function of the fading coefficients, where

$$f(x, y) := \frac{xy}{x + y + 1} \qquad (7.7)$$

The outage event for spectral efficiency R is given by $I_{AF} < R$ and the outage probability is given by

$$P_O^{AF} = \Pr\left[|h_{s,d}|^2 + \frac{1}{\text{SNR}} f\left(\text{SNR}|h_{s,r}|^2, \text{SNR}|h_{r,d}|^2\right) < \frac{2^{2R}-1}{\text{SNR}}\right] \quad (7.8)$$

When the relay is equipped with multiple antennas, and there is CSI about the source-relay and relay-destination links, the AF strategy can attain significant gains over the direct transmission by means of optimum linear filtering of the data to be forwarded by the relay [118].

7.5.3 Fixed Decode and Forward

For the simplest DF algorithm with repetition coding, the random variable representing the maximum mutual information, given the channel fading coefficients, is given by [102]

$$I_{FDF} = \frac{1}{2} \min\left\{\log\left(1+\text{SNR}\,|h_{s,r}|^2\right), \log\left(1+\text{SNR}\left(|h_{s,d}|^2+|h_{r,d}|^2\right)\right)\right\} \quad (7.9)$$

As indicated earlier, the outage probability can be computed, defining $g(\text{SNR}) = \dfrac{2^{2R}-1}{\text{SNR}}$ as follows:

$$P_O^{FDF} = \Pr\left[|h_{s,r}|^2 < g(\text{SNR})\right]$$

$$+ \Pr\left[|h_{s,r}|^2 \geq g(\text{SNR})\right]\Pr\left[|h_{s,d}|^2 + |h_{r,d}|^2 < g(\text{SNR})\right] \quad (7.10)$$

Requiring the relay to fully and correctly decode the source information limits the performance of the fixed decode-and-forward (FDF) protocol to that of the direct transmission between the source and destination. In other words, the FDF protocol does not offer a diversity gain. In order to solve this problem, a selection DF or simply a DF protocol was proposed.

7.5.4 Selection Decode and Forward

In this method, a relay attempts to detect and decode the partners' signals and then retransmits the decoded signals after re-encoding.

The transmission of one packet is divided into two phases. In the first phase, the source transmits its information to the relay and the destination. During this interval, the relay performs decoding for the received information. Unlike the FDF protocol, it does not send the decoded information, if it is not correctly decoded. In the case of incorrect decoding, the source repeats its information to the destination. The outage event is given as

$$
I_{DF} = \begin{cases}
\dfrac{1}{2}\log\left(1+2\text{SNR}\,|h_{s,d}|^2\right) & |h_{s,r}|^2 < g(\text{SNR}) \\[2mm]
\dfrac{1}{2}\log\left(1+\text{SNR}\,|h_{s,d}|^2+\text{SNR}\,|h_{r,d}|^2\right) & |h_{s,r}|^2 \geq g(\text{SNR})
\end{cases}
\tag{7.11}
$$

The outage probability is then calculated as

$$
P_{out}^{DF} = \Pr[I_{DF} < R]
$$

$$
= \Pr\left[|h_{s,r}|^2 < g(\text{SNR})\right]\Pr\left[2|h_{s,d}|^2 < g(\text{SNR})\right]
$$

$$
+ \Pr\left[|h_{s,r}|^2 \geq g(\text{SNR})\right]\Pr\left[|h_{s,d}|^2 + |h_{r,d}|^2 < g(\text{SNR})\right] \tag{7.12}
$$

All the random variables in the earlier expressions are independent exponential random variables, which makes the calculation of the outage probability straightforward. The diversity order of the earlier schemes was derived in [102] at high SNR values. It was proved that the DF and AF protocols achieve a diversity of order two when using one relay node.

In order to get insight into the performance of the different protocols, we have carried out simulations for the outage probability and the symbol error rate (SER) performance. Figure 7.5 shows a performance comparison between the different protocols in outage probability. The FDF protocol gives the worst performance. As indicated earlier, it is limited by the performance of the source-to-relay link, and this will be possible for small source rates. The diversity gain of the selection DF and AF protocols is clear from this figure. If this figure is plotted on a log–log scale, the slope of the curve will represent the diversity order,

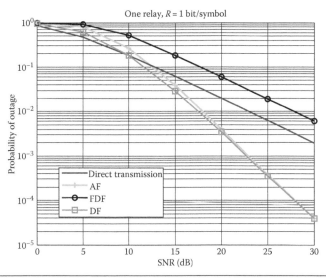

Figure 7.5 Example of outage performance of noncooperative and cooperative transmission computed via Monte Carlo simulation.

i.e., the full spatial diversity equal to the number of cooperating nodes (the source plus the number of relays). In our case, the diversity order is two. It is worth noting that this scheme uses repetition coding. In the first phase, the source broadcasts its information to the destination and also to the relays. In the second phase, the relay sends a version of the received signal at the first phase to the destination according to the protocol used, e.g., AF or DF. The source remains idle in the second phase. This causes a loss in the source rate.

In addition, we have carried out simulations to compare the SER performance of the different protocols. Figure 7.6 shows the SER variations for different cooperative protocols. For the DF protocol, the relay repeats the received information if it can decode the source information correctly. If the source information is not decoded correctly, the relay sends nothing. MRC is used at the receiver. It can be seen from this figure that a diversity of order two can be obtained using the DF and AF protocols.

From the previous discussion, we can use MRC at the output to gain a maximum diversity order. This is true, but at the expense of a reduced source rate R. The main problem with multinode DF and AF protocols is the loss in the data rate as the number of relay nodes increases.

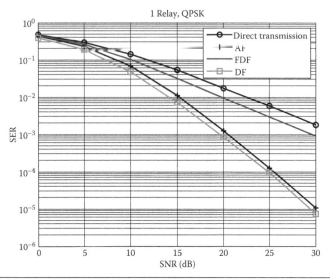

Figure 7.6   SER variations with the SNR for different cooperative protocols.

7.5.5 Compress and Forward

The compress and forward (CF) cooperative protocol differs from the DF/AF protocol in that in the latter case the relay transmits a copy of the received message, while in the CF protocol the relay transmits a quantized and compressed version of the received message. Therefore, the destination node will combine the received message from the source node and its quantized and compressed version from the relay node. The CF protocol was first described in [114]. The quantization and compression process at the relay node is a process of source encoding and is done using a set of coding techniques known as distributed source coding or Wyner–Ziv coding [119]. It should be noted that the information received at the destination from the source can be used as a side information, while decoding the message from the relay. This will allow for encoding at a lower source encoding rate [120].

7.6 Cooperative Diversity Techniques

In this section, we introduce various algorithms in the literature for cooperative diversity in which sets of wireless terminals benefit from relaying messages for each other to propagate redundant signals over multiple paths in the network. This redundancy allows the

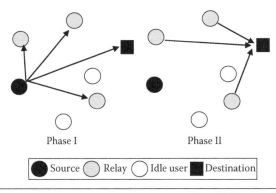

Phase I   Phase II

Source ◐ Relay ◯ Idle user ■ Destination

Figure 7.7 Two-phase cooperative diversity system with multiple relays. Phase I is the broadcasting phase and phase II is the cooperation phase.

ultimate receivers to essentially average channel variations resulting from fading. Generally, both AF and DF protocols consist of two transmission phases as shown in Figure 7.7 for multiple relay cooperative diversity systems. In the first phase, the source node transmits the information to its intended destination and all potential relays. In the second phase, the relays transmit the information to the destination. The classification of cooperative diversity schemes in the literature depends mainly on the way the relay nodes forward the received information in the first phase to the destination as will be explained in the following sections.

7.6.1 Cooperative Diversity Based on Repetition Coding

The various cooperative diversity protocols proposed in [102] are based on repetition coding using a single relay node. Consider a wireless network with a set of transmitting terminals $S = \{1, 2, \ldots, S\}$. Each user $s \in S$ has the ability to use the other terminals $S - \{s\}$ as relays to send his information to a single destination terminal denoted as $d(s) \notin S$. Figure 7.8 shows an example of channel allocation for noncooperative transmission in which the channel allocation is made across the entire frequency band, and each transmitting terminal utilizes a fraction $1/S$ of the total degrees of freedom in the channel. Furthermore, subchannel allocation for different relay nodes using repetition coding is made across the time domain as shown in Figure 7.9. Each subchannel contains a fraction of $1/S^2$ of the total degrees of freedom in the channel. For the model shown in Figure 7.9, the relays either

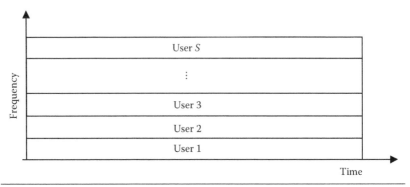

Figure 7.8   Channel allocation for *S* noncooperative users.

	Phase I		Phase II		
S Transmits	1 Repeats *S*	1 Repeats *S*	...		*S*-1 Repeats *S*
⋮	⋮	⋮	⋮		
3 Transmits	1 Repeats 3	2 Repeats 3	...		*S* Repeats 3
2 Transmits	1 Repeats 2	3 Repeats 2	...		*S* Repeats 2
1 Transmits	2 Repeats 1	3 Repeats 1	...		*S* Repeats 1

Figure 7.9   Example of channel allocation for *S* users in a repetition-based cooperative diversity. The source channel allocations across frequency and relay subchannel allocations across time.

amplify what they receive or fully decode and repeat the source signal. Similar to the noncooperative case, the transmission between the source s and destination $d(s)$ utilizes a fraction $1/S$ of the total degrees of freedom in the channel.

The fixed AF scheme proposed in [102] can be viewed as a repetition coding from two users, except that the relay amplifies its own receiver noise. The source-relay channel fading coefficient is utilized at the relay to amplify the signal. Therefore, it is important to have accurate CSI at the relay node. However, CSI may not be accurately obtainable if the fading coefficients vary quickly. Since the coherence time decreases in the fast fading case, estimation of the CSI substantially reduces the data transmission rate due to the pilot insertion. The destination receiver can use the MRC technique to decode the information. Repetition-based cooperative diversity using multiple relays and the AF protocol was presented in [121].

In the DF protocol, the relay can decode the information either by fully decoding the source message to estimate the source codeword or by estimating symbol by symbol. The performance of the DF cooperative diversity scheme was analyzed for more than two relays in [122]. In the earlier schemes, the source and relays use a common codebook, which is equivalent to repetition coding for the destination.

However, the spatial diversity achieved by cooperative diversity protocols using repetition coding comes at the price of a decreased transmission rate with the number of cooperating users. The use of orthogonal subchannels for the relay node transmissions, either time division multiple access (TDMA) or frequency division multiple access (FDMA), results in a high loss of the system spectral efficiency. This leads to the use of what is known as DSTC, which will be discussed in the following section.

7.6.2 *Cooperative Diversity Based on Space–Time Coding*

Distributed space–time coding can be achieved through node cooperation to emulate multiple antenna transmission. The term distributed comes from the fact that the virtual multiantenna transmitter is distributed between the relays. The relay nodes are allowed to transmit simultaneously on the same subchannels in the second phase as shown in Figure 7.10. Note that $D(s)$ denotes the decoding set, which represents the relay nodes that have decoded correctly

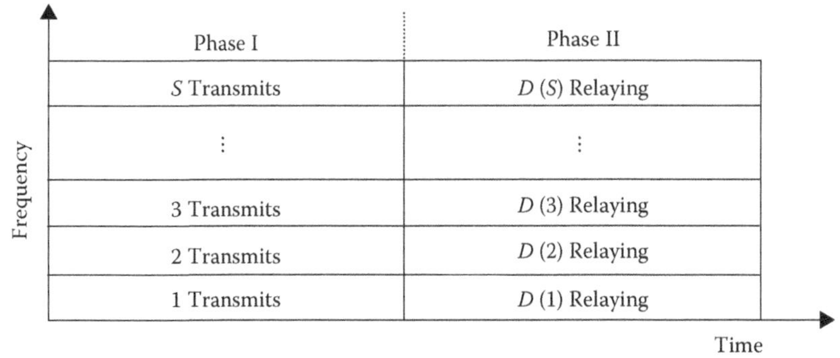

	Phase I	Phase II
	S Transmits	*D* (*S*) Relaying
	⋮	⋮
	3 Transmits	*D* (3) Relaying
	2 Transmits	*D* (2) Relaying
	1 Transmits	*D* (1) Relaying

Frequency (vertical axis) — Time (horizontal axis)

Figure 7.10 Channel allocation for *S* users in distributed space–time-coded cooperative diversity with the DF protocol.

the signal of the source s transmitted in the first phase. For the case of using the AF protocol, all the relay nodes transmit simultaneously in the second phase. Again, transmission between source s and destination $d(s)$ utilizes $1/S$ of the total degrees of freedom in the channel. However, in contrast to noncooperative transmission and repetition-based cooperative diversity transmission, each terminal employing space–time-coded cooperative diversity transmits in half of the total degrees of freedom in the channel.

From Figure 7.10, we realize that using the DSTC the user can exploit half the degrees of freedom allocated to him for transmitting his information to the destination regardless of the number of relays. In contrast, using repetition coding as shown in Figure 7.9, the source utilizes only $1/S$ of the degrees of freedom allocated to the user. So DSTC is more spectrally efficient than repetition coding.

There is a lot of work in literature to exploit the space–time codes designed for MIMO systems and apply them in a distributed manner. Suppose that we have the following space–time coding matrix:

$$C = \begin{pmatrix} c_{1,1} & c_{1,2} & \cdots & c_{1,N} \\ c_{2,1} & c_{2,2} & \cdots & c_{2,N} \\ \vdots & \vdots & \ddots & \vdots \\ c_{T,1} & c_{T,2} & \cdots & c_{T,N} \end{pmatrix}, \tag{7.13}$$

where $C_{i,j}$ is the symbol transmitted at time slot i from the transmit antenna j. In DSTC, each relay node transmits one column of the earlier matrix. So, N cooperating relay nodes represent virtual N antennas in the MIMO systems. In cooperative diversity, we deal with a single user at the destination. So the cooperative diversity system behaves like a MISO system. Several works in the literature have considered the application of existing space–time codes in a distributed fashion for the wireless relay networks [122–125].

In practice, the number of users participating in the cooperative protocol is unknown a priori. Moreover, in the case of the DF protocol, if any relay cannot decode the received source signal correctly, the corresponding column in the code matrix should be omitted (or replaced with zeros). To overcome this problem, the columns of the code matrix should be orthogonal to each other. Space–time block codes based on orthogonal design arranged for MIMO system are considered as a good candidate for coded cooperative systems. It was proved in [126] that any code designed for MIMO systems that can achieve full diversity and maximum coding gain will achieve this if it is used with the AF protocol. If it is used with the DF protocol, it will give full diversity but not necessarily maximize the coding gain.

Most of the works reported earlier consider the flat-fading channel. In [127], the authors proposed three different DSTC schemes that were originally proposed for the MIMO systems with frequency-selective fading channels. These schemes are distributed time reversal, distributed single carrier, and distributed orthogonal frequency division multiplexing.

The DSTC systems have a major disadvantage in synchronous reception in the second phase from all the relay nodes due to the different propagation delays among the relays to the destination node.

7.6.3 Cooperative Diversity Based on Relay Selection

In practice, DSTC design is quite difficult due to the distributed nature of cooperative links, as opposed to codes designed for colocated MIMO systems. Therefore, DSTC requires accurate CSI and strict symbol-level synchronization. In order to alleviate these problems and also to reduce the multiplexing loss of orthogonal transmission, relay selection schemes were proposed as a practical and efficient method of implementing cooperative diversity in distributed systems [128–132]. Instead of transmitting the data from all the relays, only one relay can be selected based on its channel quality to the source and the destination.

There are two main selection schemes proposed in the literature known as selection cooperation and opportunistic relaying schemes [128,129] as shown in Figures 7.11 and 7.12, respectively. It should be noted that in [130], the authors used the terms reactive and proactive opportunistic relaying to refer to what is termed here as selection cooperation and opportunistic relaying, respectively. In both schemes, only a single relay out of N available relay nodes is selected to forward the information to the destination. However, their modes of operation are different:

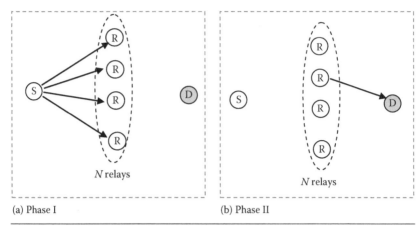

Figure 7.11   Selection cooperation scheme. (a) The source transmits its information and the nodes that decode correctly constitute a decoding set. (b) The relay is selected from the decoding set.

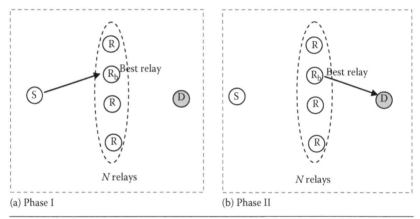

Figure 7.12   Opportunistic relaying scheme. (a) The best relay is selected and the source node transmits the information to the selected relay. (b) The selected relay forwards the information to the destination node.

Selection cooperation: This scheme is more relevant to the DF protocol. In the first phase, all relays listen to the source node, and relays that decode the signal correctly constitute the decoding set D. In the second phase, the relaying node R_b is selected from this decoding set such that it maximizes the relay-destination mutual information or equivalently the relay-destination channel gain:

$$R_b = \underset{R_i \in \mathcal{D}}{\mathrm{argmax}} \left| h_{R_i,D} \right|^2 \tag{7.14}$$

where $h_{R_i,D}$ is the channel gain between the ith relay node and the destination node. It should be noted that in order to select the best relay, the fact that an individual relay has decoded correctly must be communicated to the node making the selection, i.e., that node must have knowledge of the decoding set.

Opportunistic relaying: In this scheme, the relay is selected in the first phase before transmitting the message from the source node. The participating (best) relay is selected according to the following criteria:

$$R_b = \underset{R_i=1,2,\dots,N}{\operatorname{argmax}} \quad \min\left(\left|h_{S,R_i}\right|^2, \left|h_{R_i,D}\right|^2\right) \tag{7.15}$$

where h_{S,R_i} is the channel gain between the source node and the ith relay node. Both AF and DF protocols can be applied with this scheme. Note that in the DF opportunistic relaying, the relay is picked from the entire set of the available relays, not just from the decoding set. The constraint of successful decoding is introduced after the relay selection.

A performance comparison between the two schemes in terms of the outage probability and the BER was presented in [133]. A threshold-based DF protocol was considered in the analysis. It was concluded that the selection cooperation scheme outperforms the opportunistic relaying scheme in terms of outage probability. In terms of BER, both schemes may outperform one another, depending on the SNR threshold that determines the set of relays that participate in the forwarding process.

Another point that should be addressed is the implementation of the relay selection and decision making. In other words, we should identify which identity evaluates the metrics and selects the relay. There are two protocols in the literature called destination-driven protocol and relay-driven protocol. The destination-driven protocol is suited to DF selection cooperation, where the destination node selects the best relay depending on a feedback bit sent from each trusted relay in the first phase to the destination node [128]. In the relay-driven protocol, the relay selects itself depending on a timer and the selection procedure was described in [129].

7.6.4 Cooperative Diversity Based on Channel Coding

A different cooperation framework based on channel coding, called coded cooperation, was proposed in [134–136]. In this scheme, the relay sends incremental redundancy, which when combined at the receiver of the destination node with the codeword sent by the source results in a codeword with larger redundancy. Note that the relay node does not repeat the source information as in the previously reported schemes using the DF and AF protocols. Instead of repeating the symbols, the codeword of each user is partitioned into two parts: one part is transmitted by the source and the other by the relay. Note that the coded cooperation assumes that the relay is able to successfully decode the source message, which can be accomplished using a cyclic redundancy check (CRC) encoder at the source node. If the relay cannot decode correctly, it will not be able to generate the extra parity symbols for the source. Because of this, the weaker codeword transmitted by the source has to be a valid codeword that can be decoded at the destination node, when the relay is not able to generate the extra parity check symbols due to incorrect decoding.

8

COOPERATIVE SPACE–TIME/FREQUENCY CODING SCHEMES FOR SC-FDMA SYSTEMS

Future broadband wireless networks should meet stringent requirements such as high data rate services over the dispersive channel with high transmission reliability. Since the wireless resources such as the bandwidth and power are very scarce, they cannot cope with the increasing demand for higher data rates. Furthermore, the wireless channel suffers from many impairments such as fading, shadowing, and multiuser interference that may degrade the system performance. In order to achieve such high bit rates and at the same time meet the quality of service (QoS) requirements, orthogonal frequency division multiple access (OFDMA) is considered as a mature technique to mitigate the problem of frequency selectivity and ISI among other well-known advantages [137,138]. However, OFDMA, as a multicarrier technique, suffers from two main drawbacks. One is the high peak-to-average power ratio (PAPR) which is a critical problem especially in the uplink direction. It is required for the mobile terminal to be of low cost and high power efficiency which is difficult to achieve with the earlier constraints. In addition, OFDMA suffers from CFOs. Alternatively, SC-FDMA is a promising solution which is adopted as an uplink multiple access scheme in 3GPP-LTE [14,139,140]. Due to its single-carrier nature, it has a lower PAPR while keeping most of the advantages of OFDMA such as assigning different numbers of subcarriers to different users and applying adaptive modulation and coding.

Multiple-antenna systems offer attractive ways for enhancing the capacity and reliability of the communication over fading channels using spatial multiplexing and/or space–time coding. Applying

MIMO techniques for the SC-FDMA system was considered in [90,141]. Unfortunately, designing distributed space–time/frequency codes for SC-FDMA systems is not straightforward. The codes should not affect the low PAPR, which is a main feature of the system. In [142], the authors proposed a coded cooperation scheme to facilitate the time/frequency/space diversity for SC-FDMA.

In this chapter, we discuss relay-assisted distributed space–time/frequency codes for SC-FDMA systems in the uplink direction [143,144]. The distributed SFC is achieved in two phases and the code does not affect significantly the PAPR of the transmitted signal. This code is suitable for fast-fading channels, where the channel gain changes from block to block. The coding is applied within a single transmitted block. As the two coded symbols may experience different channel gains within a single transmitted block, direct application of Alamouti decoding using MLD over one symbol may degrade the performance especially with a large number of allocated subcarriers per user. Therefore, a MMSE decoder is used at the receiver to overcome this problem with a slight increase in the decoding complexity. Furthermore, we discuss a distributed STC for the slow-fading channel environment. This code is more spectrally efficient than the classical DSTC schemes. It achieves a transmission rate of 2/3. For the two discussed codes, the DF protocol is considered and erroneous decoding at the relay node is taken into account. If the relay cannot decode correctly in the first phase, it remains idle in the second phase.

8.1 SC-FDMA System Model

In this section, we return to the system model for the conventional SC-FDMA from a different perspective. MIMO SC-FDMA system model is also briefly discussed. I_n is an $n \times n$ identity matrix; $C\mathcal{N}(\mu,\sigma^2)$ denotes a circularly symmetric complex Gaussian random variable with mean μ and variance σ^2.

8.1.1 SISO SC-FDMA System Model

The SC-FDMA system model of user i with Single-Input–Single-Output (SISO) configuration is shown in Figure 8.1. A group of

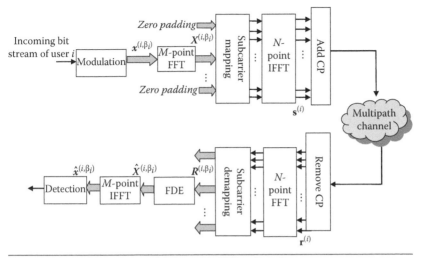

Figure 8.1 SC-FDMA system model of user i for SISO configuration.

M-modulated time-domain symbols that belong to user i, who will be given M subcarriers after the fast Fourier transform (FFT) operation, can be represented as

$$x^{(i,\beta_i)} = \left[x_{(i-1)M}^{(i,\beta_i)} \ x_{(i-1)M+1}^{(i,\beta_i)} \cdots x_{(i-1)M+M-1}^{(i,\beta_i)} \right]^T \tag{8.1}$$

where

$x_m^{(i,\beta_i)}$ is the mth time-domain symbol of user i

β_i denotes the frequency band that represents the subcarriers that belong to the ith user

The block $x^{(i,\beta_i)}$ is converted to the frequency domain

These M frequency-domain symbols of user i transmitted over the band β_i are expressed as

$$X^{(i,\beta_i)} = \left[X_{(i-1)M}^{(i,\beta_i)} \ X_{(i-1)M+1}^{(i,\beta_i)} \cdots X_{(i-1)M+M-1}^{(i,\beta_i)} \right]^T \tag{8.2}$$

Localized subcarrier allocation is assumed in the following analysis as illustrated in Figure 8.2. Each user is pre-allocated a set of M subcarriers and the rest $(N-M)$ subcarriers are padded zeros, where N denotes the total number of subcarriers in the system. Then, the signal is converted to the time domain using IFFT operation. The transmitted signal of user i without Cyclic Prefix (CP) can be expressed as

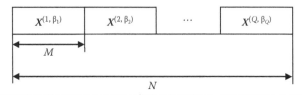

Figure 8.2   Illustration for the localized subcarriers allocations in the frequency domain for Q users.

$$s^{(i)} = \sqrt{P} F_N^{\mathcal{H}} \, T_{N,M} \, F_M \, x^{(i,\beta_i)} \tag{8.3}$$

where

P is the total average transmitted power per symbol

$F_N^{\mathcal{H}}$ is the N-point IFFT matrix

F_M is the M-point FFT matrix

The $N \times M$ matrix $T_{N,M}$ is the mapping matrix for subcarriers allocation

Each block is appended with a CP prior to transmission in order to prevent IBI and to convert the linear convolution of the channel to a circular convolution. At the receiver, the CP is first removed and the received signal in the time domain is given by

$$r^{(i)} = \sqrt{P} H s^{(i)} + n \tag{8.4}$$

where H is an $N \times N$ circulant matrix whose first column contains the channel impulse response appended by zeros. Therefore, H can be decomposed as follows:

$$H = F_N^{\mathcal{H}} \Lambda F_N \tag{8.5}$$

where Λ is an $N \times N$ diagonal matrix containing the FFT of the channel impulse response. The signal goes through the FFT operation. Then, the received signal of the ith user, after demapping, can be expressed as

$$R^{(i,\beta_i)} = \sqrt{P} \Lambda^{(\beta_i)} X^{(i,\beta_i)} + N \tag{8.6}$$

where

$\Lambda^{(\beta_i)} = \mathrm{diag}\{H_{(i-1)M}, H_{(i-1)M+1}, \ldots, H_{(i-1)M+M-1}\}$, where H_m is the frequency response of the channel for the subcarrier m

N is the additive noise vector that contains noise elements modeled as zero-mean complex Gaussian with variance $N_0/2$ per dimension

Then, single-tap equalization is performed in the frequency domain to retrieve the transmitted symbols. We will consider the MMSE equalizer, for which the output signal is given by

$$\hat{X}^{(i,\beta_i)} = W\left(\sqrt{P}\Lambda^{(\beta_i)}X^{(i,\beta_i)} + N\right)$$

$$= \sqrt{P}W\Lambda^{(\beta_i)}X^{(i,\beta_i)} + \tilde{N} \tag{8.7}$$

where

$$\tilde{N} = WN$$

W is an $M \times M$ diagonal matrix whose (k, k) element is given by

$$W(k,k) = \frac{\Lambda^{(\beta_i)*}(k,k)}{\left|\Lambda^{(\beta_i)}(k,k)\right|^2 + \dfrac{1}{\text{SNR}}} \tag{8.8}$$

where $\text{SNR} = P/N_0$ is the average SNR at the destination node. The decoding in the SC-FDMA system is achieved in the time domain, so we need an M-point IFFT in order to demodulate the transmitted signal. The signal after the IFFT operation is given by

$$\hat{x}^{(i,\beta_i)} = F_M^{\mathcal{H}}\hat{X}^{(i,\beta)} \tag{8.9}$$

The recovered data $\hat{x}^{(i,\beta_i)}$ is sent to the decision operation at the final stage of the SC-FDMA receiver.

8.1.2 MIMO SC-FDMA System Model

In this chapter, we will consider Alamouti-based distributed space–time/frequency codes. Therefore, in this section, the MIMO system model using space–time coding, which represents the basis for the subsequent analysis, is presented [145]. For the case of a 2×2 MIMO system, the transmitted codeword matrix from the ith user in the frequency domain can be represented as

$$C^{(i,\beta_i)} = \begin{bmatrix} X_1^{(i,\beta_i)}(1) & -X_2^{(i,\beta_i)*}(2) \\ X_2^{(i,\beta_i)}(1) & X_1^{(i,\beta_i)*}(2) \end{bmatrix} \tag{8.10}$$

where $X_j^{(i,\beta_i)}(t) = \left[X_{(i-1)M}^{(i,\beta_i)} \ X_{(i-1)M+1}^{(i,\beta_i)} \ \cdots \ X_{(i-1)M+M-1}^{(i,\beta_i)} \right]$ is the jth symbol vector of the ith user transmitted using the band β_i at the time slot t. We have assumed that each time slot corresponds to the transmission of a block of M symbols. Thus, the codeword matrix $C^{(i,\beta_i)}$ would span two time slots. Each column in (8.10) is transmitted from the two antennas at the same time and each row is transmitted from one antenna at two time slots. The received signals in the frequency domain after the FFT operation and carrier demapping at the destination can be represented as

$$
\begin{bmatrix} R_{B_1}^{(i,\beta_i)}(1) & R_{B_1}^{(i,\beta_i)}(2) \\ R_{B_2}^{(i,\beta_i)}(1) & R_{B_2}^{(i,\beta_i)}(2) \end{bmatrix} = \begin{bmatrix} \Lambda_{i_1,B_1}^{(\beta_i)} & \Lambda_{i_2,B_1}^{(\beta_i)} \\ \Lambda_{i_1,B_2}^{(\beta_i)} & \Lambda_{i_2,B_2}^{(\beta_i)} \end{bmatrix} C^{(i,\beta_i)} + \begin{bmatrix} N_1 & N_3 \\ N_2 & N_4 \end{bmatrix}
$$

$$(8.11)$$

where

$R_{B_n}^{(i,\beta_i)}(t)$ is the received vector of the ith user at the nth antenna of the destination node at time slot t

$\Lambda_{i_m,B_n}^{(\beta_i)}$ is the channel frequency response from the mth transmit antenna of the ith user to the nth receive antenna of the destination node over the band β_i

N_1 and N_3 are the noise vectors received at the first receive antenna at time slots 1 and 2, respectively

Similarly, N_2 and N_4 are the received noise at antenna 2

Each element of the noise vector is modeled as a zero-mean additive white Gaussian vector with variance N_0. Note that we have assumed in Equation 8.11 that the channel is constant over the two time slots. It is clear from Equations 8.10 and 8.11 that a diversity of order four can be obtained using Alamouti decoding. For the purpose of analysis in the following sections, we will consider the nth receive antenna, when the matrix $C^{(i,\beta_i)}$ is transmitted. The two received vectors at two consecutive time slots are given by

$$
R_{B_n}^{(i,\beta_i)}(1) = \Lambda_{i_1,B_n}^{(\beta_i)} X_1^{(i,\beta_i)}(1) + \Lambda_{i_2,B_n}^{(\beta_i)} X_2^{(i,\beta_i)}(1) + N_1 \qquad (8.12)
$$

$$
R_{B_n}^{(i,\beta_i)}(2) = -\Lambda_{i_1,B_n}^{(\beta_i)} X_2^{(i,\beta_i)*}(2) + \Lambda_{i_2,B_n}^{(\beta_i)} X_1^{(i,\beta_i)*}(2) + N_2 \qquad (8.13)
$$

The decoding is performed in the frequency domain in similar steps as in [9] assuming that the channel is perfectly known at the receiver side. The decision variables are constructed by combining $R_{B_n}^{(i,\beta_i)}(1)$ and $R_{B_n}^{(i,\beta_i)}(2)$ for $n = 1, 2$ in the frequency domain as follows:

$$\hat{X}_1^{(i,\beta_i)}(1) = \Lambda_{i_1,B_n}^{(\beta_i)^*} R_{B_n}^{(i,\beta_i)}(1) + \Lambda_{i_2,B_n}^{(\beta_i)} R_{B_n}^{(i,\beta_i)^*}(2) \tag{8.14}$$

$$\hat{X}_2^{(i,\beta_i)}(1) = \Lambda_{i_2,B_n}^{(\beta_i)^*} R_{B_n}^{(i,\beta_i)}(1) - \Lambda_{i_1,B_n}^{(\beta_i)} R_{B_n}^{(i,\beta_i)^*}(2) \tag{8.15}$$

Substituting Equations 8.12 and 8.13 into Equations 8.14 and 8.15, we get

$$\hat{X}_1^{(i,\beta_i)}(1) = (|\Lambda_{i_1,B_n}^{(\beta_i)}|^2 + |\Lambda_{i_2,B_n}^{(\beta_i)}|^2) X_1^{(i,\beta_i)}(1) + \tilde{N}_1 \tag{8.16}$$

$$\hat{X}_2^{(i,\beta_i)}(1) = (|\Lambda_{i_1,B_n}^{(\beta_i)}|^2 + |\Lambda_{i_2,B_n}^{(\beta_i)}|^2) X_2^{(i,\beta_i)}(1) + \tilde{N}_2 \tag{8.17}$$

where $\tilde{N}_1$ and $\tilde{N}_2$ are the new modified noise terms. It is clear from Equations 8.16 and 8.17 that a diversity of order two is achieved for both vectors transmitted from the two antennas.

8.2 Cooperative Space–Frequency Coding for SC-FDMA System

In this section, we present a cooperative diversity scheme for SC-FDMA based on distributed space–frequency coding. The discussed protocols utilize the space–frequency coding in the uplink SC-FDMA systems. We will consider Alamouti-like coding in which two users cooperate to send their information to the destination node. We will consider the DF protocol at the relay node.

8.2.1 Motivation and Cooperation Strategy

Consider an uplink transmission using the SC-FDMA system. The users in the system help each other to transmit cooperatively to the destination node in the second phase. Consider an example of two users sharing a total resource of N subcarriers that is divided equally between them. Specifically, user 1 is allocated N_1 subcarriers which are defined earlier as band β_1, and user 2 is allocated N_2 subcarriers

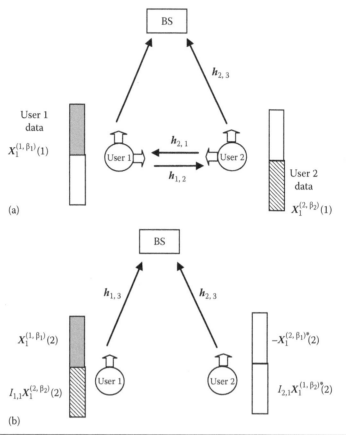

Figure 8.3 Two cooperative users in the uplink SC-FDMA network: (a) Listening Phase I and (b) cooperation Phase II.

(band β_2), where $N = N_1 + N_2$. The system has two phases as illustrated in Figure 8.3. In the first phase, both users 1 and 2 send their own information on their corresponding subcarriers to the destination node (BS). Meanwhile, each user overhears his partner's information. The transmitted signals from user 1 and user 2 are denoted by the vectors $X_1^{(1,\beta_1)}$ and $X_1^{(2,\beta_2)}$, respectively. The transmitted signals in the frequency domain are also illustrated in Table 8.1, where $I_{i,j}$ is a random variable that represents the state of the ith relay, whether it has decoded correctly at time slot j or not.

In the second phase, user 1 sends his own data on the band β_1 and his partner's data (if he received it correctly) on the band β_2. Similarly, in phase 2, user 2 sends the *processed* information of his own data on the band β_1 and his partner's processed data (if he received it correctly)

Table 8.1 Transmitted signals for the distributed SFC using DF protocol in the frequency domain

TIME SLOT (t)	USER 1		USER 2	
	β_1	β_2	β_1	β_2
1	$X_1^{(1,\beta_1)}(1)$	—	—	$X_1^{(2,\beta_2)}(1)$
2	$X_1^{(1,\beta_1)}(2)$	$l_{1,1}X_1^{(2,\beta_2)}(2)$	$X_1^{(2,\beta_1)*}(2)$	$l_{2,1}X_1^{(1,\beta_2)*}(2)$

on the band β_2 as shown in Figure 8.3. This arrangement is done so as to constitute the modified version of the Alamouti coding scheme. In other words, the transmitted signals from the two users on both bands β_1 and β_2 in the second phase constitute an Alamouti-like coding scheme in which two vectors are coded instead of two symbols. As will be shown later, the code construction does not affect the PAPR of the transmitted signal at user 1. There is only a slight increase in the PAPR for the transmitted signals from user 2 but still less than that for other space–frequency coding schemes mentioned in the literature.

As mentioned earlier, the signals transmitted in the second phase from both users constitute an Alamouti-like space–frequency code, which achieves a diversity of order two. Furthermore, the signal transmitted in the first phase can also be exploited and combined. The transmitted signals for the distributed SFC using DF protocol in the frequency domain can be combined with the signal received in the second phase at the receiver of the destination node. In the following analysis, we distinguish between two protocols regarding the received signal at the destination node in the first phase:

Protocol I: In the first phase, each user transmits his information to his partner. The destination node ignores the signals in this phase. In the second phase, in the case of correct decoding, both users communicate with the destination node to form the space–frequency code on the bands β_1 and β_2. This is motivated by protocol III proposed in [146]. As indicated in [146], the destination terminal may be engaged in data transmission to another terminal during the first time slot. Hence, the transmitted signal is received only at the relay terminal in the first phase.

Protocol II: This protocol differs from protocol I in that the signals transmitted in the first time slot are received by the destination node and combined with the coded signals in the second time slot. In the case of a fast-fading channel, the channel state changes from block to block. Therefore, a diversity of order three can be obtained as will be shown later. A further investigation of this type of non-orthogonal transmission to increase the spatial diversity was presented in [147].

8.2.2 Cooperative Space–Frequency Code for SC-FDMA with the DF Protocol

In this section, the distributed SFC with the DF protocol is considered, where each terminal is equipped with a single antenna. Table 8.1 summarizes the transmitted signals during the two phases. In the following analysis, we will consider only a one-way cooperation cycle in which user 1 is the source node and user 2 acts as a relay node. The coding is applied only over the band β_1 allocated to user 1. Note that the same procedure explained in Section 8.2.1 and Table 8.1 to constitute the modified version of the Alamouti scheme is applied here but using only the band β_1 at both users. The relay node performs decoding for the received vector in the first phase. The relay is assumed to be able to decide whether it has decoded correctly or not. This can be achieved through using error detecting codes such as CRC codes [148,149].

In the first time slot ($t=1$), the source node broadcasts its information to the destination and relay nodes. The received signals at the destination and the relay are given by

$$R_B^{(1,\beta_1)}(1)=\sqrt{P_1}\,\Lambda_{1,3}^{(\beta_1)}X_1^{(1,\beta_1)}(1)+N_{1,3}(1) \tag{8.18}$$

$$R_{U2}^{(1,\beta_1)}(1)=\sqrt{P_1}\,\Lambda_{1,2}^{(\beta_1)}X_1^{(1,\beta_1)}(1)+N_{1,2}(1) \tag{8.19}$$

where

P_1 denotes the average transmitted power per symbol at the source node

$X_1^{(1,\beta_1)}(1)$ is the frequency-domain vector that belongs to user 1 transmitted in the first time slot on the band β_1

$\Lambda_{1,j}^{(\beta_1)}, j=2,3$ are $M\times M$ diagonal matrices representing the frequency response of the channels between the source node and both the relay and destination, respectively

$$N_{i,j}(1)\sim\mathcal{CN}(0,N_0 I_{M\times M})$$

In the second time slot ($t=2$), the source node transmits the first column ($X_1^{(1,\beta_1)}$) of the distributed SFC on the frequency band β_1. Meanwhile, if the relay has decoded correctly, it will forward the processed information to the destination on the frequency band β_1. Therefore, the received signal at the destination node during the second time slot is given by

$$R_B^{(1,\beta_1)}(2)=\sqrt{P_1}\,\Lambda_{1,3}^{(\beta_1)}X_1^{(1,\beta_1)}(2)+\sqrt{P_2}\,I_{2,1}\Lambda_{2,3}^{(\beta_1)}\tilde{X}_1^{(1,\beta_1)}(2)+N(2)$$

$$(8.20)$$

where

P_2 is the average transmitted power from the relay node

$\tilde{X}_1^{(1,\beta_1)}$ is the processed user 1 information received in the first phase

$I_{i,j}$ is a Bernoulli random variable representing the state of the relay node i (user 2 in this case) at time slot j. In other words, it indicates whether the relay has decoded correctly or not and it is given by

$$I_{i,j}=\begin{cases}1 & \text{correct decoding}\\ 0 & \text{error decoding}\end{cases} \qquad i,j=1,2 \qquad (8.21)$$

Such scheme guarantees the decoding at the destination node even if one of the channels was in deep fade. This is attributed to the orthogonality of the code matrix constructed at the receiver of the destination node. In other words, a maximum diversity gain of two is achieved even if the relay cannot decode correctly in the first phase [150]. Let us partition the vector $X_1^{(1,\beta_1)}(2)$ into two subvectors such that

$$X_1^{(1,\beta_1)}(2)=[X_{1,1}^{(1,\beta_1)T}\ \ X_{1,2}^{(1,\beta_1)T}]^T \qquad (8.22)$$

where

$$X_{1,1}^{(1,\beta_1)}=\left[X_0^{(1,\beta_1)}\ X_1^{(1,\beta_1)}\ \cdots\ X_{M/2-1}^{(1,\beta_1)}\right]^T$$

$$X_{1,2}^{(1,\beta_1)}=\left[X_{M/2}^{(1,\beta_1)}\ X_{M/2+1}^{(1,\beta_1)}\ \cdots\ X_{M-1}^{(1,\beta_1)}\right]^T$$

The processed data at the relay node is performed as follows:

$$\tilde{X}_1^{(1,\beta_1)}(2)=[-X_{1,2}^{(1,\beta_1)\mathcal{H}}\ \ X_{1,1}^{(1,\beta_1)\mathcal{H}}]^T \qquad (8.23)$$

From Equations 8.22 and 8.23, it can be seen that the transmitted codeword matrix from the source and relay nodes over the band β_1 in the second phase constitutes an Alamouti-like space–frequency code. Let us divide the received vector in Equation 8.20 at the destination node into two subvectors as follows:

$$R_{B,1}^{(1,\beta_1)}(2)=\sqrt{P_1}\,\Lambda_{1,3}^{(\beta_{1,1})}X_{1,1}^{(1,\beta_1)}(2)-\sqrt{P_2}\,I_{2,1}\Lambda_{2,3}^{(\beta_{1,1})}X_{1,2}^{(1,\beta_1)*}(2)+N_1(2)$$

(8.24)

$$R_{B,2}^{(1,\beta_1)}(2)=\sqrt{P_1}\,\Lambda_{1,3}^{(\beta_{1,2})}X_{1,2}^{(1,\beta_1)}(2)+\sqrt{P_2}\,I_{2,1}\Lambda_{2,3}^{(\beta_{1,2})}X_{1,1}^{(1,\beta_1)*}(2)+N_2(2)$$

(8.25)

Note that the decoding of the received subvectors in Equations 8.24 and 8.25 is similar to the steps presented in [9], if we assume approximately equal channel gains over the two coded subcarriers, i.e., $\Lambda_{i,j}^{(\beta_{1,1})}\approx\Lambda_{i,j}^{(\beta_{1,2})}$ [151]. However, the channel response is no longer constant over the two subcarriers, especially if the two symbols encoded together are far separated. The two symbols that are coded together are separated by $M/2$ subcarriers as shown in Figure 8.4c. In the classical SFC, the coding is performed over two adjacent subcarriers as shown in Figure 8.4a. The scheme shown in Figure 8.4b was presented in [90] and it was shown that this coding scheme has a lower PAPR than the classical scheme. The rationale behind the code structure, shown in Figure 8.4c, is that the structure of the code at the relay node does not break significantly the single-carrier nature, and hence we get a low PAPR, which is close to the uncoded SC-FDMA as will be shown in the simulation examples. The code construction at the relay node can be considered as a circular shift of the original signal with simple conjugate and minus operations. So, it is expected that the resulting PAPR is close to the uncoded SC-FDMA. The performance of the code is expected to be degraded with large M using Alamouti decoding with MLD for single-symbol detection [145]. Therefore, we use a slightly more complex MMSE decoding at the receiver of the destination node. From Equations 8.24 and 8.25, the received signal can be represented as

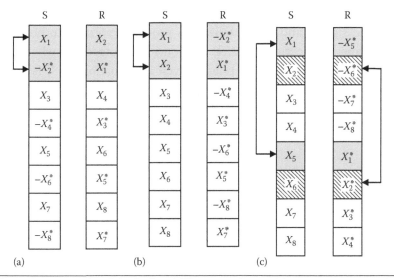

Figure 8.4   Illustration of the distributed SFC structure for user 1 for *one-way cooperation* on the band β_1 with $M = 8$ (c) compared to the classical SFC (a) and the coding scheme presented in [7] (b).

$$
\underbrace{\begin{bmatrix} R_{B,1}^{(1,\beta_1)}(2) \\ R_{B,2}^{(1,\beta_1)*}(2) \end{bmatrix}}_{R} = \sqrt{\mathcal{P}} \underbrace{\begin{bmatrix} \mathbf{\Lambda}_{1,3}^{(\beta_{1,1})} & -I_{2,1}\mathbf{\Lambda}_{2,3}^{(\beta_{1,1})} \\ I_{2,1}\mathbf{\Lambda}_{2,3}^{(\beta_{1,2})*} & \mathbf{\Lambda}_{1,3}^{(\beta_{1,2})*} \end{bmatrix}}_{H_{\mathrm{DF}}} \underbrace{\begin{bmatrix} X_{1,1}^{(1,\beta_1)} \\ X_{1,2}^{(1,\beta_1)*} \end{bmatrix}}_{X} + \begin{bmatrix} N_1(2) \\ N_2^*(2) \end{bmatrix}
$$

$$(8.26)$$

where we have defined $\mathcal{P} \triangleq P_1 = P_2$. The output of MMSE equalization can be written as

$$
\hat{X} = \left(H_{\mathrm{DF}}^{\mathcal{H}} H_{\mathrm{DF}} + 1/\mathrm{SNR} \right)^{-1} H_{\mathrm{DF}}^{\mathcal{H}} R
$$

$$(8.27)$$

where

$$\mathrm{SNR} = P/N_0$$

$$P = P_1 + P_2$$

Note that H_{DF} is not an orthogonal matrix, because $\mathbf{\Lambda}_{i,j}^{(\beta_{1,1})} \neq \mathbf{\Lambda}_{i,j}^{(\beta_{1,2})}$, $(i, j) = (1, 2)$. However, as far as we are concerned with complexity, an inversion of matrices of order two, which contains the two coded symbols, is needed as can be seen from Equation 8.26. The equalized signal $\hat{X}$ is then transformed to the time domain for the final decoding.

It should be noted that Equations 8.26 and 8.27 represent the received signals and the equalization process according to protocol I described in Section 8.2.1. If we consider protocol II, where the received signal in the first phase, given in Equation 8.18, is combined with the received signal from the second phase, which is given in Equation 8.26, the combined signal is given as

$$
\underbrace{\begin{bmatrix} R_{B,1}^{(1,\beta_1)}(2)+R_{B,1}^{(1,\beta_1)}(1) \\ R_{B,2}^{(1,\beta_1)^*}(2)+R_{B,2}^{(1,\beta_1)^*}(1) \end{bmatrix}}_{\bar{R}} = \sqrt{\mathcal{P}} \underbrace{\begin{bmatrix} \Lambda_{1,3}^{(\beta_{1,1})}+\Lambda_{1,3}^{(\beta_{1,1})}(1) & -I_{2,1}\Lambda_{2,3}^{(\beta_{1,1})} \\ I_{2,1}\Lambda_{2,3}^{(\beta_{1,2})^*} & \Lambda_{1,3}^{(\beta_{1,2})^*}+\Lambda_{1,3}^{(\beta_{1,2})^*}(1) \end{bmatrix}}_{\bar{H}_{DF}}
$$

$$
\times \underbrace{\begin{bmatrix} X_{1,1}^{(1,\beta_1)} \\ X_{1,2}^{(1,\beta_1)^*} \end{bmatrix}}_{\hat{X}} + \begin{bmatrix} \tilde{N}_1 \\ \tilde{N}_2 \end{bmatrix}
$$

$$(8.28)$$

where we have used the time index for the channel frequency response in the first time slot to distinguish it from the received signals in the second time slot. $\tilde{N}_1$ and $\tilde{N}_2$ are the new noise vectors. The MMSE criterion can be used and the equalized signal is given by

$$
\hat{X} = \left(\bar{H}_{DF}^{\mathcal{H}} \bar{H}_{DF} + 1/\text{SNR} \right)^{-1} \bar{H}_{DF}^{\mathcal{H}} \bar{R} \qquad (8.29)
$$

8.2.2.1 Peak-to-Average Power Ratio As the main concern of the discussed schemes is to achieve the cooperative diversity and at the same time preserve the low PAPR of the transmitted signals, we return back here to the definition of the PAPR and the importance of keeping this ratio as low as possible. Recall that the SC-FDMA system has a lower PAPR than the OFDMA system. However, designing coding schemes for SC-FDMA systems affect the PAPR, and hence it is important for the code to take into account this constraint.

The PAPR for the user i is defined as the ratio of the maximum power of the signal to the average power and it is given by

$$\text{PAPR}^{(i)} = \frac{\max\left(\left|s_n\right|^2\right)}{\frac{1}{N}\sum_{n=0}^{N-1}\left|s_n\right|^2}, \quad n=0,1,\ldots,N-1 \quad (8.30)$$

where $s_n, n = 0,1,\ldots,N-1$ are the elements of the transmitted time-domain vector given in Equation 8.3.

8.3 Cooperative Space–Time Code for SC-FDMA

In this section, we discuss a bandwidth-efficient distributed STC that achieves a better spectral efficiency than the classical scheme. It is illustrated in Table 8.2. The two-way cooperation cycle will last for three time slots. The channel is assumed to be constant over the three time slots.

In the first time slot, user 1 uses both his own subcarriers and his partner's subcarriers to transmit his information to the destination node. We argue that there is no practical limitation for the user to use both the subcarriers allocated to him and to his partner. At the same time ($t=1$), user 2 is in the listening mode. At the second time slot ($t=2$), user 2 uses both β_1 and β_2 to transmit his information to the destination node, while user 1 listens. At the third time slot, both users transmit on the two bands if they have decoded correctly in the previous time slots as shown in Table 8.2.

Due to the symmetry in cooperation in the two bands (β_1 and β_2), we will consider only one cooperation cycle on one band (say β_1). Note that the transmitted signals on the band β_1 in Table 8.2 (columns 1 and 3) constitute an Alamouti-like space–time code. It is clear that

Table 8.2 Transmitted signals in the frequency domain for the distributed STC with a DF protocol

TIME SLOT (t)	USER 1		USER 2	
	β_1	β_2	β_1	β_2
1	$X_1^{(1,\beta_1)}(1)$	$X_2^{(1,\beta_2)}(1)$	—	—
2	—	—	$X_1^{(2,\beta_1)}(2)$	$X_2^{(2,\beta_2)}(2)$
3	$-l_{1,2}\hat{X}_1^{(2,\beta_1)*}(3)$	$-l_{1,2}\hat{X}_2^{(2,\beta_2)*}(3)$	$l_{2,1}\hat{X}_1^{(1,\beta_1)*}(3)$	$l_{2,1}\hat{X}_2^{(1,\beta_2)*}(3)$

every user sends two signal vectors in three time slots. Therefore, a transmission rate of 2/3 is achieved. There are four possibilities in this scenario.

Case 1: Both users decode correctly, i.e., $I_{2,1} = I_{1,2} = 1$. In this case, the received signals at the destination node can be expressed as

$$R_B^{(\beta_1)}(1,2) = \sqrt{P_1}\, \Lambda_{1,3}^{(\beta_1)}\, X_1^{(1,\beta_1)}(1) + \sqrt{P_2}\, \Lambda_{2,3}^{(\beta_1)}\, X_1^{(2,\beta_1)}(2) + N(1) + N(2)$$

$$(8.31)$$

$$R_B^{(\beta_1)}(3) = -\sqrt{P_1}\, \Lambda_{1,3}^{(\beta_1)}\, X_1^{(2,\beta_1)*}(3) + \sqrt{P_2}\, \Lambda_{2,3}^{(\beta_1)}\, X_1^{(1,\beta_1)*}(3) + N(3)$$

$$(8.32)$$

where $R_B^{(\beta_1)}(1,2)$ is the linear combination of the received signals at the receiver of the destination node at time slots 1 and 2. It should be noted that the received signals in Equations 8.31 and 8.32 contain the transmitted information of both users. The receiver performs Alamouti-like combining and MLD as in [145]. The only difference here is the additional noise term in (8.31). We will investigate the effect of this variation compared to the classical scheme, where a single noise term is present in the simulation section (Section 8.4).

Case 2: User 2 decodes correctly and user 1 not, i.e., $I_{2,1} = 1$ and $I_{1,2} = 0$. In this case, user 1 does not send a signal in the third time slot in the band β_1. Therefore, the received signals at the destination node are given by

$$R_B^{(\beta_1)}(1) = \sqrt{P_1}\, \Lambda_{1,3}^{(\beta_1)}\, X_1^{(1,\beta_1)}(1) + N(1) \qquad (8.33)$$

$$R_B^{(\beta_1)}(2) = \sqrt{P_2}\, \Lambda_{2,3}^{(\beta_1)}\, X_1^{(2,\beta_1)}(2) + N(2) \qquad (8.34)$$

$$R_B^{(\beta_1)}(3) = \sqrt{P_2}\, \Lambda_{2,3}^{(\beta_1)}\, X_1^{(1,\beta_1)*}(3) + N(3) \qquad (8.35)$$

The receiver combines the received signals in Equations 8.33 and 8.35 using Equations MRC and makes a decision for the signal vector $X_1^{(1,\beta_1)}$ of user 1. Equation 8.34 is decoded separately.

Case 3: User 2 fails to decode whereas user 1 decodes correctly, i.e., $I_{2,1}=0$ and $I_{1,2}=1$. In this case, the received signals at the destination node during the first and second time slots are similar to those in Equations 8.33 and 8.34. The received signal at the third time slot is given by

$$R_B^{(\beta_1)}(3) = -\sqrt{P_1}\,\Lambda_{1,3}^{(\beta_1)}\,X_1^{(2,\beta_1)^*}(3) + N(3) \qquad (8.36)$$

Similar to Case 2, the receiver combines the signals in Equations 8.34 and 8.36 using MRC and makes a decision for the vector $X_1^{(2,\beta_1)}$ that belongs to user 2. Equation 8.33 is decoded separately.

Case 4: Both users 1 and 2 fail to decode correctly, i.e., $I_{2,1}=I_{1,2}=0$. In this case, the received signals at the destination node in the first and second time slots are similar to Equations 8.33 and 8.34. The two users send nothing in the third time slot. The receiver decodes these two received signals separately, and the decision variables in the frequency domain for users 1 and 2 are given by

$$\hat{X}_1^{(1,\beta_1)} = \Lambda_{1,3}^{(\beta_1)^*}\,R_B^{(1,\beta_1)}(1) \qquad (8.37)$$

$$\hat{X}_1^{(2,\beta_1)} = \Lambda_{2,3}^{(\beta_1)^*}\,R_B^{(1,\beta_1)}(2) \qquad (8.38)$$

In the earlier analysis, we have assumed symmetric interuser SNR, i.e., $SNR_{1,2}=SNR_{2,1}=P/N_0$. As will be shown in the simulation section (Section 8.4), the discussed scheme achieves a diversity of order two and the performance highly depends on the interuser channel quality.

It should be noted that for the receiver of the destination node to combine the received signals, the state of each node whether it has decoded correctly or not should be forwarded to the destination node. This can be accomplished by sending one bit from each user to the destination to indicate whether the user could decode correctly the received signal from his partner or not.

8.4 Simulation Examples

In this section, we verify the discussed cooperative diversity schemes via computer simulations. The uplink direction of the SC-FDMA system has been considered. A total of $N=512$ subcarriers has been used in the simulations. The signal constellation is QPSK. Channel modeling is based on an exponentially decaying power delay profile. The operating bandwidth is 20 MHz and the maximum excess delay of the channel (τ_{max}) has been chosen to produce a channel with 10 paths. The power of the channel paths is normalized to one. The CP is assumed to eliminate the ISI; it also converts the linear convolution of the channel impulse response to a circular convolution. The power is divided equally between the source node and the relay node. The SNR used in the simulation is defined as $SNR = \dfrac{P_1 + P_2}{N_0}$, where $P_1 = P_2$.

Figure 8.5 shows the SER performance of the SFC using two decoding schemes: Alamouti MLD and the MMSE schemes. As shown from this figure, the performance of the MLD scheme starts to degrade at high SNR values due to the error floor, because the two symbols coded together experience different channel responses. As the number of the allocated subcarriers M increases, the SER performance gets worse. Therefore, in the following simulations, we will use the MMSE criterion.

Figure 8.6 shows the SER performance of the distributed SFC with the DF protocol. For the case of the relay located at the midpoint between the source and the destination, all channels ($S \rightarrow D$, $S \rightarrow R$, $R \rightarrow D$) have been assumed to be of unit variance. For the case the relay is close to the source, the source-relay channel variance $\sigma_{s,r}^2$ has

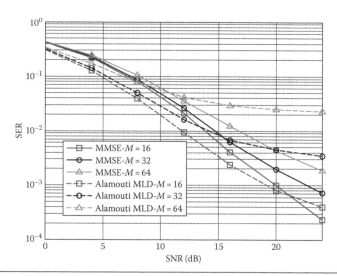

Figure 8.5 Performance comparison between MMSE decoding and Alamouti MLD for the distributed SFC with different numbers of subcarriers allocated to each user using protocol I.

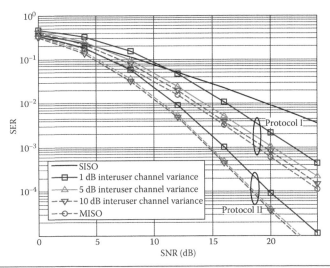

Figure 8.6 SER vs. SNR for the distributed SFC for user 1 considering both protocol I and protocol II.

been increased in the simulation. Figure 8.6 shows that as the relay node becomes closer to the source node, the performance is greatly improved. With protocol I and high interuser channel variance, the performance of the distributed SFC scheme approaches the performance of the MISO system with two antennas at the transmitter side. Note also that the DF signaling at the relay is very sensitive to the channel conditions of the source-relay link. As can be seen from this figure, a diversity of order three can be achieved using protocol II. There is a trade-off between the complexity and the performance for the two protocols. Protocol II achieves a better performance at the expense of the delay and the processing time required at the destination node to combine the two signals received at both phases of transmission.

The complementary cumulative distribution functions (CCDFs) of the PAPR of the transmitted signals at the relay node for the distributed SFC scheme, the classical scheme, and scheme II in [90] are shown in Figure 8.7. The CCDF is defined as the probability that the PAPR is larger than a certain value $PAPR_0$. The number of subcarriers allocated to each user is $M = 64$. It is clear from this figure that the distributed SFC scheme achieves about 1.5 dB gain in PAPR compared to the classical scheme at a CCDF value of 10^{-3}. Moreover, it achieves about a 0.6 dB gain compared to scheme II in [90]. Note

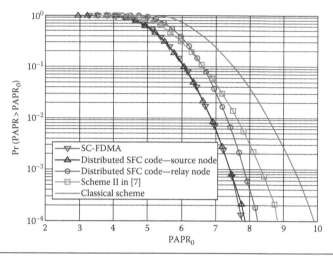

Figure 8.7 Performance comparison of the CCDF of the PAPR between the distributed SFC, classical schemes, and uncoded SC-FDMA at the relay node.

that the transmitted signal of the distributed SFC scheme at the source node achieves the same PAPR as in the case of uncoded SC-FDMA.

Figure 8.8 compares the SER performance of the distributed SFC with the classical SFC (scheme II) for different numbers of subcarriers. As can be seen from this figure, the performance of the two schemes is the same at low SNR values. At high SNR values, there is about 0.2 dB loss for the distributed SFC scheme compared to scheme II. This small performance loss is reasonable compared to the PAPR gain, which is more important for the small mobile terminal transmitting in the uplink direction.

The SER performance comparison between the distributed STC and the classical coding scheme is shown in Figure 8.9. The cooperative protocol of the classical scheme needs four time slots to transmit two blocks. Therefore, the transmission rate is 1/2. Note that the distributed STC scheme offers a better performance than the direct transmission. The performance gain is further improved with higher interuser channel variances. It can be seen that the slope of the curve of the distributed STC scheme is the same as that of the MISO case, and thereby it achieves a diversity of order two. It can also be seen that the distributed STC scheme has a better performance than the classical distributed STC, when all channel variances are equal to

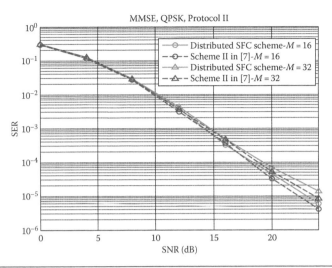

Figure 8.8 Performance comparison between the SFC scheme and the scheme in [7] with different numbers of subcarriers M.

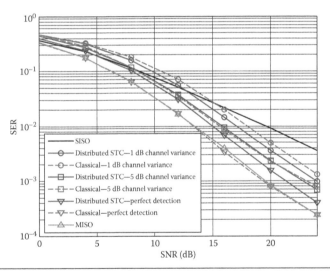

Figure 8.9 SER performance comparison between the distributed STC and the classical scheme with different interuser channel variances.

one. As the interuser channel variance increases and other channel variances are equal to one, the performance loss of the distributed STC scheme increases compared to the classical scheme, because case 1 will be the dominant case and the additional noise component in Equation 8.31 makes this performance loss.

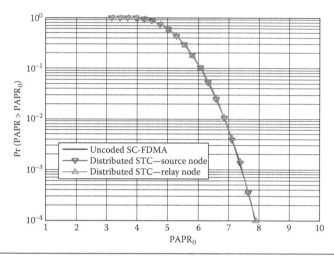

Figure 8.10   Performance comparison of the CCDF of the PAPR between the distributed STC and uncoded SC-FDMA at the relay node.

Finally, the CCDF of the PAPR of the distributed STC is shown in Figure 8.10. It can be observed that the PAPR of the transmitted signals from both the source node and the relay node are equal to that of the uncoded SC-FDMA signal. Therefore, the distributed STC scheme does not affect the PAPR of the transmitted signals at the third time slot, where a modification is made to the original signal.

9

Relaying Techniques for Improving the Physical Layer Security

Security issues for traditional wireless networks have been considered as a very important research area, and have taken a considerable attention for several years due to the broadcast nature of the wireless medium, which makes attacks more likely to happen than in wireline systems. Most of the studies in this area focus on the security in the higher layers in the protocol stack via the use of cryptography and key management. The main idea of these security approaches is based on the limited computational power of the eavesdroppers. With the advent of the infrastructureless networks such as mobile ad hoc networks (MANETs), further challenges have appeared which make the nodes more vulnerable to attacks. Among these challenges are the absence of centralized hardware for security problems, the multihop routes that the transmitted information should follow, and the power-limited terminals [152]. Therefore, to cope with these limitations, physical layer security (or information-theoretic security) has gained a considerable attention in the last few years. The goal of physical layer security is to allow the source to confidentially communicate with the legitimate destination, while the eavesdropper cannot interpret the source information, without any assumption for the eavesdropper nodes such as the computational power, or the available information. From an information-theoretic standpoint, the achieved source–destination confidential rate is called the achievable secrecy rate.

Physical layer security in the context of information theory was first introduced in [105]. It was shown that a source can communicate with its intended destination with a nonzero rate if the source–eavesdropper (wiretapper) channel is a degraded version of the source–destination channel. An extension of this work to the case of a broadcast channel with confidential messages was proposed in [106]. In [153], the authors proposed a joint consideration of reliability, physical layer security, and stability for wireless broadcast networks. Joint optimal power control and optimal scheduling schemes were proposed to enhance the secrecy rate of the intended receiver against cooperative and noncooperative eavesdropping models. In the earlier mentioned schemes, however, there is a nonzero probability of communicating with zero rate between the source and the destination if the source–eavesdropper channel is the same or better than the source–destination channel [105,154]. To alleviate this problem, and to exploit the degrees of freedom available in the MIMO systems, some recent work has been proposed in this area [155–158]. However, it is difficult to equip the small mobile units with more than one antenna with uncorrelated fading due to size and cost constraints. In such case, the benefits of MIMO systems can only be achieved through user cooperation [100,101].

The application of cooperative techniques to improve the secrecy capacity has been recently proposed [159–164]. In [159], an uplink secure transmission strategy for cellular systems was proposed. The basic idea is to schedule the downlink transmission of the cooperative BS in order to send a jamming signal to confound the possible eavesdropper. This jamming signal is assumed to be conveyed to the home BS, and hence this intentional interference is assumed to be neglected at the intended receiver. In [160], the authors proposed a variety of cooperative schemes to achieve secure data transmission with the help of multiple relays in the presence of one or multiple eavesdroppers. In the second phase of these protocols (cooperative phase), the relays transmit simultaneously a weighted version of the original message depending on the signaling strategy used at the relay node. Suboptimal weights were obtained in order to maximize the secrecy rate subject to a total power constraint or to minimize the transmitted power under the secrecy-rate constraint. An extension for these schemes with an optimal solution for the weights was proposed in [161]. In [162], a four-node model was presented, and it was shown that if the relay is closer to the destination than the eavesdropper, a positive secrecy rate can be achieved even if the source–destination rate is zero. Moreover, a noise-forwarding scheme was presented. In this scheme, the relay sends a noise signal that is independent of the source signals in order to jam the eavesdropper. On the other hand, relay selection can ensure an efficient secrecy rate for networks with several potential relays by keeping the complexity relatively low. The main idea of these schemes is that a single trusted relay is selected to assist the source message by taking into account the existence of the eavesdroppers. The strategies of relay selection introduced in [128] and [132] were first considered for improving the secrecy rate in [163], where the authors assumed different levels of channel side information. In [164], the authors extended the work presented in [163] for cooperative networks with a jamming protection, without taking into account a direct link.

In this chapter, we discuss relay selection schemes that improve the physical layer security, when there are multiple eavesdroppers in the system, and the source has also a direct link with the destination. We present optimal and suboptimal relay and jammer selection schemes in order to maximize the achievable secrecy rate. In the first phase of the adopted DF relaying protocol, a "friendly" jammer is selected from the set of potential relays to send an intentional interference to the eavesdropper nodes. In the second phase of the protocol, two relay nodes are selected: one node is selected to assist (forward) the source to deliver its information to its legitimate receiver using the DF protocol, while the other node behaves as a jammer node in order to confuse the eavesdroppers. This selection process is analyzed for scenarios without eavesdropper cooperation as well as for scenarios with eavesdropper cooperation, where the eavesdropper nodes cooperate (create a malicious network) in order to overhear the source message. The proposed selection schemes are analyzed in terms of the achievable secrecy rate and secrecy outage probability under the assumption of global instantaneous channel state information [160–162].

The rest of this chapter is organized as follows. In Section 9.1, the system and channel models are introduced. In Section 9.2, the relay selection schemes are presented. Noncooperative and cooperative eavesdroppers models and also the effect of jamming for each case are investigated. Finally, Section 9.3 presents the simulation examples.

9.1 System and Channel Models

Throughout this chapter, the following notations will be adopted. I_n is an $n \times n$ identity matrix. $\mathcal{CN}(\mu, \sigma^2)$ denotes a circularly symmetric complex Gaussian random variable with mean μ and variance σ^2; $E\{\cdot\}$ denotes the expectation operation; $[x]^+$ denotes $\max(0, x)$; and $\triangleq$ denotes equal by definition. $|x|$ represents the cardinality of the vector x.

We consider a two-phase cooperative protocol as shown in Figure 9.1. The source node S communicates with the destination node D with the help of N relay nodes. There are M eavesdroppers that try to overhear the source information. We define the following sets: S_{relays}, the set of all relay nodes; S_{evs}, the set of the eavesdroppers; and D the decoding set, which contains the relays that have decoded correctly the received messages during the first phase of the transmission.

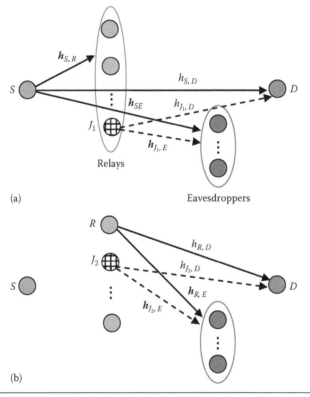

(a)

(b)

Figure 9.1   System model. (a) Broadcasting phase. (b) Cooperative phase.

In the first (broadcasting) phase, the source node sends its information to the destination node, and due to the broadcasting nature of the transmission, the relays and eavesdroppers overhear the transmitted information. Meanwhile, a jammer $J_1 \in S_{\text{relays}}$ is selected to send intentional interference to jam the eavesdroppers. The received signal at the destination D for the first phase of the protocol is given by

$$y_d^{(1)} = \sqrt{P_S}\, h_{S,D}\, x + \sqrt{P_{J_1}}\, h_{J_1,D}\, \tilde{x} + n_d^{(1)} \tag{9.1}$$

where P_S and P_{J_1} denote the average transmitted power per symbol at the source node and at the selected jammer in the first phase, respectively. The noise component $n_d^{(1)} \sim \mathcal{CN}(0, N_0)$. x denotes the transmitted symbol from the source node, and $\tilde{x}$ denotes the transmitted signal by the jammer. $h_{S,D}$ and $h_{J_1,D}$ represent the Rayleigh fading coefficient of the source–destination channel and the first selected jammer–destination channel, respectively.

The received signals at the relays and the eavesdroppers are represented in vector form as follows:

$$\mathbf{y}_r = \sqrt{P_S}\, \mathbf{h}_{S,R}\, x + \mathbf{n}_r \tag{9.2}$$

$$\mathbf{y}_e^{(1)} = \sqrt{P_S}\, \mathbf{h}_{S,E}\, x + \sqrt{P_{J_1}}\, \mathbf{h}_{J_1,E}\, \tilde{x} + \mathbf{n}_e^{(1)} \tag{9.3}$$

where

$\mathbf{h}_{S,R}$ is an $N \times 1$ vector, which represents the channel between the source and the relays

$\mathbf{h}_{S,E}$ is an $M \times 1$ vector, which represents the channel between the source and the eavesdroppers

$\mathbf{h}_{J_1,E}$ is an $M \times 1$ vector, which represents the channel between the selected jammer and the eavesdroppers

$$\mathbf{n}_r \sim \mathcal{CN}(0, N_0 \mathbf{I}_N)$$

$$\mathbf{n}_e^{(1)} \sim \mathcal{CN}(0, N_0 \mathbf{I}_M)$$

Furthermore, we have assumed unit-variance noise. Note that the jammer node J_1 that is selected in the first phase of the protocol does not participate in the decoding process in order to obey the half-duplex constraint (a node cannot transmit and receive simultaneously). Moreover, it is assumed that the jamming signal is known at the relay nodes (a jamming sequence that is predefined before transmission), and therefore can be removed from the relay nodes. Equation 9.2 assumes that the jamming signal is removed from the received signal. This assumption can also be justified by the fact that the relay nodes are located close to each other in the same cluster. Therefore, it is easy to distribute this jamming sequence among the relay nodes with low transmit power so that it cannot be overheard by the eavesdroppers. A similar assumption is used in [159] and [165], where an infinite-capacity backbone between the receiving and jamming antennas is leveraged in order to remove the jamming signal at the receiving antenna.

In the second (cooperative) phase, two nodes are selected: one as a relay $R \in D$ to help the source node to transmit its information to the destination node. At the same time, a jammer $J_2 \in \{S_{\text{relays}} - R\}$ is selected to confound the eavesdroppers as in the first phase. The received signals at the destination and the eavesdroppers are given as

$$y_d^{(2)} = \sqrt{P_R}\, h_{R,D} x + \sqrt{P_{J_2}}\, h_{J_2,D} \tilde{x} + n_d^{(2)} \tag{9.4}$$

$$\mathbf{y}_e^{(2)} = \sqrt{P_R}\, \mathbf{h}_{R,E} x + \sqrt{P_{J_2}}\, \mathbf{h}_{J_2.E} \tilde{x} + \mathbf{n}_e^{(2)} \tag{9.5}$$

where

$h_{R,D}$ denotes the fading coefficient between the selected relay in the second phase of the protocol and the destination node

$h_{J_2,D}$ denotes the fading coefficient between the selected jammer in the second phase and the destination node

$\mathbf{h}_{R,E}$ and $\mathbf{h}_{J_2.E}$ represent the channel vectors for the relay–eavesdroppers and jammer–eavesdroppers links, respectively

$$n_d^{(2)} \sim \mathcal{CN}(0, N_0)$$

$$\mathbf{n}_e^{(2)} \sim \mathcal{CN}(0, N_0 \mathbf{I}_M)$$

Maximum ratio combining (MRC) is used to combine the two received signals represented in Equations 9.1 and 9.4 at the destination.

The instantaneous SNRs for the source–destination link, the source–eavesdropper link, the selected relay–destination link, the selected relay–eavesdropper link, the jammer–destination link, and the jammer–eavesdropper link are denoted by $\gamma_{S,D} = |h_{S,D}|^2 P_S$, $\gamma_{S,E_m} = |h_{S,E_m}|^2 P_S$, $\gamma_{R,D} = |h_{R,D}|^2 P_R$, $\gamma_{R,E_m} = |h_{R,E_m}|^2 P_R$, $\gamma_{J_i,D} = |h_{J_i,D}|^2 P_{J_i}$, and $\gamma_{J_i,E_m} = |h_{J_i,E_m}|^2 P_{J_i}$, respectively, where $m = 1, 2, \ldots, M$ and $i = 1, 2$. The distribution of the channel coefficients between the nodes i and j ($h_{i,j}$) is given as $h_{i,j} \sim \mathcal{CN}(0, \sigma_{i,j}^2)$, where $\sigma_{i,j}^2 \triangleq \mathrm{E}\{|h_{i,j}|^2\} = d_{i,j}^{-\alpha}$, $d_{i,j}$ is the Euclidian distance between terminals i and j, and α is the path-loss exponent. Furthermore, a block Rayleigh fading channel is assumed, where the channel remains constant over each transmitted block and changes independently from block to block. As the channel gains change from phase to phase, the selection of the jammer J_1 is based on the channel gain during the first phase. The global CSI is assumed to be known. This assumption is also considered in [160–162]. Moreover, for several cases such as military applications, this continuous tracking of the channels is justified, especially when multiple eavesdroppers are available in the network, and the possibility of cooperation between the eavesdroppers is high. The average value of the instantaneous SNR $\gamma_{i,j}$ between the nodes i and j is given by $\overline{\gamma}_{i,j} \triangleq \sigma_{i,j}^2 P_i$ and its probability density function (PDF) is given by $p_{\gamma_{i,j}}(\gamma) = \lambda \exp(-\lambda \gamma)$, where $\lambda = 1 / \overline{\gamma}_{i,j}$.

9.2 Relay and Jammers Selection Schemes

In this section, relay and jammers selection schemes are presented. The objective is to select the nodes R, J_1, and J_2 in order to maximize the achievable secrecy rate. First, we will consider a cooperative network with multiple eavesdroppers that can individually decipher the source message. Then, we will consider the case where the eavesdroppers have the ability to cooperate in order to decipher the transmitted source message.

9.2.1 Selection Schemes with Noncooperative Eavesdroppers

In this section, we consider noncooperative eavesdroppers that try to decode the source information individually. In this case, the instantaneous achievable secrecy rate for the network shown in Figure 9.1, with a decoding set $\mathcal{D}$, is given as follows [153,162]:

$$
C_S^{|\mathcal{D}|}(R, J_1, J_2)
$$

$$
= \begin{cases}
\left[\left[\dfrac{1}{2}\log_2\left(1 + \dfrac{\gamma_{S,D}}{1 + \gamma_{J_1,D}}\right)\right. \\
\left. - \dfrac{1}{2}\log_2\left(\max_{E \in \mathcal{E}_{\text{evs}}}\left\{1 + \dfrac{\gamma_{S,E_m}}{1 + \gamma_{J_1,E_m}}\right\}\right)\right]^+ & |\mathcal{D}| = 0 \\[2em]
\left[\left[\dfrac{1}{2}\log_2\left(1 + \dfrac{\gamma_{S,D}}{1 + \gamma_{J_1,D}} + \dfrac{\gamma_{R,D}}{1 + \gamma_{J_2,D}}\right)\right. \\
\left. - \dfrac{1}{2}\log_2\left(\max_{E_m \in \mathcal{S}_{\text{evs}}}\left\{1 + \dfrac{\gamma_{S,E_m}}{1 + \gamma_{J_1,E_m}} + \dfrac{\gamma_{R,E_m}}{1 + \gamma_{J_2,E_m}}\right\}\right)\right]^+ & |\mathcal{D}| > 0
\end{cases}
$$

$$(9.6)$$

Note that in (9.6), we have considered the worst case, which represents the eavesdropper that can achieve the maximum rate. In other words, the secrecy rate achieved at the destination node is limited by the maximum rate achieved at the eavesdroppers.

The secrecy outage probability is defined as the probability that the secrecy rate is less than a given target secrecy rate $R_s > 0$ [154]. For the secrecy rate given in Equation 9.6, the secrecy outage probability is given by

$$
P_O = \sum_{n=1}^{N} \Pr\left\{C_S^n(R, J_1, J_2) < R_s \,\big|\, |\mathcal{D}| = n\right\} \Pr\left\{|\mathcal{D}| = n\right\} \quad (9.7)
$$

The objective is to obtain the following optimization problem:

$$
\left(J_1^*, R^*, J_2^*\right) = \underset{\substack{J_1 \in \mathcal{S}_{\text{relays}} \\ R \in |\mathcal{D}| \\ J_2 \in \{\mathcal{S}_{\text{relays}} - R^*\}}}{\arg\max}\left\{C_S^{|\mathcal{D}|}(J_1, R, J_2)\right\} \quad (9.8)
$$

The selection scheme in Equation 9.8 that maximizes the instantaneous secrecy rate also minimizes the secrecy outage probability:

$$\left(J_1^*, R^*, J_2^*\right) = \operatorname*{argmin}_{\substack{J_1 \in S_{\text{relays}} \\ R \in |\mathcal{D}| \\ J_2 \in \left\{S_{\text{relays}} - R^*\right\}}} \left\{ \Pr\left\{C_S^{|\mathcal{D}|}\left(J_1, R, J_2\right)\right\} < R_S \right\} \tag{9.9}$$

where $J_1^*, R^*,$ and J_2^* denote the selected first jammer, relay, and second jammer, respectively. In this chapter, we will consider two performance metrics: the secrecy rate and the secrecy outage probability. It should be noted here that the secrecy rate metric is suitable for some applications, when the wireless links are deterministic [159]. For fading channels, no instantaneous nonzero secrecy rate can be achieved, and therefore, the most appropriate performance metric is the secrecy outage probability. As we deal with Rayleigh fading channels, the main performance metric will be the secrecy outage probability. Based on Equation 9.6, we will distinguish between three cases in order to investigate the effect of jamming: no jammer selection, conventional jamming where the jamming signal is not known at the destination, and controlled jamming where the jamming signal is known at the destination. The selection process associated with each case is given based on the general optimization problem of Equations 9.8 and 9.9.

9.2.1.1 Noncooperative Eavesdroppers without Jamming (NC) In this case, only one relay node is selected in the second phase. Therefore, ignoring the effect of the jamming signals in Equation 9.6 and assuming $|\mathcal{D}| > 0$, the instantaneous secrecy rate is given by

$$C_S^{\text{NC}}(R) = \left[\frac{1}{2} \log_2 \left(\frac{1 + \gamma_{S,D} + \gamma_{R,D}}{\max\limits_{E_m \in S_{\text{evs}} \forall m} \left\{1 + \gamma_{S,E_m} + \gamma_{R,E_m}\right\}} \right) \right]^+ \tag{9.10}$$

Note from Equation 9.10 that a positive secrecy rate is achieved only when the rate of the source node with respect to the destination is larger than the maximal rate over the eavesdroppers with respect to the source. This case is similar to the wiretap channel with multiple eavesdroppers presented in [166]. The relay selection process that maximizes the secrecy capacity in Equation 9.10 is given as

$$R^* = \operatorname*{argmax}_{R \in D} \left[\frac{1 + \gamma_{S,D} + \gamma_{R,D}}{\max\limits_{E_m \in S_{\text{evs}} \forall m} \left\{1 + \gamma_{S,E_m} + \gamma_{R,E_m}\right\}} \right] \tag{9.11}$$

9.2.1.1.1 Asymptotic Secrecy Outage Probability for the Symmetric NC Case   In this section, we derive an asymptotic approximation for the secrecy outage probability at high SNRs. In order to simplify the analysis, we consider the symmetric case, where the source–destination, source–eavesdropper, relay–destination, and relay–eavesdropper distances are equal. This configuration simplifies the analysis and is a guideline for the general asymptotic case (a similar approach has been used in [164]). Simulation results in Section 9.3 are concerned with both the symmetric and asymmetric cases. The average SNRs are then given as $\mathrm{E}\{\gamma_{S,D}\} = \mathrm{E}\{\gamma_{S,E_m}\} = \mathrm{E}\{\gamma_{R,D}\} = \mathrm{E}\{\gamma_{R,E_m}\} = \mathcal{P}\sigma^2$, where $m = 1, 2,..., M$, $\mathcal{P} \triangleq P_S = P_R$, and σ^2 is the channel variance. The average SNRs due to the jammers power are given as $\mathrm{E}\{\gamma_{J_i,D}\} = \mathrm{E}\{\gamma_{J_i,E_m}\} = P_{J_i}\sigma^2, j = 1,2$ Moreover, at high SNR values, all the relays are assumed to decode the source signal correctly, so that $|\mathcal{D}| = N$. Therefore, the secrecy outage probability in (9.7) is now given by $\Pr\{C_S^{\mathrm{NC}}(J_1, R, J_2)||\mathcal{D}| = N\} < R_S$. For the analysis, we use the following two propositions:

Proposition 9.1

If X_1 and X_2 are two i.i.d exponential random variables with parameter λ, the PDF and cumulative distribution function (CDF) of the new random variable $X = X_1 + X_2$ are given by

$$p_X(x) = \lambda^2 x e^{-\lambda x} \tag{9.12}$$

$$P_X(x) = 1 - e^{-\lambda x}(1 + \lambda x) \tag{9.13}$$

∎

Proposition 9.2

If (X_m, Y_m), $m = 1, 2, ..., M$ are M pairs of i.i.d exponential random variables with parameter λ, the CDF and PDF of the new random variable $Z \triangleq \max_{m}(X_m + Y_m)$, using proposition 1 and order statistics [167,168], are given by

$$P_Z(z) = \left(1 - e^{-\lambda z}(1 + \lambda z)\right)^M \tag{9.14}$$

$$p_Z(z) = M\left(\lambda^2 z e^{-\lambda z}\right)\left[1 - e^{-\lambda z}(1 + \lambda z)\right]^{M-1} \tag{9.15}$$

■

According to Appendix E.1, the secrecy outage probability for the case of noncooperative eavesdroppers without jamming is given by

$$P_O^{NC} \approx \left[M \int_0^\infty \left[1 - e^{-\lambda \rho z}(1 + \lambda \rho z)\right]\left(\lambda^2 z e^{-\lambda z}\right)\left[1 - e^{-\lambda z}(1 + \lambda z)\right]^{M-1} dz \right]^N \tag{9.16}$$

where $\rho = 2^{2R_s}$. For the special case of $M = 2$, and considering unit-variance channels, the outage probability in Equation 9.16 can be solved to get a closed form as

$$P_O^{NC} = \left[\frac{M}{2} - \frac{M}{(\rho+1)} + \frac{M\left(3\rho^2 + 16\rho + 8\right)}{(\rho+2)^4} \right]^N \tag{9.17}$$

9.2.1.2 Noncooperative Eavesdroppers with Jamming (NCJ)   In this case, we introduce the effect of jamming on the secrecy rate and the secrecy outage probability. Assuming that $|\mathcal{D}| > 0$, the instantaneous secrecy capacity in Equation 9.6 is given as

$$C_S^{NCJ}(J_1, R, J_2) = \left[\frac{1}{2}\log_2\left(\frac{1 + \dfrac{\gamma_{S,D}}{1 + \gamma_{J_1,D}} + \dfrac{\gamma_{R,D}}{1 + \gamma_{J_2,D}}}{1 + \max_{E_m \in \mathcal{S}_{evs} \; \forall m}\left\{ \dfrac{\gamma_{S,E_m}}{1 + \gamma_{J_1,E_m}} + \dfrac{\gamma_{R,E_m}}{1 + \gamma_{J_2,E_m}} \right\}} \right) \right]^+ \tag{9.18}$$

The objective of the selection process is to maximize the following optimization problem:

$$
\left(J_1^*, R^*, J_2^*\right) = \underset{\substack{J_1 \in \mathcal{S}_{\text{relays}} \\ R \in \mathcal{D} \\ J_2 \in \left\{\mathcal{S}_{\text{relays}} - R^*\right\}}}{\arg\max} \left(\frac{1 + \dfrac{\gamma_{S,D}}{1 + \gamma_{J_1,D}} + \dfrac{\gamma_{R,D}}{1 + \gamma_{J_2,D}}}{1 + \underset{E_m \in \mathcal{S}_{\text{evs}} \ \forall m}{\max} \left[\dfrac{\gamma_{S,E_m}}{1 + \gamma_{J_1,E_m}} + \dfrac{\gamma_{R,E_m}}{1 + \gamma_{J_2,E_m}} \right]} \right)
$$

$$(9.19)$$

The selection process in Equation 9.19 requires a large number of comparisons and algebraic computations. Therefore, the earlier optimization can be divided into suboptimal selections for the three nodes. Considering a high SNR and the assumption that the jammer-transmitted power is much lower than the source and relay powers [164], it can be shown that

$$
\left(J_1^*, R^*, J_2^*\right) = \underset{\substack{J_1 \in \mathcal{S}_{\text{relays}} \\ R \in \mathcal{D} \\ J_2 \in \left\{\mathcal{S}_{\text{relays}} - R^*\right\}}}{\arg\max} \left(\frac{\dfrac{\gamma_{S,D}}{\gamma_{J_1,D}} + \dfrac{\gamma_{R,D}}{\gamma_{J_2,D}}}{\underset{E_m \in \mathcal{S}_{\text{evs}} \ \forall m}{\max} \left[\dfrac{\gamma_{S,E_m}}{\gamma_{J_1,E_m}} + \dfrac{\gamma_{R,E_m}}{\gamma_{J_2,E_m}} \right]} \right) \qquad (9.20)
$$

In the light of Equation 9.20, we discuss suboptimal selection schemes in order to maximize the secrecy rate in Equation 9.18. The jammer in the first phase is selected through the following optimization problem:

$$
J_1^* = \underset{J_1 \in \mathcal{S}_{\text{relays}}}{\arg\min} \left(\frac{\gamma_{J_1,D}}{\underset{\forall m \in \mathcal{S}_{\text{evs}}}{\min} \gamma_{J_1,E_m}} \right) \qquad (9.21)
$$

The relay and the jammer in the second phase are selected as follows:

$$
R^* = \underset{R \in \mathcal{S}_{\text{relays}}}{\arg\max} \left(\frac{\gamma_{R,D}}{\underset{\forall m \in \mathcal{S}_{\text{evs}}}{\max} \gamma_{R,E_m}} \right) \quad \text{and}
$$

$$
J_2^* = \underset{J_2 \in \left\{\mathcal{S}_{\text{relays}} - R^*\right\}}{\arg\min} \left(\frac{\gamma_{J_2,D}}{\underset{\forall m \in \mathcal{S}_{\text{evs}}}{\min} \gamma_{J_2,E_m}} \right) \qquad (9.22)
$$

Note that in Equations 9.21 and 9.22, the jammers that represent the worst scenario correspond to a minimum jamming for the eavesdroppers. In the case that the individual selection of the two nodes in the second phase using Equation 9.22 is expected to degrade the performance, a slightly more complex joint selection for the relay and the jammer in the second phase is given by

$$
\left(R^*, J_2^*\right) = \underset{\substack{R \in D \\ J_2 \in \left\{S_{\text{relays}} - R^*\right\}}}{\operatorname{argmax}} \left(\frac{\dfrac{\gamma_{R,D}}{\gamma_{J_2,D}}}{\underset{E_m \in S_{\text{evs}}}{\max} \; \forall m \left(\dfrac{\gamma_{R,E_m}}{\gamma_{J_2,E_m}} \right)} \right)
\tag{9.23}
$$

The suboptimal solutions in Equation 9.22 can be used for some applications with critical computation constraints.

9.2.1.2.1 Asymptotic Secrecy Outage Probability for NCJ Symmetric Case   As in the previous non-jamming case, we try to find the asymptotic secrecy outage probability for the symmetric case for simplicity of presentation. Let us introduce the following lemma, which will be used in the subsequent analysis.

Lemma 9.1

Let X_1, X_2, X_3, and X_4 be i.i.d exponential random variables. Define $Y_1 \triangleq X_1 / X_2$ and $Y_2 \triangleq X_3 / X_4$. The PDF and CDF of the new random variable $Z \triangleq Y_1 + Y_2$ are given by

$$
p_Z(z) = \frac{z}{(z+1)(z+2)^2} + \frac{4\ln(z+1)}{(z+2)^3}
\tag{9.24}
$$

$$
P_Z(z) = \frac{-2}{(z+2)} - \frac{2\ln(z+1)}{(z+2)^2} + 1
\tag{9.25}
$$

Proof: See Appendix E.2.   ■

As a double check, the PDF given in Equation 9.24 is verified using Monte Carlo simulation as shown in Figure 9.2. It is clear from this figure that the simulation results agree with the analytical results.

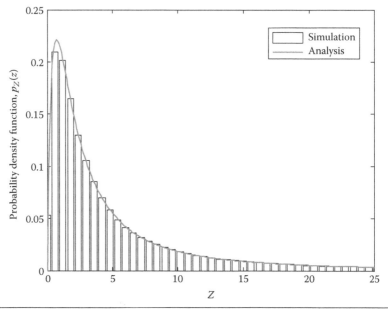

Figure 9.2 Comparison between the analytical and simulation results of the PDF given in Equation 9.24.

For the symmetric case discussed earlier and high SNR, where the decoding set D contains the same elements as the S_{relays}, the asymptotic secrecy outage probability is given by

$$
P_O^{\text{NCJ}} \approx \left(M \int_0^\infty \left[\frac{-2}{(\rho z + 2)} - \frac{2\ln(\rho z + 1)}{(\rho z + 2)^2} + 1 \right] \cdot \left[\frac{-2}{(z+2)} - \frac{2\ln(z+1)}{(z+2)^2} + 1 \right]^{M-1} \right.
$$

$$
\left. \left[\frac{2}{(z+2)^2} - \frac{2}{(z+1)(z+2)^2} + \frac{4\ln(z+1)}{(z+2)^3} \right] dz \right)^{2N\binom{N}{2}}
\tag{9.26}
$$

where the proof is given in Appendix E.3. Since the integral in (9.26) is intractable, it can be solved numerically.

9.2.1.3 Noncooperative Eavesdroppers with Controlled Jamming (NCCJ) In the previous case, we have assumed that the jamming signals sent by the selected jammers in both phases are not known at the destination. In the cases, where the computational complexity is not of much concern, like the cases we have assumed here with multiple eavesdroppers trying to decode the information with cooperation between them, we can relax the security constraints by assuming that the destination can decode the jamming signal. This can be accomplished by sending initial sequences to the destination at the beginning of the transmission. In this case, with nonzero elements decoding set, the secrecy rate in Equation 9.6 is given by

$$
C_S^{\text{NCCJ}}\left(R, J_1, J_2\right) = \left[\frac{1}{2}\log_2\left(\frac{1+\gamma_{S,D}+\gamma_{R,D}}{1+\max_{E_m \in S_{evs}\,\forall m}\left\{\frac{\gamma_{S,E_m}}{1+\gamma_{J_1,E_m}}+\frac{\gamma_{R,E_m}}{1+\gamma_{J_2,E_m}}\right\}}\right)\right]^{+}
$$

(9.27)

As previously mentioned, the selection process can be performed in two steps, one in each phase. Assuming a high SNR value, and following the same assumptions in obtaining Equation 9.20, the selection policy that maximizes the secrecy rate given in Equation 9.27 is given by

$$
J_1^* = \operatorname*{argmax}_{J_1 \in S_{\text{realys}}}\left(\min_{\forall m \in S_{evs}}\gamma_{J_1,E_m}\right)
$$

(9.28)

$$
R^* = \operatorname*{argmax}_{R \in S_{\text{realys}}}\left(\frac{\gamma_{R,D}}{\max_{\forall m \in S_{evs}}\gamma_{R,E_m}}\right), J_2^* = \operatorname*{argmax}_{J_2 \in \left\{S_{\text{realys}} - R^*\right\}}\left(\min_{\forall m \in S_{evs}}\gamma_{J_2,E_m}\right)
$$

(9.29)

9.2.1.3.1 Asymptotic Secrecy Outage Probability for Symmetric NCCJ Case According to Appendix E.4, the asymptotic secrecy outage probability for this case converges to

$$
P_O^{\text{NCCJ}} \approx \left(M\int_0^\infty \left[1-e^{-\lambda\bar{\rho}z}\left(1+\lambda\bar{\rho}z\right)\right]\left[\frac{-2}{(z+2)}-\frac{2\ln(z+1)}{(z+2)^2}+1\right]^{M-1}\right.
$$

$$
\left.\left[\frac{z}{(z+1)(z+2)^2}+\frac{4\ln(z+1)}{(z+2)^3}\right]dz\right)^{2N\binom{N}{2}}
$$

(9.30)

where $\bar{\rho} = \rho/P_{J_i}$ and $i = 1,2$. The earlier integration can be solved numerically.

9.2.2 Selection Schemes with Cooperative Eavesdroppers

In this section, we consider the cooperative eavesdropping model in which the eavesdroppers have the ability to exchange their obtained information in order to decode the source information. The instantaneous secrecy rate for this model with decoding set $\mathcal{D}$ is given by [153,162]

$$
C_S^{|\mathcal{D}|}(R, J_1, J_2)
$$

$$
= \begin{cases}
\left[\left[\dfrac{1}{2}\log_2\left(1+\dfrac{\gamma_{S,D}}{1+\gamma_{J_1,D}}\right)\right.\right. \\
\qquad\left.\left. -\dfrac{1}{2}\log_2\left(1+\sum_{m=1}^{M}\dfrac{\gamma_{S,E_m}}{1+\gamma_{J_1,E_m}}\right)\right]^{+} & |\mathcal{D}|=0 \\[4ex]
\left[\left[\dfrac{1}{2}\log_2\left(1+\dfrac{\gamma_{S,D}}{1+\gamma_{J_1,D}}+\dfrac{\gamma_{R,D}}{1+\gamma_{J_2,D}}\right)\right.\right. \\
\qquad\left.\left. -\dfrac{1}{2}\log_2\left(1+\sum_{m=1}^{M}\left(\dfrac{\gamma_{S,E_m}}{1+\gamma_{J_1,E_m}}+\dfrac{\gamma_{R,E_m}}{1+\gamma_{J_2,E_m}}\right)\right)\right]^{+} & |\mathcal{D}|>0
\end{cases}
$$

$$\tag{9.31}$$

It is clear from Equation 9.31 that the cooperation between the eavesdroppers adds more constraints on the achievable secrecy rate by the source node. In the following discussion, a set of selection schemes is proposed in order to maximize the secrecy capacity in Equation 9.31. Again, the objective is to maximize the secrecy rate or equivalently minimize the secrecy outage probability. The optimization problems in Equations 9.8 and 9.9 are analyzed here for the case of cooperative eavesdroppers. We will consider first the relay selection without jamming, and then the effects of the jamming operation.

9.2.2.1 Cooperative Eavesdroppers without Jamming (Cw/oJ)   Assuming that $|\mathcal{D}| > 0$, the instantaneous achievable secrecy rate in Equation 9.31, ignoring the effect of jamming signals, is given by

$$C_S^{Cw/oJ}(R) = \left[\frac{1}{2} \log_2 \left(\frac{1 + \gamma_{S,D} + \gamma_{R,D}}{1 + \sum_{m=1}^{M} \left(\gamma_{S,E_m} + \gamma_{R,E_m} \right)} \right) \right]^+$$ (9.32)

It can be seen from Equation 9.32 that a nonzero secrecy rate can be achieved only if the rate of the source to the destination is larger than the sum of the rates of the source to all eavesdroppers. The selection of the relay node in the second phase can be achieved by the following selection scheme:

$$R^* = \underset{R \in D}{\arg\max} \left\{ \frac{1 + \gamma_{S,D} + \gamma_{R,D}}{1 + \sum_{m=1}^{M} \left(\gamma_{S,E_m} + \gamma_{R,E_m} \right)} \right\}$$ (9.33)

9.2.2.2 Cooperative Eavesdroppers with Jamming (CJ)   In this case, we consider the effects of the jamming signals transmitted by the selected jammers. For a decoding set with $|D| > 0$, the achievable secrecy rate in Equation 9.31 can be written as follows:

$$C_S^{CJ}(R, J_1, J_2) = \left[\frac{1}{2} \log_2 \left(\frac{1 + \dfrac{\gamma_{S,D}}{\gamma_{J_1,D}} + \dfrac{\gamma_{R,D}}{\gamma_{J_2,D}}}{1 + \sum_{m=1}^{M} \left(\dfrac{\gamma_{S,E_m}}{1 + \gamma_{J_1,E_m}} + \dfrac{\gamma_{R,E_m}}{1 + \gamma_{J_2,E_m}} \right)} \right) \right]^+$$ (9.34)

The objective of the selection process is to solve the following optimization problem:

$$\left(J_1^*, R^*, J_2^* \right) = \underset{\substack{J_1 \in S_{relays} \\ R \in D \\ J_2 \in \{S_{relays} - R^*\}}}{\arg\max} \left(\frac{1 + \dfrac{\gamma_{S,D}}{1 + \gamma_{J_1,D}} + \dfrac{\gamma_{R,D}}{1 + \gamma_{J_2,D}}}{1 + \sum_{m=1}^{M} \left(\dfrac{\gamma_{S,E_m}}{1 + \gamma_{J_1,E_m}} + \dfrac{\gamma_{R,E_m}}{1 + \gamma_{J_2,E_m}} \right)} \right)$$ (9.35)

Similar to the case of noncooperative eavesdroppers with jamming, we can follow the following suboptimal selection schemes: one jammer is selected in the first phase from the set of the relays and two nodes are selected in the second phase as follows:

$$J_1^* = \underset{J_1 \in S_{\text{realys}}}{\arg\min} \left(\frac{\gamma_{J_1,D}}{\sum_{\forall m \in S_{\text{evs}}} \gamma_{J_1,E_m}} \right) \tag{9.36}$$

$$R^* = \underset{R \in \mathcal{D}}{\arg\max} \left(\frac{\gamma_{R,D}}{\sum_{\forall m \in S_{\text{evs}}} \gamma_{R,E_m}} \right) \text{ and}$$

$$J_2^* = \underset{J_2 \in \{S_{\text{relays}} - R^*\}}{\arg\min} \left(\frac{\gamma_{J_2,D}}{\sum_{\forall m \in S_{\text{evs}}} \gamma_{J_2,E_m}} \right) \tag{9.37}$$

The individual selection for the nodes in the second phase with Equation 9.37 may degrade the performance. If a high degradation is expected or when the computational complexity is not of major concern, a suboptimal joint selection of the two nodes is given as

$$\left(R^*, J_2^* \right) = \underset{\substack{R \in \mathcal{D} \\ J_2 \in \{S_{\text{relays}} - R^*\}}}{\arg\max} \left(\frac{\dfrac{\gamma_{R,D}}{\gamma_{J_2,D}}}{\displaystyle\sum_{m=1}^{M} \left(\dfrac{\gamma_{R,E_m}}{\gamma_{J_2,E_m}} \right)} \right) \tag{9.38}$$

9.2.2.3 Cooperative Eavesdroppers with Controlled Jamming (CCJ) When the eavesdroppers have the ability to cooperate in order to decode the source information, a higher constraint is added to the achievable secrecy rate than in the case of noncooperative eavesdroppers. Therefore, removing the jamming signals at the destination node is expected to greatly enhance the secrecy rate and the secrecy outage performance. In this case, the instantaneous secrecy rate given in Equation 9.31 is modified as

$$C_S^{CCJ}(J_1, R, J_2) = \left[\frac{1}{2}\log_2 \left(\frac{1 + \gamma_{S,D} + \gamma_{R,D}}{\sum_{m=1}^{M}\left(1 + \frac{\gamma_{S,E_m}}{1 + \gamma_{J_1,E_m}} + \frac{\gamma_{R,E_m}}{1 + \gamma_{J_2,E_m}}\right)} \right) \right]^+$$

$$(9.39)$$

The selection policy that maximizes the secrecy rate in Equation 9.39 is given by the following selection schemes:

$$J_1^* = \underset{J_1 \in S_{\text{relays}}}{\operatorname{argmax}} \left(\sum_{m=1}^{M} \gamma_{J_1,E_m} \right) \qquad (9.40)$$

$$R^* = \underset{R \in S_{\text{relays}}}{\operatorname{argmax}} \left(\frac{\gamma_{R,D}}{\sum_{m=1}^{M} \gamma_{R,E_m}} \right) \quad \text{and} \quad J_2^* = \underset{J_2 \in (S_{\text{relays}} - R^*)}{\operatorname{argmax}} \left(\sum_{m=1}^{M} \gamma_{J_2,E_m} \right)$$

$$(9.41)$$

It should be noted that the asymptotic secrecy outage probability for the cases of cooperative eavesdroppers, specifically for Cw/oJ, CJ, and CCJ schemes, can be derived in a similar way to the cases considered for the noncooperative eavesdroppers.

9.3 Simulation Examples

We have investigated the effectiveness of the discussed selection schemes via computer simulations. We have considered the system model shown in Figure 9.3 and followed the model described in Section 9.2. We have assumed that $N = 4$ relays and $M = 4$ eavesdroppers are deployed in a 2D unit-square area. The locations of the source and the relays are fixed at $\{x_S, y_S\} = \{0, 0.5\}$ and $\{x_{R_i}, y_{R_i}\}_{i=1}^{4} = \{(0.5, 0.2), (0.5, 0.4), (0.5, 0.6), (0.5, 0.8)\}$, respectively, as shown in Figure 9.3. Three scenarios have been considered and each scenario has distinct locations of the destination and eavesdroppers [169].

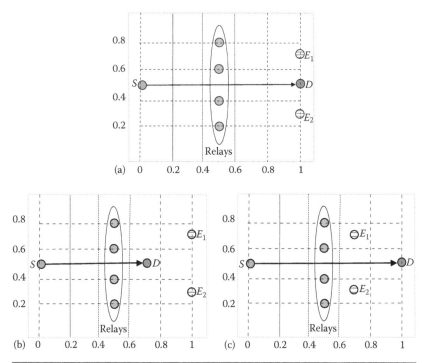

Figure 9.3 The simulation environment with a square unit area, $N=4$, $M=2$, and $\alpha=3$. (a) Scenario A: The destination and the eavesdroppers are of equal distances from the relays. (b) Scenario B: The destination is closer to the relays than the eavesdroppers. (c) Scenario C: The eavesdroppers are closer to the relays.

Scenario A: As shown in Figure 9.3a, the destination is located at $\{x_D,\ y_D\}=\{1,0.5\}$, and the eavesdroppers are located at $\{x_{E_i},\ y_{E_i}\}_{i=1}^2=\{(1,0.3),(1,0.7)\}$, i.e., the destination and the eavesdroppers are at the same distance from the relays.

Scenario B: As shown in Figure 9.3b, the destination is located at $\{x_D,\ y_D\}=\{0.7,0.5\}$, and the eavesdroppers are located at $\{x_{E_i},\ y_{E_i}\}_{i=1}^2=\{(1,0.3),(1,0.7)\}$. Hence, in this case the destination is closer to the relays than the eavesdroppers.

Scenario C: In this scenario, the destination is located at $\{x_D,\ y_D\}=\{1,0.5\}$, and the eavesdroppers are located at $\{x_{E_i},\ y_{E_i}\}_{i=1}^2=\{(0.7,0.3),(0.7,0.7)\}$, as shown in Figure 9.3c. Hence, the eavesdroppers are closer to the relays than the destination in this case.

Furthermore, the path loss exponent has been taken as $\alpha = 3$, the transmission rate is equal to $R_0 = 2$ bits/s/Hz, and the target secrecy rate is equal to 0.1 bits/s/Hz. Note that for the relay to belong to the decoding set, the transmission rate R_0 must be less than the capacity of the source–relay channel. We have adopted the secrecy outage probability and the achievable secrecy rate as performance metrics for the discussed schemes. The SNR used in the simulation is defined as $SNR = P/N_0$, where $P \triangleq P_S = P_R$. In order to justify the assumption used to obtain Equation 9.20 that the power allocated to the jammers should be much lower than the power of the source and the selected relay, we have considered the power ratio $P/P_{j_i} = 100$ and $i = 1, 2$.

First, we considered Scenario A along with the previously described selection schemes, namely *NC*, *NCJ*, *NCCJ*, *Cw/oJ*, *CJ*, and *CCJ*. As shown in Figure 9.4, the outage performance for the cases with noncooperative eavesdroppers is better than that of the cases with cooperative eavesdroppers. The effect of jamming in reducing the secrecy outage probability can be observed in

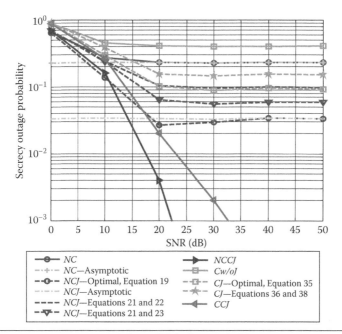

Figure 9.4 Secrecy outage probability vs. SNR for Scenario A with $N=4$ and $M=2$.

Figure 9.4. For instance, the case of *NCJ* (optimal selection based on (9.19)) reduces the secrecy outage probability from 0.22 to 3×10^{-2}. The performance of the two *NCJ* suboptimal selection schemes can also be observed. For the first scheme, J_1 is selected using Equation 9.21, and R and J_2 are selected individually using Equation 9.22. For the second scheme, J_1 is selected using Equation 9.21, and the two nodes in the second phase are jointly selected using Equation 9.23. As shown, the two suboptimal schemes still give better performance than the *NC* scheme. The second scheme with joint selection in the second phase gives a better performance than that of the first suboptimal scheme. It should be noted that there is a trade-off between the computational complexity and the performance with the two schemes as indicated in Section 9.3. We also note that the *NCCJ* and the *CCJ* schemes give the best performance, because the effect of jamming signals is removed at the main destination node. Moreover, Figure 9.4 shows the asymptotic secrecy outage probability for the case of noncooperative eavesdroppers. Note that the asymptotic approximations match well with the simulation results of *NC* and *NCJ* schemes at high SNR values.

Figure 9.5 depicts the achievable secrecy rate for the various schemes. The obtained results for the various schemes confirm the results of the secrecy outage probability shown in Figure 9.4. It can be seen that the jamming techniques significantly improve the secrecy rate for both cooperative and noncooperative eavesdroppers models. An improvement of about 1 bit/s/Hz is achieved using the jamming techniques. Note also that the difference between the optimal and suboptimal selection schemes is small (about 0.1 bit/s/Hz).

After that, we have considered the configuration of Scenario B shown in Figure 9.3b. The secrecy outage probability of Scenario B is shown in Figure 9.6. It is interesting to note that the non-jamming selection schemes achieve a much better performance than those of the cases with jamming, because for the case of jamming the destination is closer to the relay nodes and the jamming signals are strong at the destination node. This in turn increases the secrecy outage probability and decreases the secrecy rate. We also note that the secrecy outage probability for the cases with jamming starts to decrease and then increase again at high SNR values, because at high

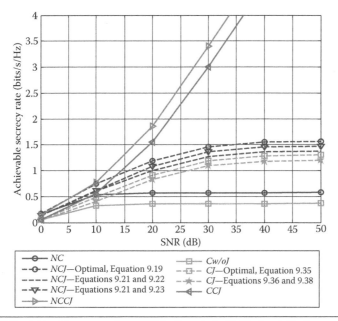

Figure 9.5 Achievable secrecy rate vs. SNR for Scenario A with $N=4$ and $M=2$.

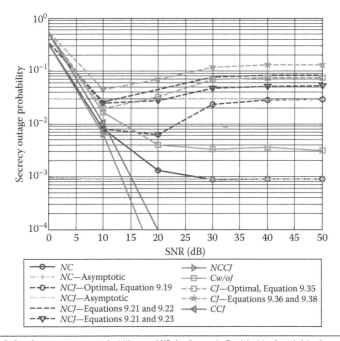

Figure 9.6 Secrecy outage probability vs. SNR for Scenario B with $N=4$ and $M=2$.

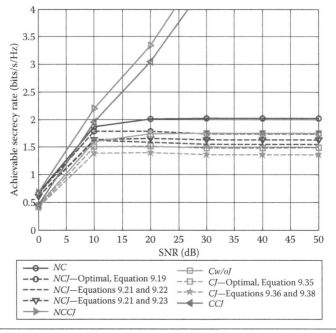

Figure 9.7 Achievable secrecy rate vs. SNR for Scenario B with $N=4$ and $M=2$.

SNRs the jamming signals become stronger and decrease the outage performance. The secrecy rate of this configuration, averaged over all channel realizations, is shown in Figure 9.7. Note that the secrecy rate is higher in the cases of non-jamming techniques. Comparing Figures 9.5 and 9.7, it is interesting to notice that the secrecy rate achieved with Scenario B is better than that with Scenario A even with jamming techniques. This reflects the importance (effect) of the location of the eavesdroppers relative to the location of the destination node.

The third configuration we have considered is the Scenario C shown in Figure 9.3c. The secrecy outage probability for this case is shown in Figure 9.8. It is clear that the non-jamming schemes are not effective in this case, because strong information signals are received by the eavesdroppers either from the source node or from the selected relay. Note that the outage probability is one even in the case of noncooperative eavesdroppers. Figure 9.9 confirms the secrecy outage probability results, where an approximately zero secrecy rate is achieved with the non-jamming schemes. Obviously, as shown

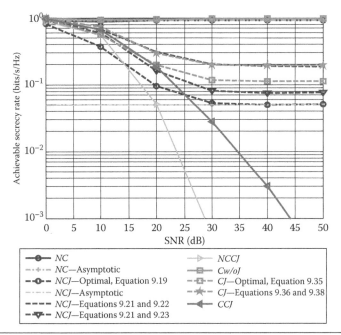

Figure 9.8 Secrecy outage probability vs. SNR for Scenario C with $N=4$ and $M=2$.

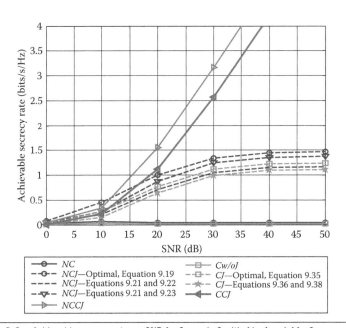

Figure 9.9 Achievable secrecy rate vs. SNR for Scenario C with $N=4$ and $M=2$.

in Figures 9.8 and 9.9, the jamming schemes improve significantly the secrecy outage probability and the secrecy rate. It can be seen from Figure 9.8 that the performance difference between the *NCCJ* and the *CCJ* schemes is larger than that in Figure 9.6 for Scenario B, where the destination is closer to the relay nodes, because in the latter case (Scenario B) the source–destination and the relay–destination links are very strong compared to the source–eaves-droppers and relay–eavesdroppers links. Therefore, the cooperation between the eavesdroppers does not give a significant decrease in the secrecy rate achieved at the destination node.

Finally, we have tested the impact of the eavesdroppers' locations on the achievable secrecy rate. We have fixed the destination loca-tion at (1, 0.5) and moved the location of the eavesdroppers on the *x*-axis from 0.6 to 1.8. Figure 9.10 illustrates the relation between the eavesdroppers' locations and the achievable secrecy rate. It can be seen that for the non-jamming schemes (*NC* and *Cw/oJ*), as the eavesdroppers are closer to the relay nodes than the destination node, the secrecy rate is zero. However, as they move away from the relays

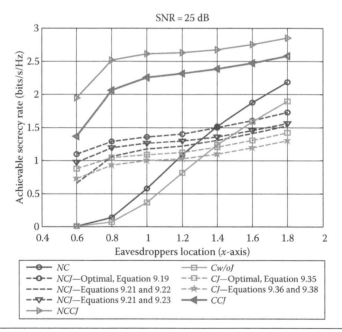

Figure 9.10   Achievable secrecy rate vs. eavesdroppers location along the *x*-axis with *N*=4 and *M*=2.

(and the source node), the achievable secrecy rate increases rapidly and for farther distances, it exceeds the achievable secrecy rate of the jamming schemes. Note that the secrecy rate of the jamming schemes does not improve rapidly as the eavesdroppers move away from the relay nodes. This conclusion recommends the use of non-jamming schemes in this case. The *NCCJ* and *CCJ* schemes achieve the best performance.

Appendix A: Channel Models

Vehicular-A outdoor and Stanford University Interim (SUI) channel models have been used to simulate the slow-fading frequency-selective Rayleigh channel. In the following sections, the characteristics of these models are described.

A.1 Vehicular-A Outdoor Channel Model

The multipath profiles of this model are tabulated in Table A.1. This model has a mobile speed of 120 km/h. This speed corresponds to a Doppler spread of 223 Hz for a carrier frequency of 2 GHz. $L = 14$ paths have been considered in the simulations. These paths correspond to a sampling rate of 5 mega samples per second.

A.2 SUI-3 Channel Model

SUI channel models are six models. These models can be used for simulations, design, development, and testing of technologies suitable for fixed broadband wireless applications. The SUI-3 channel is one of these models adopted by the IEEE 802.16a standard for evaluating the performance of broadband wireless systems in the 2–11 GHz band [64]. It is suitable to model the hilly terrain with light tree density or

Table A.1   Multipath Profiles of the Vehicular-A Outdoor Channel Model

	TAPE 1	TAPE 2	TAPE 3	TAPE 4	TAPE 5	TAPE 6
Delay (ns)	0	310	710	1090	1730	2510
Power (dB)	0	−1	−9	−10	−15	−20

Table A.2   Multipath Profiles of the SUI-3 Channel Model

	TAPE 1	TAPE 2	TAPE 3
Delay (μs)	0	0.5	1
Power (dB)	0	−5	−10
Doppler	0.4	0.4	0.4

flat terrain with moderate to heavy tree density. The multipath profiles of the SUI-3 channel model are shown in Table A.2. $L = 6$ paths have been considered. These paths correspond to a sampling rate of 5 mega samples per second.

Appendix B: Derivation of the Interference Coefficients for the DFT-SC-FDMA System over an AWGN Channel

B.1 Derivation of the Interference Coefficients

This appendix considers a DFT-SC-FDMA system operating over an AWGN channel in the presence of CFOs. At the transmitter side, the signal after the DFT can be expressed as follows:

$$X^u(k) = \sum_{n=0}^{N-1} x^u(n) e^{-\frac{j2\pi nk}{N}} \tag{B.1}$$

where

$x^u(n)$ is the modulated data symbol of the uth user

N is the DFT length

The signal after the IDFT can be formulated as follows:

$$\bar{x}^u(m) = \frac{1}{M} \sum_{l=0}^{M-1} \bar{X}^u(l) e^{\frac{j2\pi ml}{M}} \tag{B.2}$$

where

M is the IDFT length

$\bar{X}^u(l)$ is the lth sample after the demapping process

At the receiver side, the received signal is given by

$$r(m) = \sum_{u=1}^{U} \bar{x}^u(m) e^{j\frac{2\pi m \varepsilon_u}{M}} + n(m) \tag{B.3}$$

where

$\varepsilon_u = \Delta f T$ is the normalized CFO, Δf is the CFO, and T is the symbol period

$n(m)$ is the mth symbol of the AWGN

Then, the received signal is transformed into the frequency domain via a DFT as follows:

$$R(k) = \sum_{m=0}^{M-1} r(m) e^{-j\frac{2\pi m k}{M}} \tag{B.4}$$

Substituting Equation B.3 into Equation B.4, we get

$$R(k) = \sum_{u=1}^{U} \sum_{m=0}^{M-1} \bar{x}^u(m) e^{\left(j\frac{2\pi m \varepsilon_u}{M}\right)} e^{\left(-j\frac{2\pi m k}{M}\right)} + N(k)$$

$$= \sum_{u=1}^{U} \frac{1}{M} \sum_{l=0}^{M-1} \bar{X}^u(l) \sum_{m=0}^{M-1} e^{\left(j\frac{2\pi m(l+\varepsilon_u-k)}{M}\right)} + N(k) \tag{B.5}$$

Using the geometric series as $\dfrac{1}{M} \sum_{m=0}^{M-1} e^{\left(\frac{j 2\pi m(l+\varepsilon_i-k)}{N}\right)}$, we can expand

$$\frac{1}{M} \sum_{m=0}^{M-1} e^{\left(\frac{j 2\pi m(l+\varepsilon_i-k)}{M}\right)} = \frac{1}{M} \frac{1 - e^{(j 2\pi(l+\varepsilon_i-k))}}{1 - e^{\left(\frac{j 2\pi(l+\varepsilon_i-k)}{M}\right)}} = \frac{1}{M} e^{(j\pi(l+\varepsilon-k)(1-1/M))}$$

$$\times \frac{\sin(\pi(l-k+\varepsilon_i))}{\sin(\pi(l-k-\varepsilon_i)/M)} \tag{B.6}$$

Substituting Equation B.6 into Equation B.5, we get

$$R(k) = \sum_{u=1}^{U} \sum_{l=0}^{M-1} \bar{X}^u(l) S^u(l,k,\varepsilon_u) + N(k) \qquad \text{(B.7)}$$

where $S^u(l,k,\varepsilon_u)$ are the complex coefficients of the interference components in the received signal due to the uth user. $S^u(l,k,\varepsilon_u)$ is given by

$$S^u(l,k,\varepsilon_u) = e^{(j\pi(l+\varepsilon_u-k)(1-1/M))} \frac{\sin(\pi(l-k+\varepsilon_u))}{M \sin(\pi(l-k+\varepsilon_u)/M)} \qquad \text{(B.8)}$$

Appendix C: Derivation of the Interference Coefficients for the DCT-SC-FDMA System over an AWGN Channel

C.1 Derivation of the Interference Coefficients

This appendix considers a DCT-SC-FDMA system operating over an AWGN channel in the presence of CFOs.

At the transmitter side, the signal after the DCT can be expressed as follows:

$$X^u(k) = \sqrt{\frac{2}{N}}\beta_k \sum_{n=0}^{N-1} x^u(n)\cos\left(\frac{\pi k(2n+1)}{2N}\right) \qquad (\text{C.1})$$

where

$x^u(n)$ is the modulated data symbol of the uth user

N is the DCT length

β_k can be expressed as in Equation 3.4

The signal after the IDCT can be formulated as follows:

$$\bar{x}^u(m) = \sqrt{\frac{2}{M}} \sum_{l=0}^{M-1} \bar{X}^u(l)\beta_l \cos\left(\frac{\pi l(2m+1)}{2M}\right) \tag{C.2}$$

where M is the IDCT length (number of subcarriers). At the receiver side, the received signal can be written as follows:

$$r(m) = \sum_{u=1}^{U} \bar{x}^u(m) e^{\left[j\frac{\pi\zeta_u(2m+1)}{2M} + \phi_u\right]} + n(m)$$

$$= \sum_{u=1}^{U} \bar{x}^u(m)\left(\cos\left[\frac{\pi\zeta_u(2m+1)}{2M} + \phi_u\right]\right.$$

$$\left. + j\sin\left[\frac{\pi\zeta_u(2m+1)}{2M} + \phi_u\right]\right) + n(m) \tag{C.3}$$

where

$\zeta_u = 2T\Delta f$ is the normalized CFO of the uth user

ϕ_u is the phase error

The effect of the phase error has not been considered, i.e., $\phi_u = 0$. Thus, Equation C.3 can be written as follows:

$$r(m) = \sum_{u=1}^{U} \left(r_I^u(m) + jr_Q^u(m)\right) + n(m) \tag{C.4}$$

where

$$r_I^u(m) = \bar{x}^u(m)\cos\left[\frac{\pi\zeta_u(2m+1)}{2M}\right] \tag{C.5}$$

$$r_Q^u(m) = \bar{x}^u(m)\sin\left[\frac{\pi\zeta_u(2m+1)}{2M}\right] \tag{C.6}$$

Then the received signal is transformed into the frequency domain via the DCT. Based on Equations C.2 and C.3, the frequency-domain in-phase component of the received signal can be derived as follows:

$$R_I(k) = \sum_{u=1}^{U} \sqrt{\frac{2}{M}} \beta_k \sum_{m=0}^{M-1} x^u(m) \cos\left[\frac{\pi k(2m+1)}{2M}\right] \cos\left[\frac{\pi \zeta_u(2m+1)}{2M}\right]$$

$$= \sum_{u=1}^{U} \frac{2}{M} \beta_k \sum_{l=0}^{M-1} \bar{X}^u(l) \beta_l \sum_{m=0}^{M-1} \cos\left[\frac{\pi l(2m+1)}{2M}\right] \cos\left[\frac{\pi k(2m+1)}{2M}\right]$$

$$\cos\left[\frac{\pi \zeta_u(2m+1)}{2M}\right] \tag{C.7}$$

Equation C.7 can be written as follows:

$$R_I(k) = \sum_{u=1}^{U} \frac{1}{2M} \beta_k \sum_{l=0}^{M-1} \bar{X}^u(l) \beta_l \sum_{m=0}^{M-1} \left(\cos\left[\frac{\pi(2m+1)(l+k+\zeta_u)}{2M}\right] \right.$$

$$+ \cos\left[\frac{\pi(2m+1)(l+k-\zeta_u)}{2M}\right] + \cos\left[\frac{\pi(2m+1)(l-k+\zeta_u)}{2M}\right]$$

$$\left. + \cos\left[\frac{\pi(2m+1)(l-k-\zeta_u)}{2M}\right] \right) \tag{C.8}$$

We can simplify Equation C.8 as follows:

$$R_I(k) = \sum_{u=1}^{U} \frac{1}{2M} \sum_{l=0}^{M-1} \bar{X}^u(l) \beta_k \beta_l \left[\varphi^u(l+k-\zeta_u) + \varphi^u(l-k-\zeta_u) \right.$$

$$\left. + \varphi^u(l+k+\zeta_u) + \varphi^u(l-k+\zeta_u) \right] + N(k)$$

$$= \sum_{u=1}^{U} \sum_{l=0}^{M-1} \bar{X}^u(l) S_I^u(l,k,\zeta_u) \tag{C.9}$$

where $S_I^u(l,k,\zeta_u)$ is the in-phase component of the interference coefficients due to the uth user. $S_I^u(l,k,\zeta_u)$ and $\varphi(x)$ are given by

$$S_I^u(l,k,\zeta_u) = \frac{1}{2M} \beta_k \beta_l \left[\varphi^u(l+k-\zeta_u) + \varphi^u(l-k-\zeta_u) \right.$$

$$\left. + \varphi^u(l+k+\zeta_u) + \varphi^u(l-k+\zeta_u) \right] \tag{C.10}$$

$$\varphi(x) = \frac{\sin\left(\dfrac{\pi x}{2}\right)\cos\left(\dfrac{\pi x}{2}\right)}{\sin\left(\dfrac{\pi x}{2M}\right)} \tag{C.11}$$

Using the same method, the frequency-domain quadrature-phase component of the received signal can be derived as follows:

$$R_Q(k) = \sum_{u=1}^{U} \sqrt{\frac{2}{M}} \beta_k \sum_{m=0}^{M-1} x^u(m) \cos\left[\frac{\pi k(2m+1)}{2M}\right] \sin\left[\frac{\pi \zeta_u(2m+1)}{2M}\right]$$

$$= \sum_{u=1}^{U} \frac{2}{M} \beta_k \sum_{l=0}^{M-1} \overline{X}^u(l) \beta_l \sum_{m=0}^{M-1} \cos\left[\frac{\pi l(2m+1)}{2M}\right]$$

$$\cos\left[\frac{\pi k(2m+1)}{2M}\right] \sin\left[\frac{\pi \zeta_u(2m+1)}{2M}\right] \tag{C.12}$$

Equation C.12 can also be simplified as follows:

$$R_Q(k) = \sum_{u=1}^{U} \frac{1}{2M} \sum_{l=0}^{M-1} X^u(l)\beta_k\beta_l \Big[\Gamma^u(l+k-\zeta_u) + \Gamma^u(l-k-\zeta_u)$$

$$- \Gamma^u(l+k+\zeta_u) - \Gamma^u(l-k+\zeta_u) \Big]$$

$$= \sum_{u=1}^{U} \sum_{l=0}^{M-1} X^u(l) S_Q^u(l,k,\zeta_u) \tag{C.13}$$

where $S_Q^u(l,k,\zeta_u)$ is the quadrature-phase component of the interference coefficients due to the uth user. $S_Q^u(l,k,\zeta_u)$ and $\Gamma(x)$ are given by

$$S_Q^u(l,k,\zeta_u) = \frac{1}{2M}\beta_k\beta_l \Big[\Gamma^u(l+k-\zeta_u) + \Gamma^u(l-k-\zeta_u)$$

$$- \Gamma^u(l+k+\zeta_u) - \Gamma^u(l-k+\zeta_u) \Big] \tag{C.14}$$

$$\Gamma(x) = \frac{\sin\left(\dfrac{\pi x}{2}\right)^2}{\sin\left(\dfrac{\pi x}{2M}\right)} \tag{C.15}$$

Now, we can rewrite Equation C.4 as follows:

$$R(k) = \sum_{u=1}^{U}\left[R_I^u(k) + jR_Q^u(k)\right] + N(k)$$

$$= \sum_{u=1}^{U}\sum_{l=0}^{M-1}\bar{X}^u(l)\left(S_I^u(l,k,\zeta_u) + jS_Q^u(l,k,\zeta_u)\right) + N(k) \tag{C.16}$$

$$= \sum_{u=1}^{U}\sum_{l=0}^{M-1}\bar{X}^u(l)S^u(l,k,\zeta_u) + N(k)$$

Using Equations C.10 and C.14, $S^u(l,k,\zeta_u)$ can be expressed as follows:

$$S^u(l,k,\zeta_u) = \frac{1}{2M}\beta_k\beta_l\Big[\varphi^u(l+k-\zeta_u) + \varphi^u(l-k-\zeta_u)$$

$$+ \varphi^u(l+k+\zeta_u) + \varphi^u(l-k+\zeta_u)\Big] + j\Big[\Gamma^u(l+k-\zeta_u)$$

$$+ \Gamma^u(l-k-\zeta_u) - \Gamma^u(l+k+\zeta_u) - \Gamma^u(l-k+\zeta_u)\Big] \tag{C.17}$$

C.2 Derivation of Equation C.11

Let $x = l + k + \zeta$. From Equation C.8, we get the following:

$$\varphi(x) = \sum_{m=0}^{M-1}\cos\left[\frac{\pi(2m+1)x}{2M}\right] = \sum_{m=0}^{M-1}\left[\frac{e^{\left[\frac{j\pi(2m+1)x}{2M}\right]} + e^{\left[\frac{-j\pi(2m+1)x}{2M}\right]}}{2}\right] \tag{C.18}$$

Equation C.18 can also be simplified as follows:

$$\varphi(x) = \frac{e^{\left[\frac{j\pi x}{2M}\right]}}{2}\sum_{m=0}^{M-1}e^{\left[\frac{j\pi mx}{M}\right]} + \frac{e^{\left[\frac{-j\pi x}{2M}\right]}}{2}\sum_{m=0}^{M-1}e^{\left[\frac{-j\pi mx}{M}\right]} \tag{C.19}$$

Using the geometric series, the first term of Equation C.19 can be expressed as follows:

$$\frac{e^{\left[\frac{j\pi x}{2M}\right]}}{2}\sum_{m=0}^{M-1}e^{\left[\frac{j\pi mx}{M}\right]} = \frac{e^{\left[\frac{j\pi x}{2M}\right]}}{2}\frac{1-e^{\left[j\pi x\right]}}{1-e^{\left[\frac{j\pi x}{M}\right]}} = \frac{e^{\left[\frac{j\pi}{2}x\right]}}{2}\frac{\sin\left(\frac{\pi x}{2}\right)}{\sin\left(\frac{\pi x}{2M}\right)} \tag{C.20}$$

Similar to that in Equation C.20, the second term of Equation C.19 can also be expressed as follows:

$$\frac{e^{\left[\frac{-j\pi x}{2M}\right]}}{2}\sum_{m=0}^{M-1}e^{\left[\frac{-j\pi mx}{M}\right]} = \frac{e^{\left[\frac{-j\pi}{2}x\right]}}{2}\frac{\sin\left(\frac{\pi x}{2}\right)}{\sin\left(\frac{\pi x}{2M}\right)} \tag{C.21}$$

Substituting Equations C.20 and C.21 into Equation C.19, we get

$$\varphi(x)=\frac{e^{\left[\frac{j\pi}{2}x\right]}}{2}\frac{\sin\left(\frac{\pi x}{2}\right)}{\sin\left(\frac{\pi x}{2M}\right)}+\frac{e^{\left[\frac{-j\pi}{2}x\right]}}{2}\frac{\sin\left(\frac{\pi x}{2}\right)}{\sin\left(\frac{\pi x}{2M}\right)} = \frac{\sin\left(\frac{\pi x}{2}\right)\cos\left(\frac{\pi x}{2}\right)}{\sin\left(\frac{\pi x}{2M}\right)}$$

$$\tag{C.22}$$

C.3 Derivation of Equation C.15

Let $x = l + k + \zeta$. From Equation C.13,

$$\Gamma(x)=\sum_{m=0}^{M-1}\sin\left[\frac{\pi(2m+1)x}{2M}\right]=\sum_{m=0}^{M-1}\left[\frac{e^{\left[\frac{j\pi(2m+1)x}{2M}\right]}-e^{\left[\frac{-j\pi(2m+1)x}{2M}\right]}}{2j}\right] \tag{C.23}$$

$$\Gamma(x) = \frac{e^{\left[\frac{j\pi x}{2M}\right]}}{2j} \sum_{m=0}^{M-1} e^{\left[\frac{j\pi mx}{M}\right]} - \frac{e^{\left[\frac{-j\pi x}{2M}\right]}}{2j} \sum_{m=0}^{M-1} e^{\left[\frac{-j\pi mx}{M}\right]} \tag{C.24}$$

Using the geometric series, the first term of Equation C.24 can be expressed as follows:

$$\frac{e^{\left[\frac{j\pi x}{2M}\right]}}{2j} \sum_{m=0}^{M-1} e^{\left[\frac{j\pi mx}{M}\right]} = \frac{e^{\left[\frac{j\pi x}{2M}\right]}}{2j} \frac{1 - e^{[j\pi x]}}{1 - e^{\left[\frac{j\pi x}{M}\right]}} = \frac{e^{\left[\frac{j\pi}{2}x\right]}}{2j} \frac{\sin\left(\frac{\pi x}{2}\right)}{\sin\left(\frac{\pi x}{2M}\right)} \tag{C.25}$$

Similar to that in Equation C.25, the second term of Equation C.24 can also be expressed as follows:

$$\frac{e^{\left[\frac{-j\pi x}{2M}\right]}}{2j} \sum_{m=0}^{M-1} e^{\left[\frac{-j\pi mx}{M}\right]} = \frac{e^{\left[\frac{-j\pi}{2}x\right]}}{2j} \frac{\sin\left(\frac{\pi x}{2}\right)}{\sin\left(\frac{\pi x}{2M}\right)} \tag{C.26}$$

Substituting Equations C.25 and C.26 into Equation C.24, we get

$$\Gamma(x) = \frac{e^{\left[\frac{j\pi}{2}x\right]}}{2j} \frac{\sin\left(\frac{\pi x}{2}\right)}{\sin\left(\frac{\pi x}{2M}\right)} - \frac{e^{\left[\frac{-j\pi}{2}x\right]}}{2j} \frac{\sin\left(\frac{\pi x}{2}\right)}{\sin\left(\frac{\pi x}{2M}\right)} = \frac{\sin\left(\frac{\pi x}{2}\right)^2}{\sin\left(\frac{\pi x}{2M}\right)} \tag{C.27}$$

Appendix D: Derivation of the Optimum Solution of the JLRZF Scheme in Chapter 6

The received signal can be written as follows:

$$R_M^k = \Pi_p^k X_M^k + N_P \tag{D.1}$$

The structures of all components of Equation D.1 are given as follows:

$$\bar{\Pi}_P^k = \begin{bmatrix} \bar{\Pi}_{11}^k & \bar{\Pi}_{12}^k \\ \bar{\Pi}_{21}^k & \bar{\Pi}_{22}^k \end{bmatrix} \tag{D.2}$$

$$N_P = \sum_{\substack{u=1 \\ u \neq k}}^{U} \bar{\Pi}_P^u X_M^k + N_M = \begin{bmatrix} N_{P1} \\ N_{P2} \end{bmatrix} = \begin{bmatrix} \displaystyle\sum_{\substack{u=1 \\ u \neq k}}^{U} \left(\Pi_{11}^u X_1^k + \Pi_{12}^u X_2^k \right) + N_{M1} \\ \displaystyle\sum_{\substack{u=1 \\ u \neq k}}^{U} \left(\Pi_{21}^u X_1^k + \Pi_{22}^u X_2^k \right) + N_{M2} \end{bmatrix} \tag{D.3}$$

$$X_M^k = \begin{bmatrix} X_1^k \\ X_2^k \end{bmatrix}$$

(D.4)

where

   N_{M1} and N_{M2} represent the noise at the first and the second receive antennas after the demapping process, respectively

   N_{P1} and N_{P2} represent the noise and the MAI at the first and the second receive antennas, respectively

   X_1^k and X_2^k are the transmitted signals from the first and the second antennas, respectively

After the first step, the received signal can be written as follows:

$$\breve{X}_M^k = W_1^k R_M^k$$

(D.5)

W_1^k can be formulated as follows:

$$W_1^k = \begin{bmatrix} I_N & -C_2^k \\ -C_1^k & I_{NN} \end{bmatrix}$$

(D.6)

where

$$C_1^k = \mathbf{\Pi}_{21}^k \left(\mathbf{\Pi}_{11}^k \right)^{-1}$$

$$C_2^k = \mathbf{\Pi}_{12}^k \left(\mathbf{\Pi}_{22}^k \right)^{-1}$$

Equation D.5 can be rewritten as follows:

$$\breve{X}_M^k = \begin{bmatrix} \breve{X}_1^k \\ \breve{X}_2^k \end{bmatrix} = \begin{bmatrix} \mathbf{\Phi}_1^k X_1^k \\ \mathbf{\Phi}_2^k X_2^k \end{bmatrix} + \begin{bmatrix} N_{P1} - C_2^k N_{P2} \\ -C_1^k N_{P1} + N_{P2} \end{bmatrix}$$

(D.7)

where

$$\mathbf{\Phi}_1^k = \mathbf{\Pi}_{11}^k - C_2^k \mathbf{\Pi}_{21}^k$$

$$\mathbf{\Phi}_2^k = \mathbf{\Pi}_{22}^k - C_1^k \mathbf{\Pi}_{12}^k$$

After the first step, MIMO signals for the kth user are separated. Thus, we can write each signal at each antenna separately as follows:

$$\breve{X}_1^k = \mathbf{\Phi}_1^k X_1^k + \left(N_{P1} - C_2^k N_{P2} \right) = \mathbf{\Phi}_1^k X_1^k + \bar{N}_{P1}$$

(D.8)

$$\breve{X}_2^k = \Phi_2^k X_2^k + \left(-C_1^k N_{P1} + N_{P2}\right) = \Phi_2^k X_2^k + \bar{N}_{P2} \qquad (D.9)$$

Due to the similarity between Equations D.8 and D.9, we will only derive the solution of Equation D.8. After the second step, the received signal at the first antenna can be expressed as follows:

$$\hat{X}_1^k = W_{21}^k \breve{X}_1^k \qquad (D.10)$$

Then, we define the error e between the estimated symbols $\hat{X}_1^k = W_{21}^k \breve{X}_1^k$ and the transmitted symbols X_1^k as follows:

$$e^k = W_{21}^k \breve{X}_M^k - X_1^k \qquad (D.11)$$

The equalization matrix W_{21}^k is determined by the minimization of the following MSE cost function:

$$J^k = E\left\{\left\|e^k\right\|^2\right\} = E\left\{\left\|W_{21}^k \breve{X}_1^k - X_1^k\right\|^2\right\} \qquad (D.12)$$

where $E\{\cdot\}$ is the expectation. Solving $\partial J^k / \partial W_{21}^k = 0$, we obtain

$$W_{21}^k = \left(\Phi_1^{k^H} R_{X_1^k}^{-1} \Phi_1^k + R_{\bar{N}_{P1}}^{-1}\right)^{-1} \Phi_1^{k^H} R_{X_1^k}^{-1} \qquad (D.13)$$

where

$$R_{X_1^k}^{-1} = E\left\{X_1^k X_1^{k^H}\right\} \qquad (D.14)$$

$$R_{\bar{N}_{P1}}^{-1} = E\left\{\bar{N}_{P1} \bar{N}_{P1}^H\right\} \qquad (D.15)$$

$R_{X_1^k}^{-1}$ and $R_{\bar{N}_{P1}}^{-1}$ are the data and the overall noise covariance matrices of the kth user. If all users have the same power and the average powers of the received signals on all subcarriers are the same, we can have $R_{X_1^k}^{-1} = \sigma_{X_1^k}^2 I_N$. Thus, Equation D.13 can be simplified as follows:

$$W_{21}^k = \left(\Phi_1^{k^H} \Phi_1^k + \frac{R_{\bar{N}_{P1}}^{-1}}{\sigma_{X_1^k}^2}\right)^{-1} \Phi_1^{k^H} \qquad (D.16)$$

After the second step, the received signal at the second antenna can be expressed as follows:

$$\hat{X}_2^k = W_{22}^k \breve{X}_2^k \tag{D.17}$$

Using the MSE criteria, W_{22}^k can be expressed as follows:

$$W_{22}^k = \left(\Phi_2^{k^H} \Phi_2^k + \frac{R_{\bar{N}_{P2}}^{-1}}{\sigma_{X_2^k}^2} \right)^{-1} \Phi_2^{k^H} \tag{D.18}$$

where

$$R_{\bar{N}_{P2}}^{-1} = E\left\{ \bar{N}_{P2} \bar{N}_{P2}^H \right\} \tag{D.19}$$

Now, we can write the received signal as follows:

$$\hat{X}_M^k = W_2^k W_1^k R_M^k \tag{D.20}$$

where

$$W_2^k = \begin{bmatrix} \left(\Phi_1^{k^H} \Phi_1^k + \dfrac{R_{\bar{N}_{P1}}^{-1}}{\sigma_{X_1^k}^2} \right)^{-1} \Phi_1^k & 0_{N \times N} \\[2em] 0_{N \times N} & \left(\Phi_2^{k^H} \Phi_2^k + \dfrac{R_{\bar{N}_{P2}}^{-1}}{\sigma_{X_2^k}^2} \right)^{-1} \Phi_2^{k^H} \end{bmatrix} \tag{D.21}$$

Appendix E: Derivations for Chapter 9

E.1 Asymptotic Secrecy Outage Probability of *NC* Scheme

As indicated in Section 9.2, the analysis is restricted to the symmetric case, where the source–destination, the source–eavesdropper, the relay–destination, and the relay–eavesdropper channels have the same variance σ^2. We consider also the high SNR values where all the relay nodes can decode the source information, correctly. From Equations 9.10 and 9.11, the secrecy outage probability, considering a high SNR, can be written as

$$
P_O^{NC} = \Pr\left\{ \max_{R \in S_{relays}} \left[\frac{|h_{S,D}|^2 + |h_{R,D}|^2}{\max_{E_m \in S_{evs} \, \forall m} \left\{ |h_{S,E_m}|^2 + |h_{R,E_m}|^2 \right\}} \right] < \rho \right\} \qquad \text{(E.1)}
$$

Where $\rho = 2^{2R_S}$ and $h_{i,j}$ denotes the channel coefficient between nodes i and j. Suppose that we have one realization of the possible selections with relay R and the two jammers J_1 and J_2. Define $X \triangleq |h_{S,D}|^2 + |h_{R,D}|^2$, $Z \triangleq \max_{E_m \in S_{evs} \, \forall m} \left\{ |h_{S,E_m}|^2 + |h_{R,E_m}|^2 \right\}$, and $\mathcal{Z} \triangleq X/Z$. Note that in Equation E.1, according to our previous assumptions, $|h_{i,j}|^2 \, \forall i,j$ are exponential random variables with parameter λ. Moreover, the distribution of the random variables X and Z are given by Propositions 9.1 and 9.2, respectively. Let us compute the following probability:

$$\Pr\{Z < \rho\} = \Pr\left\{\frac{X}{Z} < \rho\right\}$$

$$= \int_0^\infty P_X(Z\rho) p_Z(z) dz$$

$$= M \int_0^\infty \left[1 - e^{-\lambda\rho z}(1 + \lambda\rho z)\right]\left(\lambda^2 z e^{-\lambda z}\right)$$

$$\times \left[1 - e^{-\lambda z}(1 + \lambda z)\right]^{M-1} dz \qquad \text{(E.2)}$$

where we have used Propositions 9.1 and 9.2 in the last equality. Note that there are N different realizations of the random variable Z, which correspond to N different selections of the relay node. It is clear from Equation E.1 that these random variables are dependent. Finding the order statistics of dependent random variables requires the joint PDF of the random variables [167], which is intractable in our case. Therefore, we assume independent random variables, and the obtained asymptotic secrecy outage probability will represent a useful lower bound. The outage probability, then, using order statistics [166], is given by

$$P_O^{NC} \approx \left(\Pr\{Z < \rho\}\right)^N$$

$$= \left[M \int_0^\infty \left[1 - e^{-\lambda\rho z}(1 + \lambda\rho z)\right]\left(\lambda^2 z e^{-\lambda z}\right)\left[1 - e^{-\lambda z}(1 + \lambda z)\right]^{M-1} dz\right]^N$$

$$\text{(E.3)}$$

E.2 Proof of Lemma 9.1

Let X_1, X_2, X_3, and X_4 be i.i.d exponential random variables with parameter λ. Define $Y_1 \triangleq X_1/X_2$ and $Y_2 \triangleq X_3/X_4$. The PDF of the random variable $Z \triangleq Y_1 + Y_2$ can be obtained by convolving the PDFs of the Y_1 and Y_2 as follows:

$$p_Z(z) = \int_{-\infty}^{\infty} p_{Y_1}(z - y_2) p_{Y_2}(y_2) dy_2$$

$$= \int_0^z \frac{1}{\left(1 + z - y_2\right)^2} \frac{1}{\left(1 + y_2\right)^2} dy_2$$

$$= \int_0^z \frac{1}{\left(y_2^2 - zy_2 - (z+1)\right)^2} dy_2 \qquad \text{(E.4)}$$

From [168, p. 67],

$$\int \frac{dx}{\left(A + 2Bx + Cx^2\right)^2} = \frac{1}{2} \frac{1}{AC - B^2} \frac{B + Cx}{A + 2Bx + Cx^2}$$

$$+ \frac{1}{2} \frac{C}{AC - B^2} \int \frac{dx}{A + 2B + Cx^2} \qquad \text{(E.5)}$$

In our case, $A = -(z+1), B = \frac{-z}{2}, C = 1, \quad B^2 - AC = \frac{z^2}{4} + z + 1$

$= \left(\frac{z}{2} + 1\right)^2 \geq 0.$ Assume that $z \neq -2$ (which is the case as z has only positive values required for the convolution process). Therefore, $B^2 - AC > 0$ or $AC < B^2$. Thus, from [168, p. 67], we have

$$\int \frac{dx}{A + 2Bx + Cx^2} = \frac{1}{2\sqrt{B^2 - AC}} \ln \left| \frac{Cx + B - \sqrt{B^2 - AC}}{Cx + B + \sqrt{B^2 - AC}} \right| \qquad \text{(E.6)}$$

Therefore, substituting Equation E.6 into Equation E.5 and substituting for A, B, and C in Equation E.5, the PDF of Z in Equation E.4 is given by

$$p_Z(z) = \int_0^z \frac{1}{\left(y_2^2 - zy_2 - (z+1)\right)^2} dy_2 = \frac{2z}{(z+1)(z+2)^2} + \frac{4\ln(z+1)}{(z+2)^3}$$

$$\text{(E.7)}$$

Finally, the CDF of the random variable Z is given by

$$P_Z(z) = \int_0^z \left[\frac{2x}{(x+1)(x+2)^2} + \frac{4\ln(x+1)}{(x+2)^3} \right] dx$$

$$= \frac{-2}{(z+2)} - \frac{2\ln(z+1)}{(z+2)^2} + 1 \qquad (E.8)$$

E.3 Asymptotic Secrecy Outage Probability of NCJ

The same assumptions, of high SNR and symmetric configuration, used in Section E.1 will be considered here. From Equations 9.18 to 9.19, the outage probability considering a high SNR is given by

$$P_O^{NCJ} = \Pr\left\{ \max_{\substack{J_1 \in S_{relays} \\ R \in S_{relays} \\ J_2 \in \{S_{relays} - R^*\}}} \left\{ \frac{\frac{|h_{S,D}|^2}{|h_{J_1,D}|^2} + \frac{|h_{R,D}|^2}{|h_{J_2,D}|^2}}{\max\limits_{E_m \in S_{evs} \; \forall m} \left\{ \frac{|h_{S,E_m}|^2}{|h_{J_1,E_m}|^2} + \frac{|h_{R,E_m}|^2}{|h_{J_2,E_m}|^2} \right\}} \right\} < \rho \right\} \qquad (E.9)$$

where $\rho = 2^{2R_S}$. Suppose that we have one realization of the selection process and the selected relay, first jammer, and second jammer are R, J_1 and J_2, respectively. Define the random variables

$$Z \triangleq \frac{|h_{S,D}|^2}{|h_{J_1,D}|^2} + \frac{|h_{R,D}|^2}{|h_{J_2,D}|^2}, \quad \mathcal{Z} \triangleq \max_{E_m \in S_{evs} \; \forall m} \left\{ \frac{|h_{S,E_m}|^2}{|h_{J_1,E_m}|^2} + \frac{|h_{R,E_m}|^2}{|h_{J_2,E_m}|^2} \right\}. \quad \text{Let us}$$

compute the following probability:

$$\Pr\left\{ \frac{Z}{\mathcal{Z}} < \rho \right\}$$

$$= \Pr\{Z < \rho \mathcal{Z}\}$$

$$= \int_0^\infty P_Z(\rho z) p_{\mathcal{Z}}(z) dz = M \int_0^\infty \left[\frac{-2}{(\rho z + 2)} - \frac{2\ln(\rho z + 1)}{(\rho z + 2)^2} + 1 \right]$$

$$\times \left[\frac{-2}{(z+2)} - \frac{2\ln(z+1)}{(z+2)^2} + 1 \right]^{M-1} \left[\frac{2z}{(z+1)(z+2)^2} + \frac{4\ln(z+1)}{(z+2)^3} \right] dz$$

$$(E.10)$$

where we have used Lemma 9.1 and order statistics in the last equality. In order to obtain the outage probability given in Equation E.9, we must find the CDF of the maximum random variable among $2N\binom{N}{2}$ different random variables, which represent the different possible selections. These random variables are dependent and identically distributed. As in Section E.1, assuming dependent random variables, we obtain a useful lower bound for the secrecy outage probability. The secrecy outage probability is, then, given by

$$P_O^{NCJ} \approx \left(\Pr\left\{ \frac{Z}{\mathcal{Z}} < \rho \right\} \right)^{2N\binom{N}{2}} \tag{E.11}$$

Substituting Equation E.10 into Equation E.11 gives (9.26).

E.4 Asymptotic Secrecy Outage Probability of NCCJ

Similar to Sections E.1 and E.3, we need to find the asymptotic secrecy outage probability for the symmetric NCCJ case. From Equation 9.27, the outage probability considering a high SNR is given by

$$P_O^{NCCJ} = \Pr\left\{ \max_{\substack{J_1 \in S_{relays} \\ R \in S_{relays} \\ J_2 \in \{S_{relays} - R^*\}}} \left\{ \frac{|h_{S,D}|^2 + |h_{R,D}|^2}{\max\limits_{E_m \in S_{eus} \forall m} \left\{ \frac{|h_{S,E_m}|^2}{|h_{J_1,E_m}|^2} + \frac{|h_{R,E_m}|^2}{|h_{J_2,E_m}|^2} \right\}} \right\} < \bar{\rho} \right\} \tag{E.12}$$

where $\bar{\rho} = \rho / P_{J_i}, i = 1, 2$. As before, define the random variables $X \triangleq |h_{S,D}|^2 + |h_{R,D}|^2$, and $\mathcal{Z} \triangleq \max\limits_{E_m \in S_{eus} \forall m} \left\{ \frac{|h_{S,E_m}|^2}{|h_{J_1,E_m}|^2} + \frac{|h_{R,E_m}|^2}{|h_{J_2,E_m}|^2} \right\}$ for one realization of the selection process. Using the CDF of the random variable X given in Proposition 9.1 and order statistics, let us compute the following probability:

$$\Pr\left\{\frac{X}{Z} < \bar{\rho}\right\}$$

$$= \Pr\{X < \bar{\rho}Z\}$$

$$= \int_0^\infty P_X(\bar{\rho}z) p_Z(z) dz$$

$$= M \int_0^\infty \left[1 - e^{-\lambda\bar{\rho}z}(1 + \lambda\bar{\rho}z)\right]$$

$$\times \left[\frac{-2}{(z+2)} - \frac{2\ln(z+1)}{(z+2)^2} + 1\right]^{M-1} \left[\frac{z}{(z+1)(z+2)^2} + \frac{4\ln(z+1)}{(z+2)^3}\right] dz$$

$$\text{(E.13)}$$

As explained in Section E.3, assuming independent random variables for the different $2N\binom{N}{2}$ realizations in Equation E.12, the secrecy outage probability is given by

$$P_O^{NCJ} \approx \left(\Pr\left\{\frac{X}{Z} < \bar{\rho}\right\}\right)^{2N\binom{N}{2}} \qquad \text{(E.14)}$$

Substituting Equation E.13 in Equation E.14 gives Equation 9.30. Note that the outage probability in Equation E.14 represents a useful lower bound.

Appendix F: MATLAB® Simulation Codes for Chapters 2 through 6

```
%= = = = = = = = = = = = = = = = = = = = = = = = = = = = = = = = = = = =
%This is the main run routine for the DFT-SC-FDMA and the DCT-SC-FDMA.
% The parameters for the simulation are set here and the simulation%functions
%are called.
%= = = = = = = = = = = = = = = = = = = = = = = = = = = = = = = = = = = =

function runSimS_DFT_and_DCT_SC_FDMA()

clear all
tic;

%%%%%%%%% Set the values of the FFT and IFFT%%%%%%%%%%%
SP.inputBlockSize = 128;
SP.FFTsize = 512;

%%%%%%%%% Choose the rang of the SNR in dB%%%%%%%%%%%%
SP.SNR = [0:3:21];

%%%%% Choose the number of simulation iterations %%%%%
SP.numRun = 200;

%%%% Choose the coding rate and the modulation Type%%%
SP.cod_rate = '1/2';
%SP.cod_rate = '1';
SP.modtype = 'QPSK1';
% SP.modtype = '16QAM';

%%%%%%%%% Choose the Equalization Type%%
% SP.equalizerType = 'ZERO';
SP.equalizerType = 'MMSE';

%%%%%%%%%Set the Cyclic prefix length %%
SP.CPsize = 20;

%%%%%%%%% Run the simulation for DFT_SC_FDMA %%
[BER_DFT_ifdma BER_DFT_lfdma] = DFT_SC_FDMA(SP);

%%%%%%%%% Run the simulation for DCT_SC_FDMA %%
[BER_DCT_ifdma BER_DCT_lfdma] = DCT_SC_FDMA(SP);

%%%%%%%%% Plot the Results %%
semilogy(SP.SNR,BER_DFT_ifdma,'r-s',SP.SNR,BER_DFT_lfdma,'r-o');
hold on
```

```
semilogy(SP.SNR,BER_DCT_ifdma,'b- s',SP.SNR,BER_DCT_lfdma,'b- o');

legend('DFT-IFDMA','DFT-LFDMA','DCT-IFDMA','DCT-LFDMA')
xlabel('SNR (dB)'); ylabel('BER');
axis([0 21 1e-4 1])
grid
toc

% = = = = = = = = = = = = = = = = = = = = = = = = = = =
% This is the simulator for the DFT-SC-FDMA System.
% = = = = = = = = = = = = = = = = = = = = = = = = = = =

function [BER_DFT_ifdma BER_DFT_lfdma] = DFT_SC_FDMA(SP)

numSymbols = SP.FFTsize;
Q = numSymbols/SP.inputBlockSize;
modtyp = SP.modtype;
for n = 1:length(SP.SNR),

rand('state',0)
randn('state',0)
noe_ifdma = zeros(1,SP.numRun);
noe_lfdma = zeros(1,SP.numRun);
for k = 1:SP.numRun,
    k

%%%%%%%%%%%%%%%%%%%%%%%%% Channel Type %%%%%%%%%%%%%%%%%%%%%%%%%
% Uniform channel

SP.channel = (1/sqrt(7))*(randn(1,7)+sqrt(-1)*randn(1,7))/sqrt(2);

%AWGN channel

% SP.channel = 1;

H_channel = fft(SP.channel,SP.FFTsize);

%%%%%%%%%%%%%%%%%%%%%%% Modulation and Encoding%%%%%%%%%%%%%%%%%%%%
    if modtyp = ='16QAM'
    switch (SP.cod_rate)
    case '1'
    m1 = 4;
    data1 = round(rand(1,SP.inputBlockSize*m1));
    mod_Symbols = mod_(data1,'16QAM');
    case '1/2'
    m1 = 4;
    trel = poly2trellis(7,[171 133]);% Trellis
    data1 = round(rand(1,(SP.inputBlockSize*m1)/2-6));
    data1 = [data1 0 0 0 0 0 0];
    tmp_coded = convenc(data1,trel);% Encode the message.
    tmp_coded = matintrlv(tmp_coded.',SP.inputBlockSize/4,16);
        %%%% Interleaving
    tmp_coded = tmp_coded.';
      mod_Symbols = mod_(tmp_coded,'16QAM');
    end

    elseif modtyp = ='QPSK1'

    switch (SP.cod_rate)
    case '1'
      tmp = round(rand(1,SP.inputBlockSize*2));
      inputSymbols = mod_(tmp,'QPSK');
    case '1/2'
      tmp = round(rand(1,SP.inputBlockSize-6));
      trel = poly2trellis(7,[171 133]);% Trellis
      tmp = [tmp zeros(1,6)];
      tmp_coded = convenc(tmp,trel);% Encode the message.
      tmp_coded = matintrlv(tmp_coded.',SP.inputBlockSize/4,8); % Matrix Interleaving
      tmp_coded = tmp_coded.';
      mod_Symbols = mod_(tmp_coded,'QPSK');
    end
    end

%%%%%%%%%%%%%%% FFT, Mapping, IFFT, and CP Insertion%%%%%%%%%%%%%
    Feq_Symbols = fft(mod_Symbols);
    Map_Samples_ifdma = zeros(1,numSymbols);
    Map_Samples_lfdma = zeros(1,numSymbols);
```

```
    Map_Samples_ifdma(1:Q:numSymbols) = Feq_Symbols;
    Map_Samples_lfdma(1:SP.inputBlockSize) = Feq_Symbols;
    OutputSamples_ifdma = ifft(Map_Samples_ifdma);
    OutputSamples_lfdma = ifft(Map_Samples_lfdma);
TxSamples_ifdma = [OutputSamples_ifdma(numSymbols- SP.CPsize+1:numSymbols)
OutputSamples_ifdma];
    TxSamples_lfdma = [OutputSamples_lfdma(numSymbols-SP.CPsize+1:numSymbols)
    OutputSamples_lfdma];

%%%%%%%%%%% Multipath Channel %%%%%%%%%%%%%%%%%%%%%%%%%
    Ch_ifdma = filter(SP.channel, 1, TxSamples_ifdma);
    Ch_lfdma = filter(SP.channel, 1, TxSamples_lfdma);

%%%%%%%%%%%%% Noise Generation %%%%%%%%%%%%%%%%%%%%%%%%
    tmpn = randn(2,numSymbols+SP.CPsize);
    complexNoise = (tmpn(1,:) + i*tmpn(2,:))/sqrt(2);
    noisePower = 10^(-SP.SNR(n)/10);

%%%%%%%%%%% Received Signal and CP removing%%%%%%%%%%%%%%
    Rx_ifdma = Ch_ifdma + sqrt(noisePower/Q)*complexNoise;
    Rx_lfdma = Ch_lfdma + sqrt(noisePower/Q)*complexNoise;
    Rx_ifdma1 = Rx_ifdma(SP.CPsize+1:numSymbols+SP.CPsize);
    Rx_lfdma1 = Rx_lfdma(SP.CPsize+1:numSymbols+SP.CPsize);

%% FFT, Frequency Domain Equalization, Demapping, IFFT%%%%
    F_ifdma = fft(Rx_ifdma1, SP.FFTsize);
    F_lfdma = fft(Rx_lfdma1, SP.FFTsize);
    H_eff = H_channel;

    if SP.equalizerType = = 'ZERO'
    F_ifdma = F_ifdma./H_eff;
    elseif SP.equalizerType = = 'MMSE'
    C = conj(H_eff)./(conj(H_eff).*H_eff + 10^(-SP.SNR(n)/10));
        F_ifdma = F_ifdma.*C;
    end
    if SP.equalizerType = = 'ZERO'
        F_lfdma = F_lfdma./H_eff;
    elseif SP.equalizerType = = 'MMSE'
        C = conj(H_eff)./(conj(H_eff).*H_eff + 10^(-SP.SNR(n)/10));
        F_lfdma = F_lfdma.*C;
    end

    Demap_ifdma1 = F_ifdma(1:Q:numSymbols);
Demap_lfdma1 = F_lfdma(1:SP.inputBlockSize);
Est_mod_ifdma = ifft(Demap_ifdma1);
Est_mod_lfdma = ifft(Demap_lfdma1);

%%%%%%%%% Demodulation and Decoding %%%%%%%%%%%%%%%%%%%%%%%
    if modtyp = ='16QAM'
    switch (SP.cod_rate)
    case '1'
    scalin_fact = sqrt(1/10);
    demod_symbol1_conv = Est_mod_ifdma/scalin_fact;
    demodulated_symbol1_conv = qamdemod(demod_symbol1_conv,16);
    symbol_size = 4;

    for y_conv = 1:size(demodulated_symbol1_conv,1)
demodulated_bit_conv = de2bi(demodulated_symbol1_conv(y_conv,:),
symbol_size,'left-msb')';
demodulated_bit1_conv(y_conv,:) = demodulated_bit_conv(:);
    end
demodata_ifdma = demodulated_bit1_conv; demod_symbol1_conv_lfdma = Est_mod_lfdma/
scalin_fact;
demodulated_symbol1_conv_lfdma = qamdemod(demod_symbol1_conv_lfdma,16);
    for y_conv = 1:size(demodulated_symbol1_conv_lfdma,1)
        demodulated_bit_conv_lfdma = de2bi(demodulated_symbol1_conv_lfdma(y_
        conv,:),symbol_size,'left-msb')';
        demodulated_bit1_conv_lfdma(y_conv,:) = demodulated_bit_conv_lfdma(:);
    end
    demodata_lfdma = demodulated_bit1_conv_lfdma;
    noe_ifdma(k) = sum(abs(demodata_ifdma-data1));
    noe_lfdma(k) = sum(abs(demodata_lfdma-data1));
```

```
    case '1/2'
      scalin_fact = sqrt(1/10);
      demod_symbol1_conv = Est_mod_ifdma/scalin_fact;
      demodulated_symbol1_conv = qamdemod(demod_symbol1_conv,16);
      symbol_size = 4;
   for y_conv = 1:size(demodulated_symbol1_conv,1)
      demodulated_bit_conv = de2bi(demodulated_symbol1_conv(y_conv,:),
      symbol_size,'left-msb)';
      demodulated_bit1_conv(y_conv,:) = demodulated_bit_conv(:);
   end
      demodata_ifdma = demodulated_bit1_conv;
      demod_symbol1_conv_lfdma = EstSymbols_lfdma/scalin_fact;
demodulated_symbol1_conv_lfdma = qamdemod(demod_symbol1_conv_lfdma,16);

   for y_conv = 1:size(demodulated_symbol1_conv_lfdma,1)

demodulated_bit_conv_lfdma = de2bi(demodulated_symbol1_conv_lfdma(y_conv,:),
symbol_size,'left-msb')';
      demodulated_bit1_conv_lfdma(y_conv,:) = demodulated_bit_conv_lfdma(:);
   end
      demodata_lfdma = demodulated_bit1_conv_lfdma;
      demodata_ifdma = matdeintrlv(demodata_ifdma,SP.inputBlockSize/4,16);
      decoded_data_ifdma = vitdec(demodata_ifdma',trel,6, 'cont','hard');% Decode.
      demodata_lfdma = matdeintrlv(demodata_lfdma,SP.inputBlockSize/4,16);
      decoded_data_lfdma = vitdec (demodata_lfdma',trel, 6,'cont','hard');% Decode.
      noe_ifdma(k) = sum(abs(data1(1:end-6)-decoded_data_ifdma(7:end).'));
      noe_lfdma(k) = sum(abs(data1(1:end-6)-decoded_data_lfdma(7:end).'));
   end
    elseif modtyp = ='QPSK1'
    switch (SP.cod_rate)
      case '1'
      EstSymbols_lfdma = sign(real(Est_mod_lfdma)) + 1i*sign(imag(Est_mod_lfdma));
      EstSymbols_lfdma = EstSymbols_lfdma/sqrt(2);
      temp_est_lfdma = QPSKDEMOD(EstSymbols_lfdma);
      noe_lfdma(k) = sum(abs(tmp-temp_est_lfdma));

      EstSymbols_ifdma = sign(real(Est_mod_ifdma)) + 1i*sign(imag(Est_mod_ifdma));
      EstSymbols_ifdma = EstSymbols_ifdma/sqrt(2);
      temp_est_ifdma = QPSKDEMOD(EstSymbols_ifdma);
      noe_ifdma(k) = sum(abs(tmp-temp_est_ifdma));
      case '1/2'
      EstSymbols_lfdma = sign(real(Est_mod_lfdma)) + 1i*sign(imag(Est_mod_lfdma));
      EstSymbols_lfdma = EstSymbols_lfdma/sqrt(2);
      temp_est_lfdma = QPSKDEMOD(EstSymbols_lfdma);
      s1_est_lfdma = matdeintrlv(temp_est_lfdma,SP.inputBlockSize/4,8);
      decoded_lfdma = vitdec(s1_est_lfdma',trel,6,'cont','hard');% Decode
      noe_lfdma(k) = sum(abs(tmp(1:end-6)- decoded_lfdma (7:end).'));
      EstSymbols_ifdma = sign(real(Est_mod_ifdma)) + 1i*sign(imag(Est_mod_ifdma));
      EstSymbols_ifdma = EstSymbols_ifdma/sqrt(2);
      temp_est_ifdma = QPSKDEMOD(EstSymbols_ifdma);
      s1_est_ifdma = matdeintrlv(temp_est_ifdma,SP.inputBlockSize/4,8);
      decoded_ifdma = vitdec(s1_est_ifdma',trel,6,'cont', 'hard');% Decode
      noe_ifdma(k) = sum(abs(tmp(1:end-6)-decoded_ifdma(7:end).'));
   end
   end
end

%%%%%%%%%%%%%%%% BER Calculation %%%%%%%%%%%%%%%%%%%%%
if modtyp = ='16QAM'
switch (SP.cod_rate)
case '1'
   BER_DFT_ifdma(n,:) = sum(noe_ifdma)/(SP.numRun*SP.inputBlockSize*m1);
   BER_DFT_lfdma(n,:) = sum(noe_lfdma)/(SP.numRun*SP.inputBlockSize*m1);

   case '1/2'
   BER_DFT_ifdma(n,:) = sum(noe_ifdma)/(SP.numRun*((SP.inputBlockSize*m1)/2-6));
   BER_DFT_lfdma(n,:) = sum(noe_lfdma)/(SP.numRun*((SP.inputBlockSize*m1)/2-6));

   end
elseif modtyp = ='QPSK1'
switch (SP.cod_rate)
   case '1'
```

```
    BER_DFT_ifdma(n,:) = sum(noe_ifdma)/(SP.numRun*SP.inputBlockSize *2);
    BER_DFT_lfdma(n,:) = sum(noe_lfdma)/(SP.numRun*SP.inputBlockSize* 2);
    case '1/2'
    BER_DFT_ifdma(n,:) = sum(noe_ifdma)/(SP.numRun*(SP.inputBlockSize-6));
    BER_DFT_lfdma(n,:) = sum(noe_lfdma)/(SP.numRun*(SP.inputBlockSize-6));
    end
end
end

% = = = = = = = = = = = = = = = = = = = = = = = = = = = =
% This is the simulator for the DCT-SC-FDMA System.
% = = = = = = = = = = = = = = = = = = = = = = = = = = = =

function [BER_DCT_ifdma BER_DCT_lfdma] = DCT_SC_FDMA(SP)

numSymbols = SP.FFTsize;
Q = numSymbols/SP.inputBlockSize;
modtyp = SP.modtype;
for n = 1:length(SP.SNR),

rand('state',0)
randn('state',0)
noe_ifdma = zeros(1,SP.numRun);
noe_lfdma = zeros(1,SP.numRun);
for k = 1:SP.numRun,
    k

% Uniform channel
SP.channel = (1/sqrt(7))*(randn(1,7)+sqrt(-1)*randn(1,7))/sqrt(2);

%AWGN channel

% SP.channel = 1;

H_channel = fft(SP.channel,SP.FFTsize);

%%%%%%%%%% Modulation and Encoding%%%%%%%%%%%%%%%%%%
    if modtyp = ='16QAM'

    switch (SP.cod_rate)
    case '1'
    m1 = 4;
    data1 = round(rand(1,SP.inputBlockSize*m1));
    modSymbols = mod_(data1,'16QAM');
    case '1/2'
    m1 = 4;
    trel = poly2trellis(7,[171 133]);% Trellis
    data1 = round(rand(1,(SP.inputBlockSize*m1)/2-6));
    data1 = [data1 0 0 0 0 0 0];
    tmp_coded = convenc(data1,trel);% Encode the message.
tmp_coded = matintrlv(tmp_coded.',SP.inputBlockSize/4,16); tmp_coded = tmp_coded.';
    modSymbols = mod_(tmp_coded,'16QAM');
    end
    elseif modtyp = ='QPSK1'

    switch (SP.cod_rate)
    case '1'
        tmp = round(rand(1,SP.inputBlockSize*2));
        modSymbols = mod_(tmp,'QPSK');
    case '1/2'
        tmp = round(rand(1,SP.inputBlockSize-6));
        trel = poly2trellis(7,[171 133]);% Trellis
        tmp = [tmp zeros(1,6)];
        tmp_coded = convenc(tmp,trel);% Encode the message.
tmp_coded = matintrlv(tmp_coded.',SP.inputBlockSize/4,8); tmp_coded = tmp_coded.';
        modSymbols = mod_(tmp_coded,'QPSK');
    end
    end

%%%%%%%%%% DCT, Mapping, IDCT, and CP Insertion%%%%%%%%%%%%%%%
    Feq_Symbols = dct(modSymbols);
    Map_Samples_ifdma = zeros(1,numSymbols);
    Map_Samples_lfdma = zeros(1,numSymbols);
```

```
Map_Samples_ifdma(1:Q:numSymbols) = Feq_Symbols;% Interleaved Mapping
Map_Samples_lfdma(1:SP.inputBlockSize) = Feq_Symbols;% Localized Mapping
   Output_Samples_ifdma = idct(Map_Samples_ifdma);
   Output_Samples_lfdma = idct(Map_Samples_lfdma);
   TxSamples_ifdma = [Output_Samples_ifdma(numSymbols-SP.CPsize+1 :numSymbols)
   Output_Samples_ifdma];
   TxSamples_lfdma = [Output_Samples_lfdma(numSymbols-SP.CPsize+1:numSymbols) Output_
   Samples_lfdma];

%%%%%%%%%%%%%%%%%%%%%%%% Multipath Channel %%%%%%%%%%%%%%%%%%%%%%%%%%%
Ch_ifdma = filter(SP.channel, 1, TxSamples_ifdma);% Multipath Channel
Ch_lfdma = filter(SP.channel, 1, TxSamples_lfdma);% Multipath Channel

%%%%%%%%%%%%%%%%%%%% Noise Generation %%%%%%%%%%%%%%%%%%
   tmpn = randn(2,numSymbols+SP.CPsize);
   complexNoise = (tmpn(1,:) + i*tmpn(2,:))/sqrt(2);
   noisePower = 10^(-SP.SNR(n)/10);

%%%%%%%%%%%% Received Signal and CP removing%%%%%%%%%%%%%%%%%%%%%%%%%
   Rx_ifdma = Ch_ifdma + sqrt(noisePower)*complexNoise;
   Rx_lfdma = Ch_lfdma + sqrt(noisePower)*complexNoise;
   Rx_ifdma1 = Rx_ifdma(SP.CPsize+1:numSymbols+SP.CPsize);
   Rx_lfdma1 = Rx_lfdma(SP.CPsize+1:numSymbols+SP.CPsize);

%%%%%%%%% FFT, Frequency Domain Equalization, IFFT%%%%%%%%%%%%%%%%%%%%
   F_ifdma = fft(Rx_ifdma1, SP.FFTsize);
   F_lfdma = fft(Rx_lfdma1, SP.FFTsize);
   H_eff = H_channel;

   if SP.equalizerType = = 'ZERO'
   F_ifdma = F_ifdma./H_eff;
   elseif SP.equalizerType = = 'MMSE'
   C = conj(H_eff)./(conj(H_eff).*H_eff + 10^(-SP.SNR(n)/10));
   F_ifdma = F_ifdma.*C;
   end

   if SP.equalizerType = = 'ZERO'
   F_lfdma = F_lfdma./H_eff;
   elseif SP.equalizerType = = 'MMSE'
   C = conj(H_eff)./(conj(H_eff).*H_eff + 10^(-SP.SNR(n)/10));
   F_lfdma = F_lfdma.*C;
   end

RxSamples_ifdma = ifft(F_ifdma);
RxSamples_lfdma = ifft(F_lfdma);

%%%%%%%%%%%%%%%%%%%%%%%% DCT, Demapping, IDCT%%%%%%%%%%%%%%%%%%%%%%%%%
Est_Map_ifdma = dct(RxSamples_ifdma);
Est_Map_lfdma = dct(RxSamples_lfdma);

Demap_ifdma = Est_Map_ifdma(1:Q:numSymbols);
Demap_lfdma = Est_Map_lfdma(1:SP.inputBlockSize);
Est_mod_ifdma = idct(Demap_ifdma);
Est_mod_lfdma = idct(Demap_lfdma);

%%%%%%%%%%%%%%%%% Demodulation and Decoding %%%%%%%%%%%%%%%%%%%%%%%%
   if modtyp = ='16QAM'
   switch (SP.cod_rate)
   case '1'
   scalin_fact = sqrt(1/10);
   demod_symbol1_conv = Est_mod_ifdma/scalin_fact;
   demodulated_symbol_conv = qamdemod(demod_symbol1_conv,16);
   symbol_size = 4;
   for y_conv = 1:size(demodulated_symbol1_conv,1)
      demodulated_bit_conv = de2bi(demodulated_symbol1_conv(y_conv,:),symbol_
      size,'left-msb')';
      demodulated_bit1_conv(y_conv,:) = demodulated_bit_conv(:);
   end
      demodata_ifdma = demodulated_bit1_conv;
      demod_symbol1_conv_lfdma = EstSymbols_lfdma/scalin_fact;
demodulated_symbol1_conv_lfdma = qamdemod(demod_symbol1_conv_lfdma,16);
```

```
      for y_conv = 1:size(demodulated_symbol1_conv_lfdma,1)
demodulated_bit_conv_lfdma = de2bi(demodulated_symbol1_conv_lfdma(y_conv,:),symbol_
size,'left-msb')';
demodulated_bit1_conv_lfdma(y_conv,:) = demodulated_bit_conv_lfdma(:);
   end
      demodata_lfdma = demodulated_bit1_conv_lfdma;
      noe_ifdma(k) = sum(abs(demodata_ifdma-data1));
      noe_lfdma(k) = sum(abs(demodata_lfdma-data1));
   case '1/2'
      scalin_fact = sqrt(1/10);
      demod_symbol1_conv = Est_mod_ifdma/scalin_fact;
      demodulated_symbol1_conv = qamdemod(demod_symbol1_conv,16);
      symbol_size = 4;
   for y_conv = 1:size(demodulated_symbol1_conv,1)
demodulated_bit_conv = de2bi(demodulated_symbol1_conv(y_conv,:),symbol_size,
'left-msb')';
      demodulated_bit1_conv(y_conv,:) = demodulated_bit_conv(:);
   end
      demodata_ifdma = demodulated_bit1_conv;
      demod_symbol1_conv_lfdma = EstSymbols_lfdma/scalin_fact;
demodulated_symbol1_conv_lfdma = qamdemod(demod_symbol1_conv_lfdma,16);
   for y_conv = 1:size(demodulated_symbol1_conv_lfdma,1)
      demodulated_bit_conv_lfdma = de2bi(demodulated_symbol1_conv_lfdma(y_
      conv,:),symbol_size,'left-msb')';
      demodulated_bit1_conv_lfdma(y_conv,:) = demodulated_bit_conv_lfdma(:);
   end
   demodata_lfdma = demodulated_bit1_conv_lfdma;

      demodata_ifdma = matdeintrlv(demodata_ifdma,SP.inputBlockSize/4,16);
      decoded_data_ifdma = vitdec(demodata_ifdma',trel,6,'cont','hard');% Decode.
      demodata_lfdma = matdeintrlv(demodata_lfdma,SP.inputBlockSize/4,16);
   decoded_data_lfdma = vitdec(demodata_lfdma',trel,6,'cont', 'hard');% Decode.
   noe_ifdma(k) = sum(abs(data1(1:end-6)-decoded_data_ifdma (7:end).'));
      noe_lfdma(k) = sum(abs(data1(1:end-6)-decoded_data_lfdma (7:end).'));
   end
   elseif modtyp = ='QPSK1'
   switch (SP.cod_rate)
      case '1'

      EstSymbols_lfdma = sign(real(Est_mod_lfdma)) + 1i*sign(imag(Est_mod_lfdma));
      EstSymbols_lfdma = EstSymbols_lfdma/sqrt(2);
      temp_est_lfdma = QPSKDEMOD(EstSymbols_lfdma);
      noe_lfdma(k) = sum(abs(tmp-temp_est_lfdma));
      EstSymbols_ifdma = sign(real(Est_mod_ifdma)) + 1i*sign(imag(Est_mod_ifdma));
      EstSymbols_ifdma = EstSymbols_ifdma/sqrt(2);
      temp_est_ifdma = QPSKDEMOD(EstSymbols_ifdma);
      noe_ifdma(k) = sum(abs(tmp-temp_est_ifdma));
      case '1/2'
      EstSymbols_lfdma = sign(real(Est_mod_lfdma)) + 1i*sign(imag(Est_mod_lfdma));
      EstSymbols_lfdma = EstSymbols_lfdma/sqrt(2);
      temp_est_lfdma = QPSKDEMOD(EstSymbols_lfdma);
      s1_est_lfdma = matdeintrlv(temp_est_lfdma,SP.inputBlockSize/4,8);
      decoded_lfdma = vitdec(s1_est_lfdma',trel,6,'cont', 'hard');% Decode
noe_lfdma(k) = sum(abs(tmp(1:end-6)-decoded_lfdma (7:end).'));
      EstSymbols_ifdma = sign(real(Est_mod_ifdma)) + 1i*sign(imag(Est_mod_ifdma));
      EstSymbols_ifdma = EstSymbols_ifdma/sqrt(2);
      temp_est_ifdma = QPSKDEMOD(EstSymbols_ifdma);
      s1_est_ifdma = matdeintrlv(temp_est_ifdma,SP.inputBlockSize/4,8);
      decoded_ifdma = vitdec(s1_est_ifdma',trel,6,'cont', 'hard');% Decode
      noe_ifdma(k) = sum(abs(tmp(1:end-6)-decoded_ifdma (7:end).'));
   end
   end
end


%%%%%%%%%%%%%%%%%%%%%%%%%%%% BER Calculation %%%%%%%%%%%%%%%%%%%%%%%%%%%%%%%%
if modtyp = ='16QAM'
switch (SP.cod_rate)
case '1'
   BER_DCT_ifdma(n,:) = sum(noe_ifdma)/(SP.numRun*SP.inputBlockSize*m1);
   BER_DCT_lfdma(n,:) = sum(noe_lfdma)/(SP.numRun*SP.inputBlockSize*m1);
```

```
    case `1/2'
    BER_DCT_ifdma(n,:) = sum(noe_ifdma)/(SP.numRun*((SP.inputBlockSize*m1)/2-6));
    BER_DCT_lfdma(n,:) = sum(noe_lfdma)/(SP.numRun*((SP.inputBlockSize*m1)/2-6));

    end
elseif modtyp = ='QPSK1'
switch (SP.cod_rate)
    case `1'
    BER_DCT_ifdma(n,:) = sum(noe_ifdma)/(SP.numRun*SP.inputBlockSize *2);
    BER_DCT_lfdma(n,:) = sum(noe_lfdma)/(SP.numRun*SP.inputBlockSize *2);
    case `1/2'

    BER_DCT_ifdma(n,:) = sum(noe_ifdma)/(SP.numRun*(SP.inputBlockSize-6));
    BER_DCT_lfdma(n,:) = sum(noe_lfdma)/(SP.numRun*(SP.inputBlockSize-6));
    end
end

end
```

```
% = = = = = = = = = = = = = = = = = = = = = = = = = = =
% This is the simulator for the mod_.
% = = = = = = = = = = = = = = = = = = = = = = = = = = =

function inputSymbols = mod_(encod_data1_conv,ModulationType)
switch (ModulationType)

case `QPSK'
symbol_size = 2;
scaling_fact = sqrt(1/2);

for e_conv = 1:symbol_size:size(encod_data1_conv,2)
mod_inp_conv(:,floor(e_conv/symbol_size)+1) = bi2de(encod_data1_conv(:,e_conv:e_
conv+symbol_size-1),`left-msb');
end

constell_qpsk_conv = qammod(mod_inp_conv,4);
%QPSK is implemented as 4 QAM
%scaled modulated data
inputSymbols = scaling_fact*constell_qpsk_conv;

case `16QAM'
symbol_size = 4;
scaling_fact = sqrt(1/10);

%convert the symbol into [0 . . . M1]
for p_conv = 1:symbol_size:size(encod_data1_conv,2)
mod_inp_conv(:,floor(p_conv/symbol_size)+1) = bi2de(encod_data1_conv(:,p_conv:p_
conv+symbol_size-1),`left-msb');
end
constell_16qpsk_conv = qammod(mod_inp_conv,16);
inputSymbols = scaling_fact*constell_16qpsk_conv;

case `64QAM'
symbol_size1 = 6;
scaling_fact = sqrt(1/42);
for p_conv = 1:symbol_size1:size(encod_data1_conv,2)
mod_inp_conv(:,floor(p_conv/symbol_size1)+1) = bi2de(encod_data1_conv(:,p_conv:p_
conv+symbol_size1-1),`left-msb');
end
constell_64qam_conv = qammod(mod_inp_conv,64);
inputSymbols = scaling_fact*constell_64qam_conv;
end
```

```
% = = = = = = = = = = = = = = = = = = = = = = = = = = =
% This is the simulator for the QPSKDEMOD.
% = = = = = = = = = = = = = = = = = = = = = = = = = = =

function temp_est = QPSKDEMOD(modulated_data1_conv)

scalin_fact = sqrt(1/2);
demod_symbol1_conv = modulated_data1_conv/scalin_fact;

%4QAM demodulation
demodulated_symbol1_conv = qamdemod(demod_symbol1_conv,4);
```

```
symbol_size = 2;
for y_conv = 1:size(demodulated_symbol1_conv)

demodulated_bit_conv = de2bi(demodulated_symbol1_conv(y_conv,:),symbol_size,'left-
msb')';
demodulated_bit1_conv(y_conv,:) = demodulated_bit_conv(:);
end
temp_est = demodulated_bit1_conv;

% = = = = = = = = = = = = = = = = = = = = = = = = = = =
% This is the simulator for the papr_DFT_SC_FDMA.
% = = = = = = = = = = = = = = = = = = = = = = = = = = =

function papr_DFT_SC_FDMA()

clear all
tic
Mod_Type = '16QAM';% Modulation Type.
totalSubcarriers = 512 ;% M.
numSymbols = 256;% N.
Q = totalSubcarriers/numSymbols;
subcarrierMapping = 'IFDMA';% Subcarrier mapping scheme.
pulseShaping = 1;% Whether to do pulse shaping or not.
filterType = 'RC';% Type of pulse shaping filter.
Alpha_ = 0.220;%Rolloff factor for the pulse shaping filter.
     %To prevent divide-by-zero, for example, use 0.099999999 instead of 0.1.
Fs = 5e6;% Sampling Frequency.
Ts = 1/Fs;% sampling rate.
Nos = 4;%Oversampling factor.
if filterType = = 'RC'% Raised-cosine filter.
psFilter = Raised_Pulse_shapping(Ts, Nos, Alpha_);
elseif filterType = = 'RR'% Root raised-cosine filter.
psFilter = Root_Raised_Pulse_shapping(Ts,Nos, Alpha_);
end
numiter = 50;% Number of iterations.
papr_DFT_SC_FDMA = zeros(1,numiter);% Initialize the PAPR results.
for n = 1:numiter,
n
% Generate random data.
if Mod_Type = = 'Q-PSK'
  tmp = round(rand(numSymbols,2));
  tmp = tmp*2 - 1;
  data = (tmp(:,1) + j*tmp(:,2))/sqrt(2);
elseif Mod_Type = = '16QAM'
  dataSet = [-3+3i -1+3i 1+3i 3+3i…
  -3+i -1+i 1+i 3+i…
  -3-i -1-i 1-i 3-i…
  -3-3i -1-3i 1-3i 3-3i];
  dataSet = dataSet/sqrt(mean(abs(dataSet).^2));
  tmp = ceil(rand(numSymbols,1)*16);
  for k = 1:numSymbols,
  if tmp(k) = = 0
    tmp(k) = 1;
  end
  data(k) = dataSet(tmp(k));
  end
  data = data.';
end

%%%%%%%%%%%%%%%% FFT%%%%%%%%%%%%%%%%%%%%%%%%.
X = fft(data);
Y = zeros(totalSubcarriers,1);

%%%%%%%%%%%%%%%% Mapping%%%%%%%%%%%%%%%%%%%%.
if subcarrierMapping = = 'IFDMA'
  Y(1:Q:totalSubcarriers) = X;
elseif subcarrierMapping = = 'LFDMA'
  Y(1:numSymbols) = X;
end

%%%%%%%%%%%%%%%%%% IFFT%%%%%%%%%%%%.
y = ifft(Y);
```

```
%%%%%%%%%%%%% Pulse Shaping%%%%%%%%%%%%%%.
if pulseShaping = = 1
y_oversampled(1:Nos:Nos*totalSubcarriers) = y;
% Perform filtering.
  y_result = filter(psFilter, 1, y_oversampled);
else
  y_result = y;
end

%%%%%%%%%%%%% PAPR Calculation%%%%%%%%%%%%%%%%%.
papr_DFT_SC_FDMA(n) = 10*log10(max(abs(y_result).^2)/mean(abs(y_result).^2));
end

%%%%%%%%%%%%%%%%% Plot CCDF%%%%%%%%%%%%%%%%%%%%%%.
[N,X] = hist(papr_DFT_SC_FDMA,1000);
semilogy(X,1-cumsum(N)/max(cumsum(N)),'b-')
% legend('a','b','C','D')
axis([0 12 1e-4 1])
xlabel('PAPR(dB)'); ylabel('CCDF');

toc

% = = = = = = = = = = = = = = = = = = = = = = = = = = =
% This is the simulator for the papr_DCT_SC_FDMA.
% = = = = = = = = = = = = = = = = = = = = = = = = = = =

function papr_DCT_SC_FDMA()

clear all
tic
Mod_Type = '16QAM';% Modulation Type.
totalSubcarriers = 512 ;% M.
numSymbols = 256;% N.
Q = totalSubcarriers/numSymbols;% Bandwidth spreading factor of IFDMA.
subcarrierMapping = 'IFDMA';% Subcarrier mapping scheme.
pulseShaping = 1;% Whether to do pulse shaping or not.
filterType = 'RC';% Type of pulse shaping filter.
Alpha_ = 0.220;% 0.3999999999;%Rolloff factor for the raised-cosine filter.
%To prevent divide-by-zero, for example, use 0.099999999 instead of 0.1.
Fs = 5e6;% Sampling Frequency.
Ts = 1/Fs;% sampling rate.
Nos = 4;%Oversampling factor.
if filterType = = 'RC'% Raised-cosine filter.
psFilter = Raised_Pulse_shapping(Ts, Nos, Alpha_);
elseif filterType = = 'RR'% Root raised-cosine filter.
psFilter = Root_Raised_Pulse_shapping(Ts,Nos, Alpha_);
end
numiter = 10;% Number of iterations.
papr_DCT_SC_FDMA = zeros(1,numiter);% Initialize the PAPR results.

for n = 1:numiter,
n
% Generate random data.
if Mod_Type = = 'Q-PSK'
  tmp = round(rand(numSymbols,2));
  tmp = tmp*2 - 1;
  data = (tmp(:,1) + j*tmp(:,2))/sqrt(2);
elseif Mod_Type = = '16QAM'
  dataSet = [-3+3i -1+3i 1+3i 3+3i…
  -3+i -1+i 1+i 3+i…
  -3-i -1-i 1-i 3-i…
  -3-3i -1-3i 1-3i 3-3i];
  dataSet = dataSet/sqrt(mean(abs(dataSet).^2));
  tmp = ceil(rand(numSymbols,1)*16);
  for k = 1:numSymbols,
  if tmp(k) = = 0
    tmp(k) = 1;
  end
  data(k) = dataSet(tmp(k));
  end
  data = data.';
end
```

```
% %%%%%%%%%%%%%% DCT%%%%%%%%%%%%%%%%%%%%%%%%%
X = dct(data);
Y = zeros(totalSubcarriers,1);


%%%%%%%%%%%%%%% Mapping%%%%%%%%%%%%%%%%%%%%%%%.
if subcarrierMapping = = 'IFDMA'
   Y(1:Q:totalSubcarriers) = X;
elseif subcarrierMapping = = 'LFDMA'
   Y(1:numSymbols) = X;
end


%%%%%%%%%%%%%%%%%% IDCT%%%%%%%%%%%%%%%%%%%%%%%%%.
y = idct(Y);


%%%%%%%%%%%%%%%%%% Pulse shaping%%%%%%%%%%%%%%%%%%.
if pulseShaping = = 1
   y_oversampled(1:Nos:Nos*totalSubcarriers) = y;
% Perform filtering.
   y_result = filter(psFilter, 1, y_oversampled);
else
   y_result = y;
end


%%%%%%%%%%%%%%%%%%%% PAPR Calculation%%%%%%%%%%%%%%%%%%%%%%.
papr_DCT_SC_FDMA(n) = 10*log10(max(abs(y_result).^2)/mean(abs(y_result).^2));
end

%%%%%%%%%%%%%%%%%%%%% Plot CCDF%%%%%%%%%%%%%%%%%%%%%%%%%%.
[N,X] = hist(papr_DCT_SC_FDMA,1000);
semilogy(X,1-cumsum(N)/max(cumsum(N)),'b-')
axis([0 12 1e-4 1])
xlabel('PAPR(dB)'); ylabel('CCDF');
toc

% = = = = = = = = = = = = = = = = = = = = = = = =
% This is the simulator for the Raised_Pulse_shapping.
% = = = = = = = = = = = = = = = = = = = = = = = =

function r = Raised_Pulse_shapping(Ts, Nos, alpha)

t1 = [-8*Ts:Ts/Nos:-Ts/Nos];
t2 = [Ts/Nos:Ts/Nos:8*Ts];

r1 = (sin(pi*t1/Ts)./(pi*t1)).*(cos(pi*alpha*t1/Ts)./(1-(4*alpha*t1/(2*Ts)).^2));
r2 = (sin(pi*t2/Ts)./(pi*t2)).*(cos(pi*alpha*t2/Ts)./(1-(4*alpha*t2/(2*Ts)).^2));

r = [r1 1/Ts r2];

% = = = = = = = = = = = = = = = = = = = = = = = =
% This is the simulator for the Root_Raised_Pulse_shapping.
% = = = = = = = = = = = = = = = = = = = = = = = =

function r = Root_Raised_Pulse_shapping(Ts, Nos, alpha)

t1 = [-6*Ts:Ts/Nos:-Ts/Nos];
t2 = [Ts/Nos:Ts/Nos:6*Ts];
r1 = (4*alpha/(pi*sqrt(Ts)))*(cos((1+alpha)*pi*t1/Ts)+(Ts./(4*alpha*t1)).* sin((1-
alpha)*pi*t1/Ts))./(1-(4*alpha*t1/Ts).^2);
r2 = (4*alpha/(pi*sqrt(Ts)))*(cos((1+alpha)*pi*t2/Ts)+(Ts./(4*alpha*t2)).* sin((1-
alpha)*pi*t2/Ts))./(1-(4*alpha*t2/Ts).^2);
r = [r1 (4*alpha/(pi*sqrt(Ts))+(1-alpha)/sqrt(Ts)) r2];

% = = = = = = = = = = = = = = = = = = = = = = = = = = = = = = = = =
%This is the main run routine for the SC-FDMA in the presence%of CFO.
% The parameters for the simulation are set here and the simulation functions
are called.
% = = = = = = = = = = = = = = = = = = = = = = = = = = = = = = = = =

function runSim_cfo_SCFDMA()

clear all
tic;
```

```
% = = Select the DFT and the IDFT length = = = = = = =
SP.FFTsize = 128;
SP.inputBlockSize = 32;
SP.subband = 0;

% = = = = Number of iterations = = = = = = = = = = =
SP.numRun = 10^4;

% = = = Select the Bandwidth of the banded Matrix = = =
SP.B = SP.inputBlockSize;%SP.inputBlockSize;%10,20,64

% = = = = = The range of the SNR values = = = = = = =
SP.SNR = [0:3:24];

% = = = = = Select the Detection Method = = = = = = =
SP.Method = 1; % single user detector = 1, Circular deconvolution = 2, MMSE = 3, and
Without CFO Compensation = 4;

% = Select the coding rate and the Modulation Format =
SP.cod_rate = '1/2';
SP.modtype = 'QPSK1';
% SP.modtype = '16QAM';

% = = = = = = = = = = = = = = = Cyclic prefix length = = = = = = = = = = = = = = = = = = =
SP.CPsize = 20;

% = = = = = = = = = = = = = = = = = = = = = = = = = = = = = = = = = = = = = = = = = =
[BER_ifdma BER_lfdma] = MULTI_CODE(SP);

%% = = = = = = = = = = = = = Plot the Results = = = = = = = = = = = = = =
semilogy(SP.SNR,BER_ifdma,'k- s',SP.SNR,BER_lfdma,'k- o');
legend('IFDMA','LFDMA')
xlabel('SNR (dB)'); ylabel('BER');
title('M = 128, N = 32,QPSK')
axis([0 24 1e-4 1])
grid

toc

% = = = = = = = = = = = = = = = = = = = = = = = = = = = = = = = = = = = = = = = = = =
% This is the simulator for the SC-FDMA System in the presence%%%of CFO.
% = = = = = = = = = = = = = = = = = = = = = = = = = = = = = = = = = = = = = = = = = =

function [BER_ifdma BER_lfdma] = MULTI_CODE(SP)

numSymbols = SP.FFTsize;
Q = numSymbols/SP.inputBlockSize;
modtyp = SP.modtype;%'16QAM';
indi_interleaved_1 = zeros(1,numSymbols);
indi_interleaved_2 = zeros(1,numSymbols);
indi_interleaved_3 = zeros(1,numSymbols);
indi_interleaved_4 = zeros(1,numSymbols);

indi_interleaved_1(1:Q:numSymbols) = ones(1,SP.inputBlockSize);
indi_interleaved_2(2:Q:numSymbols) = ones(1,SP.inputBlockSize);
indi_interleaved_3(3:Q:numSymbols) = ones(1,SP.inputBlockSize);
indi_interleaved_4(4:Q:numSymbols) = ones(1,SP.inputBlockSize);

indi_localized_1 = zeros(1,numSymbols);
indi_localized_2 = zeros(1,numSymbols);
indi_localized_3 = zeros(1,numSymbols);
indi_localized_4 = zeros(1,numSymbols);

indi_localized_1(1:SP.inputBlockSize) = ones(1,SP.inputBlockSize);
indi_localized_2((1+SP.inputBlockSize):(2*SP.inputBlockSize)) = ones(1,SP.
inputBlockSize);
indi_localized_3((2*SP.inputBlockSize+1):(3*SP.inputBlockSize)) = ones(1,SP.
inputBlockSize);
indi_localized_4((3*SP.inputBlockSize+1):(4*SP.inputBlockSize)) = ones(1,SP.
inputBlockSize);

for n = 1:length(SP.SNR),

rand('state',0)
```

```
randn('state',0)
noe_ifdma = zeros(1,SP.numRun);
noe_lfdma = zeros(1,SP.numRun);
for k = 1:SP.numRun,
  k

CFO_Values = 0.15*(2*rand(1,4)-1);

%%% = = = = = = = = = = = = = = = = = = = = = = Perfect CFO values
SP.ep1 = CFO_Values(1,1);%
SP.ep2 = CFO_Values(1,2);
SP.ep3 = CFO_Values(1,3);
SP.ep4 = CFO_Values(1,4);
A1 = exp((1*j*2*pi*SP.ep1/(numSymbols))*(0:(numSymbols-1)));
A2 = exp((1*j*2*pi*SP.ep2/(numSymbols))*(0:(numSymbols-1)));
A3 = exp((1*j*2*pi*SP.ep3/(numSymbols))*(0:(numSymbols-1)));
A4 = exp((1*j*2*pi*SP.ep4/(numSymbols))*(0:(numSymbols-1)));
A5 = exp((-1*j*2*pi*SP.ep1/(numSymbols))*(0:(numSymbols-1)));

%%% = = = = = = = = = = = = = = = = = = = = = = = = = = = = = = = = = = = =
CT1 = fft(eye(numSymbols))*diag(A1)*ifft(eye(numSymbols));
CT2 = fft(eye(numSymbols))*diag(A2)*ifft(eye(numSymbols));
CT3 = fft(eye(numSymbols))*diag(A3)*ifft(eye(numSymbols));
CT4 = fft(eye(numSymbols))*diag(A4)*ifft(eye(numSymbols));

CT_1 = CT1([1:SP.inputBlockSize],[1:SP.inputBlockSize]);
CT_2 = CT1(1:Q:numSymbols,1:Q:numSymbols);

CT5 = fft(eye(numSymbols))*diag(A5)*ifft(eye(numSymbols));
CT_12 = CT5([1:SP.inputBlockSize],[1:SP.inputBlockSize]);
CT_22 = CT5(1:Q:numSymbols,1:Q:numSymbols);
% Uniform channel
SP.channel = (1/sqrt(7))*(randn(4,7)+sqrt(-1)*randn(4,7))/sqrt(2);

% = = = = = = = = = = = = User 1 = = = = = = = = = = = = = = = = = = = = =
  SP.channel1 = SP.channel(1,:);

%% = = = = = = = = = = = = User 2 = = = = = = = = = = = = = = = = = = = = =
  SP.channel2 = SP.channel(2,:);

%% = = = = = = = = = = = = User 3 = = = = = = = = = = = = = = = = = = = = =
  SP.channel3 = SP.channel(3,:);

%% = = = = = = = = = = = = User 4 = = = = = = = = = = = = = = = = = = = = =
  SP.channel4 = SP.channel(4,:);

% = = = = = = = = = =AWGN channel = = = = = = = = = = = = = = = = = = = = =
% SP.channel1 = 1;
% SP.channel2 = 1;
% SP.channel3 = 1;
% SP.channel4 = 1;
H_channel1 = fft(SP.channel1,SP.FFTsize);
H_channel2 = fft(SP.channel2,SP.FFTsize);
H_channel3 = fft(SP.channel3,SP.FFTsize);
H_channel4 = fft(SP.channel4,SP.FFTsize);

A_channel1 = diag(fft(SP.channel1,SP.FFTsize));
A_channel2 = diag(fft(SP.channel2,SP.FFTsize));
A_channel3 = diag(fft(SP.channel3,SP.FFTsize));
A_channel4 = diag(fft(SP.channel4,SP.FFTsize));

CT1 = fft(eye(numSymbols))*diag(A1)*ifft(eye(numSymbols));
CT_1 = CT1([1:SP.inputBlockSize],[1:SP.inputBlockSize]);
CT_2 = CT1(1:Q:numSymbols,1:Q:numSymbols);

  if modtyp = ='16QAM'

  m1 = 4;

  switch (SP.cod_rate)

  case '1'
  m1 = 4;
  data1 = round(rand(1,SP.inputBlockSize*m1));
  inputSymbols = mod_(data1,'16QAM');
```

```
case '1/2'
m1 = 4;
trel = poly2trellis(7,[171 133]);% Trellis
data1 = round(rand(1,(SP.inputBlockSize*m1)/2-6));
data1 = [data1 0 0 0 0 0 0];
tmp_coded = convenc(data1,trel);% Encode the message.
tmp_coded = matintrlv(tmp_coded.',32,16);%Matrix Interleaving
tmp_coded = tmp_coded.';
inputSymbols = mod_(tmp_coded,'16QAM');
end
elseif modtyp = ='QPSK1'

switch (SP.cod_rate)
case '1'
   tmp = round(rand(1,SP.inputBlockSize*2));
   inputSymbols = mod_(tmp,'QPSK');
   tmp2 = round(rand(1,SP.inputBlockSize*2));
   inputSymbols2 = mod_(tmp2,'QPSK');
   tmp3 = round(rand(1,SP.inputBlockSize*2));
   inputSymbols3 = mod_(tmp3,'QPSK');
   tmp4 = round(rand(1,SP.inputBlockSize*2));
   inputSymbols4 = mod_(tmp4,'QPSK');
case '1/2'

   trel = poly2trellis(7,[171 133]);% Trellis

   % = = = = Coding and modulation for User 1 = = = = = = = = = = = = = = = = = = = =
   tmp = round(rand(1,SP.inputBlockSize-6));
   tmp = [tmp zeros(1,6)];
   tmp_coded = convenc(tmp,trel);% Encode the message.
   tmp_coded = matintrlv(tmp_coded.',SP.inputBlockSize/4,8); tmp_coded = tmp_
   coded.';
   inputSymbols = mod_(tmp_coded,'QPSK');

   % = = = Coding and modulation for User 2 = = = = = = = = = = = = = = = = = = = =
   tmp2 = round(rand(1,SP.inputBlockSize-6));
   tmp2 = [tmp2 zeros(1,6)];
   tmp_coded2 = convenc(tmp2,trel);% Encode the message.
   tmp_coded2 = matintrlv(tmp_coded2.',SP.inputBlockSize/4,8);
   tmp_coded2 = tmp_coded2.';
   inputSymbols2 = mod_(tmp_coded2,'QPSK');

   % = = = = = Coding and modulation for User 3 = = = = = = = = = = = = = = = = = =
   tmp3 = round(rand(1,SP.inputBlockSize-6));
   tmp3 = [tmp3 zeros(1,6)];
   tmp_coded3 = convenc(tmp3,trel);% Encode the message.
   tmp_coded3 = matintrlv(tmp_coded3.',SP.inputBlockSize/4,8);
   tmp_coded3 = tmp_coded3.';
   inputSymbols3 = mod_(tmp_coded3,'QPSK');

   % = = = = = Coding and modulation for User 4 = = = = = = = = = = = = = = = = = =
   tmp4 = round(rand(1,SP.inputBlockSize-6));
   tmp4 = [tmp4 zeros(1,6)];
   tmp_coded4 = convenc(tmp2,trel);% Encode the message.
   tmp_coded4 = matintrlv(tmp_coded4.',SP.inputBlockSize/4,8);
   tmp_coded4 = tmp_coded4.';
   inputSymbols4 = mod_(tmp_coded4,'QPSK');
end

end

% = = = = = = = = = SC-FDMA User 1 = = = = = = = = = = = = = = = = = = = = = = = = =
inputSymbols_freq = fft(inputSymbols);
inputSamples_ifdma = zeros(1,numSymbols);
inputSamples_lfdma = zeros(1,numSymbols);
inputSamples_ifdma(1+SP.subband:Q:numSymbols) = inputSymbols_freq;
inputSamples_lfdma([1:SP.inputBlockSize]+SP.inputBlockSize*SP.subband) =
inputSymbols_freq;
TxSamples_ifdma1 = ifft(inputSamples_ifdma);
TxSamples_lfdma1 = ifft(inputSamples_lfdma);

% = = = = = = = = = = SC-FDMA User 2 = = = = = = = = = = = = = = = = = = = = = = = =
inputSymbols_freq2 = fft(inputSymbols2);
```

```
  inputSamples_ifdma2 = zeros(1,numSymbols);
  inputSamples_lfdma2 = zeros(1,numSymbols);
  inputSamples_ifdma2(2+SP.subband:Q:numSymbols) = inputSymbols_freq2;
  inputSamples_lfdma2([SP.inputBlockSize+1:2*SP.inputBlockSize]+SP.inputBlockSize*SP.
  subband) = inputSymbols_freq2;
  TxSamples_ifdma2 = ifft(inputSamples_ifdma2);
  TxSamples_lfdma2 = ifft(inputSamples_lfdma2);

% = = = = = = = = = = = SC-FDMA User 3 = = = = = = = = = = = = = = = = = = = = = = =
  inputSymbols_freq3 = fft(inputSymbols3);
  inputSamples_ifdma3 = zeros(1,numSymbols);
  inputSamples_lfdma3 = zeros(1,numSymbols);
  inputSamples_ifdma3(3+SP.subband:Q:numSymbols) = inputSymbols_freq3;
  inputSamples_lfdma3([2*SP.inputBlockSize+1:3*SP.inputBlockSize]+SP.
  inputBlockSize*SP.subband) = inputSymbols_freq3;
  TxSamples_ifdma3 = ifft(inputSamples_ifdma3);
  TxSamples_lfdma3 = ifft(inputSamples_lfdma3);

% = = = = = = = = = = = SC-FDMA User 4 = = = = = = = = = = = = = = = = = = = = = = =
  inputSymbols_freq4 = fft(inputSymbols4);
  inputSamples_ifdma4 = zeros(1,numSymbols);
  inputSamples_lfdma4 = zeros(1,numSymbols);
  inputSamples_ifdma4(4+SP.subband:Q:numSymbols) = inputSymbols_freq4;
  inputSamples_lfdma4([3*SP.inputBlockSize+1:4*SP.inputBlockSize]+SP.
  inputBlockSize*SP.subband) = inputSymbols_freq4;
  TxSamples_ifdma4 = ifft(inputSamples_ifdma4);
  TxSamples_lfdma4 = ifft(inputSamples_lfdma4);

% = = = = = = = = = = = = = CP for User 1 = = = = = = = = = = = = = = = = = = = = = =
TxSamples_ifdma_u1 = [TxSamples_ifdma1(numSymbols-SP.CPsize+1:numSymbols) TxSamples_
ifdma1];
TxSamples_lfdma_u1 = [TxSamples_lfdma1(numSymbols-SP.CPsize+1:numSymbols) TxSamples_
lfdma1];

% = = = = = = = = = = = = = CP for User 2 = = = = = = = = = = = = = = = = = = = = = =
TxSamples_ifdma_u2 = [TxSamples_ifdma2(numSymbols-SP.CPsize+1:numSymbols) TxSamples_
ifdma2];
TxSamples_lfdma_u2 = [TxSamples_lfdma2(numSymbols-SP.CPsize+1:numSymbols) TxSamples_
lfdma2];

% = = = = = = = = = = = = = CP for User 3 = = = = = = = = = = = = = = = = = = = = = =
TxSamples_ifdma_u3 = [TxSamples_ifdma3(numSymbols-SP.CPsize+1:numSymbols) TxSamples_
ifdma3];
TxSamples_lfdma_u3 = [TxSamples_lfdma3(numSymbols-SP.CPsize+1:numSymbols) TxSamples_
lfdma3];

% = = = = = = = = = = = = = CP for User 4 = = = = = = = = = = = = = = = = = = = = = =
TxSamples_ifdma_u4 = [TxSamples_ifdma4(numSymbols-SP.CPsize+1:numSymbols) TxSamples_
ifdma4];
TxSamples_lfdma_u4 = [TxSamples_lfdma4(numSymbols-SP.CPsize+1:numSymbols) TxSamples_
lfdma4];

% = = = = = = = = = = = = = = Noise Generation = = = = = = = = = = = = = = = = = = = =
  tmpn = randn(2,numSymbols+SP.CPsize);
  complexNoise = (tmpn(1,:) + i*tmpn(2,:))/sqrt(2);
  noisePower = 10^(-SP.SNR(n)/10);
  noise_ = sqrt(noisePower/Q)*complexNoise;

% = = = = = = = = = Channel and CFO for User 1 = = = = = = = = = = = = = = = = = = = =
  RxSamples_ifdma_u1 = filter(SP.channel1, 1, TxSamples_ifdma_u1);%
  RxSamples_lfdma_u1 = filter(SP.channel1, 1, TxSamples_lfdma_u1);%
  RxSamples_ifdma_u1 = RxSamples_ifdma_u1.*exp((j*2*pi*SP.ep1/
(numSymbols))*(-SP.CPsize:(numSymbols-1)));
  RxSamples_lfdma_u1 = RxSamples_lfdma_u1.*exp((j*2*pi*SP.ep1/
(numSymbols))*(-SP.CPsize:(numSymbols-1)));

% = = = = = = = Channel and CFO for User 2 = = = = = = = = = = = = = = = = = = = = = =
  RxSamples_ifdma_u2 = filter(SP.channel2, 1, TxSamples_ifdma_u2);%
  RxSamples_lfdma_u2 = filter(SP.channel2, 1, TxSamples_lfdma_u2);%
  RxSamples_ifdma_u2 = RxSamples_ifdma_u2.*exp((j*2*pi*SP.ep2/
(numSymbols))*(-SP.CPsize:(numSymbols-1)));
  RxSamples_lfdma_u2 = RxSamples_lfdma_u2.*exp((j*2*pi*SP.ep2/
(numSymbols))*(-SP.CPsize:(numSymbols-1)));
```

```
% = = = = = = = Channel and CFO for User 3 = = = = = = = = = = = = = = = = = = = = =
   RxSamples_ifdma_u3 = filter(SP.channel3, 1, TxSamples_ifdma_u3);%
   RxSamples_lfdma_u3 = filter(SP.channel3, 1, TxSamples_lfdma_u3);%
   RxSamples_ifdma_u3 = RxSamples_ifdma_u3.*exp((j*2*pi*SP.ep3/
(numSymbols))*(-SP.CPsize:(numSymbols-1)));
   RxSamples_lfdma_u3 = RxSamples_lfdma_u3.*exp((j*2*pi*SP.ep3/
(numSymbols))*(-SP.CPsize:(numSymbols-1)));

% = = = = = = = Channel and CFO for User 3 = = = = = = = = = = = = = = = = = = = = =
   RxSamples_ifdma_u4 = filter(SP.channel4, 1, TxSamples_ifdma_u4);%
   RxSamples_lfdma_u4 = filter(SP.channel4, 1, TxSamples_lfdma_u4);%
   RxSamples_ifdma_u4 = RxSamples_ifdma_u4.*exp((j*2*pi*SP.ep4/
(numSymbols))*(-SP.CPsize:(numSymbols-1)));
   RxSamples_lfdma_u4 = RxSamples_lfdma_u4.*exp((j*2*pi*SP.ep4/
(numSymbols))*(-SP.CPsize:(numSymbols-1)));

% = = = = = = = = = = = = = Received signal = = = = = = = = = = = = = = = = = = =
RxSamples_ifdma1 = RxSamples_ifdma_u1+ RxSamples_ifdma_u2+RxSamples_ifdma_u3+
RxSamples_ifdma_u4+noise_;%
RxSamples_lfdma1 = RxSamples_lfdma_u1+ RxSamples_lfdma_u2+ RxSamples_lfdma_u3+
RxSamples_lfdma_u4+noise_;%

%% = = = = = = = = = Detection Method = = = = = = = = = = = = = = = = = = = =
%% = = = = = = = = = = = = = = = = = = = = = = = = = = = = = = = = = = = = = = = = =
   if SP.Method = =1

%% = = = = = = = = = = = = = = = = = = = = = = = = = = = = = = = = = = = = = = = = =
%% = = = = = = = = = = = 1. single user detector = = = = = = = = = = = = = = = = =
%% = = = = = = = = = = = = = = = = = = = = = = = = = = = = = = = = = = = = = = = = =
%

% % = = = = = = = = = = = = = CFO Compensation = = = = = = = = = = = = = =
   RxSamples_ifdma1 = RxSamples_ifdma1.*exp((-j*2*pi*SP.ep1/
(numSymbols))*(-SP.CPsize:(numSymbols-1)));;
   RxSamples_lfdma1 = RxSamples_lfdma1.*exp((-j*2*pi*SP.ep1/
(numSymbols))*(-SP.CPsize:(numSymbols-1)));;

   %%% = = = = = = = = = = = = = CP Removing = = = = = = = = = = = = =
RxSamples_ifdma11 = RxSamples_ifdma1(SP.CPsize+1:numSymbols+SP.CPsize);
RxSamples_lfdma11 = RxSamples_lfdma1(SP.CPsize+1:numSymbols+SP.CPsize);
Y_ifdma2 = fft(RxSamples_ifdma11, SP.FFTsize);
Y_lfdma2 = fft(RxSamples_lfdma11, SP.FFTsize);

   %%% = = = = = = = = = = = = = Demapping = = = = = = = = = = = = =
Y_ifdma1 = Y_ifdma2(1+SP.subband:Q:numSymbols);
Y_lfdma1 = Y_lfdma2([1:SP.inputBlockSize]+SP.inputBlockSize*SP.subband);
CT_LFDMA = diag(H_channel1([1:SP.inputBlockSize]));
CT_IFDMA = diag(H_channel1(1:Q:numSymbols));

   %%% = = = = = = = = = = = = = = Equalization = = = = = = = = = = = = =
Y_ifdma = inv(CT_IFDMA'*CT_IFDMA+10^(-SP.SNR(n)/10)*eye(SP.inputBlockSize))
*CT_IFDMA'*Y_ifdma1.';
Y_lfdma = inv(CT_LFDMA'*CT_LFDMA+10^(-SP.SNR(n)/10)*eye(SP.inputBlockSize))
*CT_LFDMA'*Y_lfdma1.';
EstSymbols_ifdma = ifft(Y_ifdma).';
EstSymbols_lfdma = ifft(Y_lfdma).';

%% = = = = = = = = = = = = = = = = = = = = = = = = = = = = = = = = = = = = = = = = =
%% = = = = = = = = = = = = = = = End of method 1 = = = = = = = = = = = = = = = = =
%% = = = = = = = = = = = = = = = = = = = = = = = = = = = = = = = = = = = = = = = = =

   elseif SP.Method = =2

%% = = = = = = = = = = = = = = = = = = = = = = = = = = = = = = = = = = = = = = = = =
%% = = = = = = = = 2. Circular Convolution Method = = = = = = = = = = = = = = = = =
%% = = = = = = = = = = = = = = = = = = = = = = = = = = = = = = = = = = = = = = = = =

RxSamples_ifdma11 = RxSamples_ifdma1(SP.CPsize+1:numSymbols+SP.CPsize);
RxSamples_lfdma11 = RxSamples_lfdma1(SP.CPsize+1:numSymbols+SP.CPsize);
Y_ifdma2 = fft(RxSamples_ifdma11, SP.FFTsize);
Y_lfdma2 = fft(RxSamples_lfdma11, SP.FFTsize);

   %%%% = = = = = = = = = = = = = Demapping = = = = = = = = = = = = =
Y_ifdma1 = Y_ifdma2(1+SP.subband:Q:numSymbols);
```

```
Y_lfdma1 = Y_lfdma2([1:SP.inputBlockSize]+SP.inputBlockSize*SP.subband);
H_eff1 = H_channel1(1:Q:numSymbols);

    %%% = = = = = = = = = = = = Deconvolution = = = = = = = = = = = = =
    Y_lfdma1 = CT_12*Y_lfdma1.';
    Y_ifdma1 = CT_22*Y_ifdma1.';

    %%% = = = = = = = = = = = = = Equalization = = = = = = = = = = = = =
CT_IFDMA = conj(H_eff1)./(conj(H_eff1).*H_eff1 + 10^(-SP.SNR(n)/10));
CT_LFDMA = conj(H_eff2)./(conj(H_eff2).*H_eff2 + 10^(-SP.SNR(n)/10));

    Y_ifdma = Y_ifdma1.*CT_IFDMA.';
    Y_lfdma = Y_lfdma1.*CT_LFDMA.';
    EstSymbols_ifdma = ifft(Y_ifdma).';
    EstSymbols_lfdma = ifft(Y_lfdma).';

%%% = = = = = = = = = = = = = = = = = = = = = = = = = = = = = = = = = =
%%% = = = = = = = = = End of method 2 = = = = = = = = = = = = = = = = =
%%% = = = = = = = = = = = = = = = = = = = = = = = = = = = = = = = = = =

    elseif SP.Method = =3

%%% = = = = = = = = = = = = = = = = = = = = = = = = = = = = = = = = = =
%%% = = = = = = = = = = 3. MMSE Method = = = = = = = = = = = = = = = = =
%%% = = = = = = = = = = = = = = = = = = = = = = = = = = = = = = = = = =

RxSamples_ifdma11 = RxSamples_ifdma1(SP.CPsize+1:numSymbols+SP.CPsize);
RxSamples_lfdma11 = RxSamples_lfdma1(SP.CPsize+1:numSymbols+SP.CPsize);
    Y_ifdma2 = fft(RxSamples_ifdma11, SP.FFTsize);
    Y_lfdma2 = fft(RxSamples_lfdma11, SP.FFTsize);

%%% = = = = = = = = = = = = = Demapping = = = = = = = = = = = = = = = = =
Y_ifdma1 = Y_ifdma2(1+SP.subband:Q:numSymbols);
Y_lfdma2([1:SP.inputBlockSize]+SP.inputBlockSize*SP.subband);
% CT_LFDMA = diag(H_channel1([1:SP.inputBlockSize]));
% CT_IFDMA = diag(H_channel1(1:Q:numSymbols));

    H_eff1 = H_channel1(1:Q:numSymbols);
    H_eff2 = H_channel1([1:SP.inputBlockSize]);

%%% = = = = = = = = = = = CFO Compensation = = = = = = = = = = = = = = = = = = = = =
Y_ifdma1 = inv(CT_2'*CT_2+10^(-SP.SNR(n)/10)*eye(SP.inputBlockSize))*CT_2'* Y_dma1.';
Y_lfdma1 = inv(CT_1'*CT_1+10^(-SP.SNR(n)/10)*eye(SP.inputBlockSize))*CT_1'* Y_
lfdma1.';

%% = = = = = = = = = = = = = = Equalization = = = = = = = = = = = = = = = = =
CT_IFDMA = conj(H_eff1)./(conj(H_eff1).*H_eff1 + 10^(-SP.SNR(n)/10));
CT_LFDMA = conj(H_eff2)./(conj(H_eff2).*H_eff2 + 10^(-SP.SNR(n)/10));
    Y_ifdma = Y_ifdma1.*CT_IFDMA.';
    Y_lfdma = Y_lfdma1.*CT_LFDMA.';
    EstSymbols_ifdma = ifft(Y_ifdma).';
    EstSymbols_lfdma = ifft(Y_lfdma).';

    else

%%% = = = = = = = = = = = = = = = = = = = = = = = = = = = = = = = = = =
%%% = = = 4. Equalization Without CFO Compensation = = = = = = = = = = = = = = =
%% = = = = = = = = = = = = = = = = = = = = = = = = = = = = = = = = = =
%

RxSamples_ifdma11 = RxSamples_ifdma1(SP.CPsize+1:numSymbols+SP.CPsize);
RxSamples_lfdma11 = RxSamples_lfdma1(SP.CPsize+1:numSymbols+SP.CPsize);
    Y_ifdma2 = fft(RxSamples_ifdma11, SP.FFTsize);
    Y_lfdma2 = fft(RxSamples_lfdma11, SP.FFTsize);

%%% = = = = = = = = = = = = = Demapping = = = = = = = = = = = = = =
    Y_ifdma1 = Y_ifdma2(1+SP.subband:Q:numSymbols);
    Y_lfdma1 = Y_lfdma2([1:SP.inputBlockSize]+SP.inputBlockSize*SP.subband);
    CT_LFDMA = diag(H_channel1([1:SP.inputBlockSize]));
    CT_IFDMA = diag(H_channel1(1:Q:numSymbols));

% = = = = = = = = = = = = = = Equalization = = = = = = = = = = = = = = = = = = = = =
Y_ifdma = inv(CT_IFDMA'*CT_IFDMA+10^(-SP.SNR(n)/10)*eye(SP.inputBlockSize))* CT_
IFDMA'*Y_ifdma1.';
```

```
Y_lfdma = inv(CT_LFDMA'*CT_LFDMA+10^(-SP.SNR(n)/10)*eye(SP.inputBlockSize))* CT_
LFDMA'*Y_lfdma1.';
EstSymbols_ifdma = ifft(Y_ifdma).';
EstSymbols_lfdma = ifft(Y_lfdma).';

%% = = = = = = = = = = = = = = = = = = = = = = = = = = = = = = = = = = = = = = = =
%% = = = = = = = = = end of method 4 = = = = = = = = = = = = = = = = = = = = = = =
%% = = = = = = = = = = = = = = = = = = = = = = = = = = = = = = = = = = = = = = = =

    end

    if modtyp = ='16QAM'
    switch (SP.cod_rate)
    case '1'
    scalin_fact = sqrt(1/10);
    demod_symbol1_conv = EstSymbols_ifdma/scalin_fact;
    demodulated_symbol1_conv = qamdemod(demod_symbol1_conv,16);
    symbol_size = 4;
    for y_conv = 1:size(demodulated_symbol1_conv,1)

       demodulated_bit_conv = de2bi(demodulated_symbol1_conv(y_conv,:),symbol_
       size,'left-msb')';
       demodulated_bit1_conv(y_conv,:) = demodulated_bit_conv(:);
    end
       demodata_ifdma = demodulated_bit1_conv;
       demod_symbol1_conv_lfdma = EstSymbols_lfdma/scalin_fact;
       demodulated_symbol1_conv_lfdma = qamdemod(demod_symbol1_conv_lfdma,16);
    for y_conv = 1:size(demodulated_symbol1_conv_lfdma,1)
       demodulated_bit_conv_lfdma = de2bi(demodulated_symbol1_conv_lfdma(y_
       conv,:),symbol_size,'left-msb')';
       demodulated_bit1_conv_lfdma(y_conv,:) = demodulated_bit_conv_lfdma(:);
    end
       demodata_lfdma = demodulated_bit1_conv_lfdma;

    case '1/2'
    scalin_fact = sqrt(1/10);
    demod_symbol1_conv = EstSymbols_ifdma/scalin_fact;
    demodulated_symbol1_conv = qamdemod(demod_symbol1_conv,16);
    symbol_size = 4;
    for y_conv = 1:size(demodulated_symbol1_conv,1)

       demodulated_bit_conv = de2bi(demodulated_symbol1_conv(y_conv,:),symbol_
       size,'left-msb')';
       demodulated_bit1_conv(y_conv,:) = demodulated_bit_conv(:);
    end
       demodata_ifdma = demodulated_bit1_conv;
       demod_symbol1_conv_lfdma = EstSymbols_lfdma/scalin_fact;
       demodulated_symbol1_conv_lfdma = qamdemod(demod_symbol1_conv_lfdma,16);

    for y_conv = 1:size(demodulated_symbol1_conv_lfdma,1)

       demodulated_bit_conv_lfdma = de2bi(demodulated_symbol1_conv_lfdma(y_
       conv,:),symbol_size,'left-msb')';
       demodulated_bit1_conv_lfdma(y_conv,:) = demodulated_bit_conv_lfdma(:);
    end
       demodata_lfdma = demodulated_bit1_conv_lfdma;
       demodata_ifdma = matdeintrlv(demodata_ifdma,8,2);
       decoded_data_ifdma = vitdec(demodata_ifdma',trel,6,'cont','hard');% Decode.
       demodata_lfdma = matdeintrlv(demodata_lfdma,8,2);
       decoded_data_lfdma = vitdec(demodata_lfdma',trel,6,'cont','hard');% Decode.
       noe_ifdma(k) = sum(abs(data1(1:end-6)-decoded_data_ifdma(7:end).'));
       noe_lfdma(k) = sum(abs(data1(1:end-6)-decoded_data_lfdma(7:end).'));
    end

    elseif modtyp = ='QPSK1'
    switch (SP.cod_rate)
       case '1'
       EstSymbols_lfdma = sign(real(EstSymbols_lfdma)) + 1i*sign(imag(EstSymbols_
       lfdma));
       EstSymbols_lfdma = EstSymbols_lfdma/sqrt(2);
       temp_est_lfdma = QPSKDEMOD(EstSymbols_lfdma);
       noe_lfdma(k) = sum(abs(tmp-temp_est_lfdma));
```

```
    EstSymbols_ifdma = sign(real(EstSymbols_ifdma)) + 1i*sign(imag(EstSymbols_
    ifdma));
    EstSymbols_ifdma = EstSymbols_ifdma/sqrt(2);
    temp_est_ifdma = QPSKDEMOD(EstSymbols_ifdma);
    noe_ifdma(k) = sum(abs(tmp-temp_est_ifdma));
    case '1/2'
    EstSymbols_lfdma = sign(real(EstSymbols_lfdma)) + 1i*sign(imag(EstSymbols_
    lfdma));
    EstSymbols_lfdma = EstSymbols_lfdma/sqrt(2);
    temp_est_lfdma = QPSKDEMOD(EstSymbols_lfdma);
    s1_est_lfdma = matdeintrlv(temp_est_lfdma,SP.inputBlockSize/4,8);
    decoded_lfdma = vitdec(s1_est_lfdma',trel,6,'cont','hard');% Decode
    noe_lfdma(k) = sum(abs(tmp(1:end-6)-decoded_lfdma(7:end).'));
    EstSymbols_ifdma = sign(real(EstSymbols_ifdma)) + 1i*sign(imag(EstSymbols_
    ifdma));
    EstSymbols_ifdma = EstSymbols_ifdma/sqrt(2);
    temp_est_ifdma = QPSKDEMOD(EstSymbols_ifdma);
    s1_est_ifdma = matdeintrlv(temp_est_ifdma,SP.inputBlockSize/4,8);
    decoded_ifdma = vitdec(s1_est_ifdma',trel,6,'cont','hard');% Decode
    noe_ifdma(k) = sum(abs(tmp(1:end-6)-decoded_ifdma(7:end).'));
  end
  end

end
if modtyp = ='16QAM'
switch (SP.cod_rate)
case '1'
  BER_ifdma(n,:) = sum(noe_ifdma)/(SP.numRun*SP.inputBlockSize*m1);
  BER_lfdma(n,:) = sum(noe_lfdma)/(SP.numRun*SP.inputBlockSize*m1);
  case '1/2'
  BER_ifdma(n,:) = sum(noe_ifdma)/(SP.numRun*((SP.inputBlockSize*m1)/2-6));
  BER_lfdma(n,:) = sum(noe_lfdma)/(SP.numRun*((SP.inputBlockSize*m1)/2-6));
end
elseif modtyp = ='QPSK1'
switch (SP.cod_rate)
case '1'
  BER_ifdma(n,:) = sum(noe_ifdma)/(SP.numRun*SP.inputBlockSize);
  BER_lfdma(n,:) = sum(noe_lfdma)/(SP.numRun*SP.inputBlockSize);
  case '1/2'

  BER_ifdma(n,:) = sum(noe_ifdma)/(SP.numRun*(SP.inputBlockSize-6));
  BER_lfdma(n,:) = sum(noe_lfdma)/(SP.numRun*(SP.inputBlockSize-6));
end
end
end


% = = = = = = = = = = = = = = = = = = = = = = = = = = = = = = = = = = = =
%This is the main run routine for BER vs. the CFO for the SC-%FDMA%system.
% The parameters for the simulation are set here and the simulation functions are
called.
% = = = = = = = = = = = = = = = = = = = = = = = = = = = = = = = = = = = =

function runSimSCFDMA()

clear all
tic;

% = = = = = = = = = = = Select the DFT and the IDFT length = = = = = = = = = = = = = = = = =
SP.FFTsize = 128;
SP.inputBlockSize = 32;

SP.subband = 0;

% SP.sigm_ = 0.01;

% = = = = = = = = = = = = = = Number of iterations = = = = = = = = = = = = = = = = = = = =
SP.numRun = 10;%10^4;

% = = = = = = = = = = = = = = SNR value = = = = = = = = = = = = = = = = = = = = = = = = =
SP.SNR = 20;

% = = = = = = = = = The range of the max. CFO = = = = = = = = = = = = = = = = = = = = =
SP.CFO = [0:0.05:0.25];
```

```
% = = = = = = = = = = = = = = Select the coding rate and the Modulation Format = = = = =
SP.cod_rate = '1/2';
SP.modtype = 'QPSK1';
% SP.modtype = '16QAM';

% = = = = = = = = = = = = = = = Cyclic prefix length = = = = = = = = = = = = = = = = = =
SP.CPsize = 20;

% = = = = = = = = = = = = = = = = = = = = = = = = = = = = = = = = = = = = = = = = = = = =
[BER_ifdma BER_lfdma] = MULTI_CODE(SP);

% = = = = = = = = = = = = = = = = = = = = Plot the Results = = = = = = = = = = = = = = = =
semilogy(SP.CFO,BER_ifdma,'b-^',SP.CFO,BER_lfdma,'b-^');
legend('IFDMA','LFDMA')
xlabel('epsilon_m_a_x'); ylabel('BER');
title('M = 128, N = 32, QPSK, SC-FDMA')
axis([0 0.25 1e-6 1])
grid

toc

% = = = = = = = = = = = = = = = = = = = = = = = = = = = = = = = = = = = = = = = = = = = =
% This is the simulator for the BER vs. the CFO for SC-FDMA.
% = = = = = = = = = = = = = = = = = = = = = = = = = = = = = = = = = = = = = = = = = = = =

function [BER_ifdma BER_lfdma] = MULTI_CODE(SP)

numSymbols = SP.FFTsize;
Q = numSymbols/SP.inputBlockSize;
modtyp = SP.modtype;%'16QAM';
indi_interleaved_1 = zeros(1,numSymbols);
indi_interleaved_2 = zeros(1,numSymbols);
indi_interleaved_3 = zeros(1,numSymbols);
indi_interleaved_4 = zeros(1,numSymbols);

indi_interleaved_1(1:Q:numSymbols) = ones(1,SP.inputBlockSize);
indi_interleaved_2(2:Q:numSymbols) = ones(1,SP.inputBlockSize);
indi_interleaved_3(3:Q:numSymbols) = ones(1,SP.inputBlockSize);
indi_interleaved_4(4:Q:numSymbols) = ones(1,SP.inputBlockSize);

indi_localized_1 = zeros(1,numSymbols);
indi_localized_2 = zeros(1,numSymbols);
indi_localized_3 = zeros(1,numSymbols);
indi_localized_4 = zeros(1,numSymbols);

indi_localized_1(1:SP.inputBlockSize) = ones(1,SP.inputBlockSize);
indi_localized_2((1+SP.inputBlockSize):(2*SP.inputBlockSize)) = ones(1,SP.
inputBlockSize);
indi_localized_3((2*SP.inputBlockSize+1):(3*SP.inputBlockSize)) = ones(1,SP.
inputBlockSize);
indi_localized_4((3*SP.inputBlockSize+1):(4*SP.inputBlockSize)) = ones(1,SP.
inputBlockSize);

for n = 1:length(SP.CFO),

rand('state',0)
randn('state',0)
noe_ifdma = zeros(1,SP.numRun);
noe_lfdma = zeros(1,SP.numRun);
for k = 1:SP.numRun,
  k
CFO_Values = SP.CFO(n)*(2*rand(1,4)-1);

%%% = = = = = = = = = = = = = = = = = = = = = = = Perfect CFO values
SP.ep1 = CFO_Values(1,1);%
SP.ep2 = CFO_Values(1,2);
SP.ep3 = CFO_Values(1,3);
SP.ep4 = CFO_Values(1,4);
A1 = exp((1*j*2*pi*SP.ep1/(numSymbols))*(0:(numSymbols-1)));
A2 = exp((1*j*2*pi*SP.ep2/(numSymbols))*(0:(numSymbols-1)));
A3 = exp((1*j*2*pi*SP.ep3/(numSymbols))*(0:(numSymbols-1)));
A4 = exp((1*j*2*pi*SP.ep4/(numSymbols))*(0:(numSymbols-1)));
A5 = exp((-1*j*2*pi*SP.ep1/(numSymbols))*(0:(numSymbols-1)));
```

```
CT1 = fft(eye(numSymbols))*diag(A1)*ifft(eye(numSymbols));
CT2 = fft(eye(numSymbols))*diag(A2)*ifft(eye(numSymbols));
CT3 = fft(eye(numSymbols))*diag(A3)*ifft(eye(numSymbols));
CT4 = fft(eye(numSymbols))*diag(A4)*ifft(eye(numSymbols));

%%% = = = = = = = = = = = = = = = = = = = = = = = = = = = = = = = = = = = =
CT1 = fft(eye(numSymbols))*diag(A1)*ifft(eye(numSymbols));
CT_1 = CT1([1:SP.inputBlockSize],[1:SP.inputBlockSize]);
CT_2 = CT1(1:Q:numSymbols,1:Q:numSymbols);
CT5 = fft(eye(numSymbols))*diag(A5)*ifft(eye(numSymbols));
CT_12 = CT5([1:SP.inputBlockSize],[1:SP.inputBlockSize]);
CT_22 = CT5(1:Q:numSymbols,1:Q:numSymbols);

% Uniform channel
  SP.channel = (1/sqrt(7))*(randn(4,7)+sqrt(-1)*randn(4,7))/sqrt(2);

% = = = = = = = = = = = = = User 1 = = = = = = = = = = = = = = = = = = = = =
  SP.channel1 = SP.channel(1,:);

%% = = = = = = = = = = = = = User 2 = = = = = = = = = = = = = = = = = = = = =
  SP.channel2 = SP.channel(2,:);

%% = = = = = = = = = = = = = User 3 = = = = = = = = = = = = = = = = = = = = =
  SP.channel3 = SP.channel(3,:);

%% = = = = = = = = = = = = = User 4 = = = = = = = = = = = = = = = = = = = = =
  SP.channel4 = SP.channel(4,:);

% = = = = = = = = = = = = AWGN channel = = = = = = = = = = = = = = = = = = = =
% SP.channel1 = 1;
% SP.channel2 = 1;
% SP.channel3 = 1;
% SP.channel4 = 1;

H_channel1 = fft(SP.channel1,SP.FFTsize);
H_channel2 = fft(SP.channel2,SP.FFTsize);
H_channel3 = fft(SP.channel3,SP.FFTsize);
H_channel4 = fft(SP.channel4,SP.FFTsize);

A_channel1 = diag(fft(SP.channel1,SP.FFTsize));
A_channel2 = diag(fft(SP.channel2,SP.FFTsize));
A_channel3 = diag(fft(SP.channel3,SP.FFTsize));
A_channel4 = diag(fft(SP.channel4,SP.FFTsize));

  if modtyp = ='16QAM'
  m1 = 4;

  switch (SP.cod_rate)

  case '1'
  m1 = 4;
  data1 = round(rand(1,SP.inputBlockSize*m1));
  inputSymbols = mod_(data1,'16QAM');

  case '1/2'
  m1 = 4;
  trel = poly2trellis(7,[171 133]);% Trellis
  data1 = round(rand(1,(SP.inputBlockSize*m1)/2-6));
  data1 = [data1 0 0 0 0 0 0];
  tmp_coded = convenc(data1,trel);% Encode the message.
  tmp_coded = matintrlv(tmp_coded.',32,16);%Matrix Interleaving
  tmp_coded = tmp_coded.';
  inputSymbols = mod_(tmp_coded,'16QAM');
  end
  elseif modtyp = ='QPSK1'

  switch (SP.cod_rate)
  case '1'
     tmp = round(rand(1,SP.inputBlockSize*2));
     inputSymbols = mod_(tmp,'QPSK');
     tmp2 = round(rand(1,SP.inputBlockSize*2));
     inputSymbols2 = mod_(tmp2,'QPSK');
     tmp3 = round(rand(1,SP.inputBlockSize*2));
     inputSymbols3 = mod_(tmp3,'QPSK');
```

```
    tmp4 = round(rand(1,SP.inputBlockSize*2));
    inputSymbols4 = mod_(tmp4,'QPSK');
  case '1/2'

    trel = poly2trellis(7,[171 133]);% Trellis

% = = = = = = = Coding and modulation for User 1 = = = = = = = = = = = = = = = = = =
    tmp = round(rand(1,SP.inputBlockSize-6));
    tmp = [tmp zeros(1,6)];
    tmp_coded = convenc(tmp,trel);% Encode the message.
    tmp_coded = matintrlv(tmp_coded.',SP.inputBlockSize/4,8);
    tmp_coded = tmp_coded.';
    inputSymbols = mod_(tmp_coded,'QPSK');

% = = = = = = = = = = = = = = Coding and modulation for User = = = = = = = = = = =
    tmp2 = round(rand(1,SP.inputBlockSize-6));
    tmp2 = [tmp2 zeros(1,6)];
    tmp_coded2 = convenc(tmp2,trel); tmp_coded2 = matintrlv(tmp_coded2.',SP.
    inputBlockSize/4,8);
    tmp_coded2 = tmp_coded2.';
    inputSymbols2 = mod_(tmp_coded2,'QPSK');

% = = = = = = = = = = Coding and modulation for User = = = = = = = = = = = = = = = =
tmp2 = round(rand(1,SP.inputBlockSize-6));
    tmp3 = round(rand(1,SP.inputBlockSize-6));
    tmp3 = [tmp3 zeros(1,6)];
    tmp_coded3 = convenc(tmp3,trel); tmp_coded3 = matintrlv(tmp_coded3.',SP.
    inputBlockSize/4,8);
    tmp_coded3 = tmp_coded3.';
    inputSymbols3 = mod_(tmp_coded3,'QPSK');

% = = = = = = = = = = = Coding and modulation for User 4 = = = = = = = = = = = = = =
    tmp4 = round(rand(1,SP.inputBlockSize-6));
    tmp4 = [tmp4 zeros(1,6)];
    tmp_coded4 = convenc(tmp2,trel); tmp_coded4 = matintrlv(tmp_coded4.',SP.
    inputBlockSize/4,8);
    tmp_coded4 = tmp_coded4.';
    inputSymbols4 = mod_(tmp_coded4,'QPSK');
  end

  end

% = = = = = = = = = = = = = SC-FDMA User 1 = = = = = = = = = = = = = = = = = = = =
  inputSymbols_freq = fft(inputSymbols);
  inputSamples_ifdma = zeros(1,numSymbols);
  inputSamples_lfdma = zeros(1,numSymbols);
  inputSamples_ifdma(1+SP.subband:Q:numSymbols) = inputSymbols_freq;
  inputSamples_lfdma([1:SP.inputBlockSize]+SP.inputBlockSize*SP.subband) =
  inputSymbols_freq;
  TxSamples_ifdma1 = ifft(inputSamples_ifdma);
  TxSamples_lfdma1 = ifft(inputSamples_lfdma);

%% = = = = = = = = = = = = = SC-FDMA User 2 = = = = = = = = = = = = = = = = = = = =
  inputSymbols_freq2 = fft(inputSymbols2);
  inputSamples_ifdma2 = zeros(1,numSymbols);
  inputSamples_lfdma2 = zeros(1,numSymbols);
  inputSamples_ifdma2(2+SP.subband:Q:numSymbols) = inputSymbols_freq2;
  inputSamples_lfdma2([SP.inputBlockSize+1:2*SP.inputBlockSize]+SP.inputBlockSize*SP.
  subband) = inputSymbols_freq2;
  TxSamples_ifdma2 = ifft(inputSamples_ifdma2);
  TxSamples_lfdma2 = ifft(inputSamples_lfdma2);
%

%% = = = = = = = = = = = = = SC-FDMA User 3 = = = = = = = = = = = = = = = = = = = =
  inputSymbols_freq3 = fft(inputSymbols3);
  inputSamples_ifdma3 = zeros(1,numSymbols);
  inputSamples_lfdma3 = zeros(1,numSymbols);
  inputSamples_ifdma3(3+SP.subband:Q:numSymbols) = inputSymbols_freq3;
  inputSamples_lfdma3([2*SP.inputBlockSize+1:3*SP.inputBlockSize]+SP.
  inputBlockSize*SP.subband) = inputSymbols_freq3;
  TxSamples_ifdma3 = ifft(inputSamples_ifdma3);
  TxSamples_lfdma3 = ifft(inputSamples_lfdma3);
```

```
%% = = = = = = = = = SC-FDMA User 4 = = = = = = = = = = = = = = = = = = = = = = = = = = = =
   inputSymbols_freq4 = fft(inputSymbols4);
   inputSamples_ifdma4 = zeros(1,numSymbols);
   inputSamples_lfdma4 = zeros(1,numSymbols);
   inputSamples_ifdma4(4+SP.subband:Q:numSymbols) = inputSymbols_freq4;
   inputSamples_lfdma4([3*SP.inputBlockSize+1:4*SP.inputBlockSize]+SP.
   inputBlockSize*SP.subband) = inputSymbols_freq4;
   TxSamples_ifdma4 = ifft(inputSamples_ifdma4);
   TxSamples_lfdma4 = ifft(inputSamples_lfdma4);
%

% = = = = = = = CP for User 1 = = = = = = = = = = = = = = = = = = = = = = = = = = = = = =
TxSamples_ifdma_u1 = [TxSamples_ifdma1(numSymbols-SP.CPsize+1:numSymbols) TxSamples_
ifdma1];
TxSamples_lfdma_u1 = [TxSamples_lfdma1(numSymbols-SP.CPsize+1:numSymbols) TxSamples_
lfdma1];

%% = = = = = = = = = = = CP for User 2 = = = = = = = = = = = = = = = = = = = = = = = = =
TxSamples_ifdma_u2 = [TxSamples_ifdma2(numSymbols-SP.CPsize+1:numSymbols) TxSamples_
ifdma2];
TxSamples_lfdma_u2 = [TxSamples_lfdma2(numSymbols-SP.CPsize+1:numSymbols) TxSamples_
lfdma2];

% = = = = = = = = = CP for User 3 = = = = = = = = = = = = = = = = = = = = = = = = = = =
TxSamples_ifdma_u3 = [TxSamples_ifdma3(numSymbols-SP.CPsize+1:numSymbols) TxSamples_
ifdma3];
TxSamples_lfdma_u3 = [TxSamples_lfdma3(numSymbols-SP.CPsize+1:numSymbols) TxSamples_
lfdma3];

% = = = = = = = = CP for User 4 = = = = = = = = = = = = = = = = = = = = = = = = = = = =
TxSamples_ifdma_u4 = [TxSamples_ifdma4(numSymbols-SP.CPsize+1:numSymbols) TxSamples_
ifdma4];
TxSamples_lfdma_u4 = [TxSamples_lfdma4(numSymbols-SP.CPsize+1:numSymbols) TxSamples_
lfdma4];

% = = = = = = = = = = = = Noise Generation = = = = = = = = = = = = = = = = = = = = = = =
   tmpn = randn(2,numSymbols+SP.CPsize);
   complexNoise = (tmpn(1,:) + i*tmpn(2,:))/sqrt(2);
   noisePower = 10^(-SP.SNR/10);
   noise_ = sqrt(noisePower/Q)*complexNoise;

% = = = = = = = = = = Channel and CFO for User 1 = = = = = = = = = = = = = = = = = = = =
   RxSamples_ifdma_u1 = filter(SP.channel1, 1, TxSamples_ifdma_u1);% Multipath Channel
   RxSamples_lfdma_u1 = filter(SP.channel1, 1, TxSamples_lfdma_u1);% Multipath Channel
   RxSamples_ifdma_u1 = RxSamples_ifdma_u1.*exp((j*2*pi*SP.ep1/(numSymbols))*(-SP.
   CPsize:(numSymbols-1)));
   RxSamples_lfdma_u1 = RxSamples_lfdma_u1.*exp((j*2*pi*SP.ep1/(numSymbols))*(-SP.
   CPsize:(numSymbols-1)));

%% = = = = = = = = = Channel and CFO for User 2 = = = = = = = = = = = = = = = = = = = =
   RxSamples_ifdma_u2 = filter(SP.channel2, 1, TxSamples_ifdma_u2);% Multipath Channel
   RxSamples_lfdma_u2 = filter(SP.channel2, 1, TxSamples_lfdma_u2);% Multipath Channel
   RxSamples_ifdma_u2 = RxSamples_ifdma_u2.*exp((j*2*pi*SP.ep2/(numSymbols))*(-SP.
   CPsize:(numSymbols-1)));
   RxSamples_lfdma_u2 = RxSamples_lfdma_u2.*exp((j*2*pi*SP.ep2/(numSymbols))*(-SP.
   CPsize:(numSymbols-1)));

% = = = = = = = = = = Channel and CFO for User 3 = = = = = = = = = = = = = = = = = = = =
   RxSamples_ifdma_u3 = filter(SP.channel3, 1, TxSamples_ifdma_u3);% Multipath Channel
   RxSamples_lfdma_u3 = filter(SP.channel3, 1, TxSamples_lfdma_u3);% Multipath Channel
   RxSamples_ifdma_u3 = RxSamples_ifdma_u3.*exp((j*2*pi*SP.ep3/(numSymbols))*(-SP.
   CPsize:(numSymbols-1)));
   RxSamples_lfdma_u3 = RxSamples_lfdma_u3.*exp((j*2*pi*SP.ep3/(numSymbols))*(-SP.
   CPsize:(numSymbols-1)));

% = = = = = = = = = = = Channel and CFO for User 3 = = = = = = = = = = = = = = = = = = =
   RxSamples_ifdma_u4 = filter(SP.channel4, 1, TxSamples_ifdma_u4);% Multipath Channel
   RxSamples_lfdma_u4 = filter(SP.channel4, 1, TxSamples_lfdma_u4);% Multipath Channel
   RxSamples_ifdma_u4 = RxSamples_ifdma_u4.*exp((j*2*pi*SP.ep4/(numSymbols))*(-SP.
   CPsize:(numSymbols-1)));
   RxSamples_lfdma_u4 = RxSamples_lfdma_u4.*exp((j*2*pi*SP.ep4/(numSymbols))*(-SP.
   CPsize:(numSymbols-1)));
%
```

```matlab
% = = = = = = = = = = = = = Received signal = = = = = = = = = = = = = = = = = = =
RxSamples_ifdma1 = RxSamples_ifdma_u1 + RxSamples_ifdma_u2 +RxSamples_ifdma_u3+
RxSamples_ifdma_u4+noise_;%
RxSamples_lfdma1 = RxSamples_lfdma_u1 + RxSamples_lfdma_u2 + RxSamples_lfdma_u3+
RxSamples_lfdma_u4+noise_;%

%% = = = = = = = = = = = = = = = = = = = = = = = = = = = = = = = = = = = = = = =
%% = = = = = = = = = = = Detection Method = = = = = = = = = = = = = = = = = = = =
%% = = = = = = = = =%%%

RxSamples_ifdma11 = RxSamples_ifdma1(SP.CPsize+1:numSymbols+SP.CPsize);
RxSamples_lfdma11 = RxSamples_lfdma1(SP.CPsize+1:numSymbols+SP.CPsize);
  Y_ifdma2 = fft(RxSamples_ifdma11, SP.FFTsize);
  Y_lfdma2 = fft(RxSamples_lfdma11, SP.FFTsize);

  %%% = = = = = = = = = = = = Demapping = = = = = = = = = = = = =
  Y_ifdma1 = Y_ifdma2(1+SP.subband:Q:numSymbols);
  Y_lfdma2([1:SP.inputBlockSize]+SP.inputBlockSize*SP.subband);
  H_eff2 = H_channel1([1:SP.inputBlockSize]);
  H_eff1 = H_channel1(1:Q:numSymbols);

  % = = = = = = = = = = = = = Equalization = = = = = = = = = = = = = = = = = = = = = =
  CT_IFDMA = conj(H_eff1)./(conj(H_eff1).*H_eff1 + 10^(-SP.SNR/10));
  CT_LFDMA = conj(H_eff2)./(conj(H_eff2).*H_eff2 + 10^(-SP.SNR/10));
  Y_ifdma = Y_ifdma1.*CT_IFDMA;
  Y_lfdma = Y_lfdma1.*CT_LFDMA;
  EstSymbols_ifdma = ifft(Y_ifdma);
  EstSymbols_lfdma = ifft(Y_lfdma);
  if modtyp = ='16QAM'
  switch (SP.cod_rate)
  case '1'
  scalin_fact = sqrt(1/10);
  demod_symbol1_conv = EstSymbols_ifdma/scalin_fact;
  demodulated_symbol1_conv = qamdemod(demod_symbol1_conv,16);
  symbol_size = 4;
  for y_conv = 1:size(demodulated_symbol1_conv,1)

     demodulated_bit_conv = de2bi(demodulated_symbol1_conv(y_conv,:),symbol_
     size,'left-msb')';
     demodulated_bit1_conv(y_conv,:) = demodulated_bit_conv(:);
  end
     demodata_ifdma = demodulated_bit1_conv;
     demod_symbol1_conv_lfdma = EstSymbols_lfdma/scalin_fact;
     demodulated_symbol1_conv_lfdma = qamdemod(demod_symbol1_conv_lfdma,16);
  for y_conv = 1:size(demodulated_symbol1_conv_lfdma,1)
     demodulated_bit_conv_lfdma = de2bi(demodulated_symbol1_conv_lfdma(y_
     conv,:),symbol_size,'left-msb')';
     demodulated_bit1_conv_lfdma(y_conv,:) = demodulated_bit_conv_lfdma(:);
  end
     demodata_lfdma = demodulated_bit1_conv_lfdma;

  case '1/2'
  scalin_fact = sqrt(1/10);
  demod_symbol1_conv = EstSymbols_ifdma/scalin_fact;
  demodulated_symbol1_conv = qamdemod(demod_symbol1_conv,16);
  symbol_size = 4;
  for y_conv = 1:size(demodulated_symbol1_conv,1)

     demodulated_bit_conv = de2bi(demodulated_symbol1_conv(y_conv,:),symbol_
     size,'left-msb')';
     demodulated_bit1_conv(y_conv,:) = demodulated_bit_conv(:);
  end
     demodata_ifdma = demodulated_bit1_conv;
     demod_symbol1_conv_lfdma = EstSymbols_lfdma/scalin_fact;
     demodulated_symbol1_conv_lfdma = qamdemod(demod_symbol1_conv_lfdma,16);

  for y_conv = 1:size(demodulated_symbol1_conv_lfdma,1)

     demodulated_bit_conv_lfdma = de2bi(demodulated_symbol1_conv_lfdma(y_
     conv,:),symbol_size,'left-msb')';
     demodulated_bit1_conv_lfdma(y_conv,:) = demodulated_bit_conv_lfdma(:);
  end
     demodata_lfdma = demodulated_bit1_conv_lfdma;
```

```
      demodata_ifdma = matdeintrlv(demodata_ifdma,8,2);
      decoded_data_ifdma = vitdec(demodata_ifdma',trel,6,'cont','hard');% Decode.
      demodata_lfdma = matdeintrlv(demodata_lfdma,8,2);
      decoded_data_lfdma = vitdec(demodata_lfdma',trel,6,'cont','hard');% Decode.
      noe_ifdma(k) = sum(abs(data1(1:end-6)-decoded_data_ifdma(7:end).'));
      noe_lfdma(k) = sum(abs(data1(1:end-6)-decoded_data_lfdma(7:end).'));
   end

   elseif modtyp = ='QPSK1'
   switch (SP.cod_rate)
      case '1'
      EstSymbols_lfdma = sign(real(EstSymbols_lfdma)) + 1i*sign(imag(EstSymbols_
      lfdma));
      EstSymbols_lfdma = EstSymbols_lfdma/sqrt(2);
      temp_est_lfdma = QPSKDEMOD(EstSymbols_lfdma);
      noe_lfdma(k) = sum(abs(tmp-temp_est_lfdma));

      EstSymbols_ifdma = sign(real(EstSymbols_ifdma)) + 1i*sign(imag(EstSymbols_
      ifdma));
      EstSymbols_ifdma = EstSymbols_ifdma/sqrt(2);
      temp_est_ifdma = QPSKDEMOD(EstSymbols_ifdma);
      noe_ifdma(k) = sum(abs(tmp-temp_est_ifdma));
      case '1/2'
      EstSymbols_lfdma = sign(real(EstSymbols_lfdma)) + 1i*sign(imag(EstSymbols_
      lfdma));
      EstSymbols_lfdma = EstSymbols_lfdma/sqrt(2);
      temp_est_lfdma = QPSKDEMOD(EstSymbols_lfdma);
      s1_est_lfdma = matdeintrlv(temp_est_lfdma,SP.inputBlockSize/4,8);
      decoded_lfdma = vitdec(s1_est_lfdma',trel,6,'cont','hard');% Decode
      noe_lfdma(k) = sum(abs(tmp(1:end-6)-decoded_lfdma(7:end).'));
      EstSymbols_ifdma = sign(real(EstSymbols_ifdma)) + 1i*sign(imag(EstSymbols_
      ifdma));
      EstSymbols_ifdma = EstSymbols_ifdma/sqrt(2);
      temp_est_ifdma = QPSKDEMOD(EstSymbols_ifdma);
      s1_est_ifdma = matdeintrlv(temp_est_ifdma,SP.inputBlockSize/4,8);
      decoded_ifdma = vitdec(s1_est_ifdma',trel,6,'cont','hard');% Decode
      noe_ifdma(k) = sum(abs(tmp(1:end-6)-decoded_ifdma(7:end).'));
   end

   end

end

if modtyp = ='16QAM'
switch (SP.cod_rate)
case '1'
   BER_ifdma(n,:) = sum(noe_ifdma)/(SP.numRun*SP.inputBlockSize*m1);
   BER_lfdma(n,:) = sum(noe_lfdma)/(SP.numRun*SP.inputBlockSize*m1);
   case '1/2'
   BER_ifdma(n,:) = sum(noe_ifdma)/(SP.numRun*((SP.inputBlockSize*m1)/2-6));
   BER_lfdma(n,:) = sum(noe_lfdma)/(SP.numRun*((SP.inputBlockSize*m1)/2-6));
end
elseif modtyp = ='QPSK1'
switch (SP.cod_rate)
case '1'
   BER_ifdma(n,:) = sum(noe_ifdma)/(SP.numRun*SP.inputBlockSize);
   BER_lfdma(n,:) = sum(noe_lfdma)/(SP.numRun*SP.inputBlockSize);
   case '1/2'
   BER_ifdma(n,:) = sum(noe_ifdma)/(SP.numRun*(SP.inputBlockSize-6));
   BER_lfdma(n,:) = sum(noe_lfdma)/(SP.numRun*(SP.inputBlockSize-6));
end
end
end

% = = = = = = = = = = = = = = = = = = = = = = = = = = = = = = = = = = = =
%This is the main run routine for the regularized equalization scheme for MIMOSC-FDMA.
% The parameters for the simulation are set here and the simulation functions
are called.
% = = = = = = = = = = = = = = = = = = = = = = = = = = = = = = = = = = = =

clear all
tic;
```

```
% = = = = = = = = = = = = Select the DFT and the IDFT length = = = = = = = = = = = =
SP.FFTsize = 128;
SP.inputBlockSize = 32;
SP.subband = 0;

% = = = = = = = = = = = = = The range of the SNR values = = = = = = = = = = = = = =
SP.SNR = [0:3:21];

% = = = = = = = = = = = = = Number of iterations = = = = = = = = = = = = = = = = = =
SP.numRun = 10%10^4;

% = = = = = = = = = = = = = Select the detection type = = = = = = = = = = = = = = =
% SP.detection_type = 'ZFE'
% SP.detection_type = 'LZF'
SP.detection_type = 'LRZ'
% SP.detection_type = 'RZF'

SP.sucariermapping = 'IFDMA'

% = = = = = = = = = = = = = = Select the coding rate and the Modulation Format = = = =
SP.cod_rate = '1/2';
SP.modtype = 'QPSK1';
% SP.modtype = '16QAM';

SP.nd = SP.FFTsize;

% = = = = = = = = = = = = = = Cyclic prefix length = = = = = = = = = = = = = = = = = =
SP.CPsize = 20;

% = = = = = = = = = = = = = = = = = = = = = = = = = = = = = = = = = = = = = = = = = = =
BER_ = MULTI_CODE(SP);

%% = = = = = = = = = = = = = = Plot the Results = = = = = = = = = = = = = = = = = =
semilogy(SP.SNR,BER_,'r-*');

toc
% legend('MIMO SM ZF','MIMO SM Proposed','MIMO SM RZF')
% xlabel('SNR (dB)'); ylabel('BER');
% axis([0 21 1e-4 1])
% grid

% = = = = = = = = = = = = = = = = = = = = = = = = = = = = = = = = = = = = = = = = = = =
% This is the simulator for the regularized equalization for%%%MIMO SC-FDMA System.
% = = = = = = = = = = = = = = = = = = = = = = = = = = = = = = = = = = = = = = = = = = =

function [BER_] = MULTI_CODE(SP)

numSymbols = SP.FFTsize;
Q = numSymbols/SP.inputBlockSize;
modtyp = SP.modtype;%'16QAM';
for n = 1:length(SP.SNR),
rand('state',0)
randn('state',0)
noe_ifdma = zeros(1,SP.numRun);
noe_lfdma = zeros(1,SP.numRun);
for k = 1:SP.numRun,
   k

%%%%%%%%%%%%%%%%%%%%%%%%%%%%%%%%%%%%%%%%%%%%%%%%%%%%%
%%%%%%%%%%%%%%%%%%%%%%%%%%%%%%%% Channels Generation
%%%%%%%%%%%%%%%%%%%%%%%%%%%%%%%%%%%%%%%%%%%%%%%%%%%%%

   % Uniform channel
   SP.channel = (1/sqrt(7))*(randn(4,7)+sqrt(-1)*randn(4,7))/sqrt(2);
%% = = = = = = = = = = = = = = = = = = = = = = = = = = = = = = = = = = = = = = = = =
   h1 = SP.channel(1,:);

%% = = = = = = = = = = = = = = = = = = = = = = = = = = = = = = = = = = = = = = = = =
   h2 = SP.channel(2,:);

%% = = = = = = = = = = = = = = = = = = = = = = = = = = = = = = = = = = = = = = = = =
   h3 = SP.channel(3,:);

%% = = = = = = = = = = = = = = = = = = = = = = = = = = = = = = = = = = = = = = = = =
   h4 = SP.channel(4,:);
```

```
%% = = = = = = = = = = = = = = = = = = = = = = = = = = = = = = = = = = = = = = = = =
    H_channel11 = fft(h1,SP.FFTsize);
    H_channel12 = fft(h2,SP.FFTsize);
    H_channel21 = fft(h3,SP.FFTsize);
    H_channel22 = fft(h4,SP.FFTsize);


%%%%%%%%%%%%%%%%%%%%%%%%%%%%%%%%%%%%%%%%%%%%%%%%%%%%%%%
%%%%%%%%%%%%%%%%%%%%%%%%%%% Encoder and Modulation
%%%%%%%%%%%%%%%%%%%%%%%%%%%%%%%%%%%%%%%%%%%%%%%%%%%%%%%

    if modtyp = ='16QAM'

    m1 = 4;
    switch (SP.cod_rate)
    case '1'
       data1 = round(rand(1,2*SP.inputBlockSize*m1));
       inputSymbols = mod_(tmp_coded,'16QAM');
    case '1/2'
       m1 = 4;
       trel = poly2trellis(7,[171 133]);% Trellis
       data1 = round(rand(1,2*(SP.inputBlockSize*m1)/2-6));
       data1 = [data1 0 0 0 0 0 0];
       tmp_coded = convenc(data1,trel);% Encode the message.
       tmp_coded = matintrlv(tmp_coded.',32,16); tmp_coded = tmp_coded.';
       inputSymbols = mod_(tmp_coded,'16QAM');
    end
    elseif modtyp = ='QPSK1'
    switch (SP.cod_rate)
    case '1'
       m = 2;
       tmp = round(rand(1,2*m*SP.inputBlockSize));
       inputSymbols = mod_(tmp,'QPSK');
    case '1/2'
       tmp = round(rand(1,2*SP.inputBlockSize-6));
       trel = poly2trellis(7,[171 133]);% Trellis
       tmp = [tmp zeros(1,6)];
       tmp_coded = convenc(tmp,trel);% Encode the message.
       tmp_coded = matintrlv(tmp_coded.',16,8);
tmp_coded = tmp_coded.';
       inputSymbols = mod_(tmp_coded,'QPSK');
    end

    end


%%%%%%%%%%%%%%%%%%%%%%%%%%%%%%%%%%%%%%%%%%%%%%%%%%%%%%%
%%%%%%%%%%%%%%%%%%%%%% SM %%%%%%%%%%%%%%%%%%%%%%%%%%%%%
%%%%%%%%%%%%%%%%%%%%%%%%%%%%%%%%%%%%%%%%%%%%%%%%%%%%%%%

    inputSymbols_freq1 = fft(inputSymbols(1:2:end));
    inputSymbols_freq2 = fft(inputSymbols(2:2:end));
    inputSamples_ifdma1 = zeros(1,numSymbols);
    inputSamples_ifdma2 = zeros(1,numSymbols);
    inputSamples_lfdma1 = zeros(1,numSymbols);
    inputSamples_lfdma2 = zeros(1,numSymbols);


%%%%%%%%%%%%%%%%%%%%%%%%%%%%%%%%%%%%%%%%%%%%%%%%%%%%%%%
%%%%%%%%%%%%%%% SC-FDMA Modulation %%%%%%%%%%%%%%%%%%
%%%%%%%%%%%%%%%%%%%%%%%%%%%%%%%%%%%%%%%%%%%%%%%%%%%%%%%

    if SP.sucariermapping = ='IFDMA'
    inputSamples_ifdma1(1+SP.subband:Q:numSymbols) = inputSymbols_freq1;
    inputSamples_ifdma2(1+SP.subband:Q:numSymbols) = inputSymbols_freq2;
    inputSamples_lfdma3 = ifft(inputSamples_ifdma1);
    inputSamples_lfdma4 = ifft(inputSamples_ifdma2);
    else
    inputSamples_lfdma1([1:SP.inputBlockSize]) = inputSymbols_freq1;
    inputSamples_lfdma2([1:SP.inputBlockSize]) = inputSymbols_freq2;
    inputSamples_lfdma3 = ifft(inputSamples_lfdma1);
    inputSamples_lfdma4 = ifft(inputSamples_lfdma2);
    end
```

```
%%%%%%%%%%%%%%%%%%%%%%%%%%%%%%%%%%%%%%%%%%%%%%%%%%%%%
%%%%%%%%%%%%%%%%%%%%% Add Cyclic Prefix %%%%%%%%%%%%%%%%
%%%%%%%%%%%%%%%%%%%%%%%%%%%%%%%%%%%%%%%%%%%%%%%%%%%%%%

   TxSamples_lfdma1 = [inputSamples_lfdma3(numSymbols-SP.CPsize+1:numSymbols)
   inputSamples_lfdma3];
   TxSamples_lfdma2 = [inputSamples_lfdma4(numSymbols-SP.CPsize+1:numSymbols)
   inputSamples_lfdma4];

%%%%%%%%%%%%%%%%%%%%%%%%%%%%%%%%%%%%%%%%%%%%%%%%%%%%%
%%%%%%%%%%%%%%%%%%%% Rayleigh channels%%%%%%%%%%%%%%%%%%%
%%%%%%%%%%%%%%%%%%%%%%%%%%%%%%%%%%%%%%%%%%%%%%%%%%%%%

RxSamples_lfdma11 = filter(h1, 1, TxSamples_lfdma1);% Multipath Channel
RxSamples_lfdma12 = filter(h2, 1, TxSamples_lfdma2);% Multipath Channel
RxSamples_lfdma21 = filter(h3, 1, TxSamples_lfdma1);% Multipath Channel
RxSamples_lfdma22 = filter(h4, 1, TxSamples_lfdma2);% Multipath Channel

   tmpn = randn(2, numSymbols+SP.CPsize);
   complexNoise = (tmpn(1,:) + i*tmpn(2,:))/sqrt(2);
   noisePower = 10^(-SP.SNR(n)/10);
   rx1 = RxSamples_lfdma11+RxSamples_lfdma12 + sqrt(noisePower/Q)*complexNoise;
   rx2 = RxSamples_lfdma21+RxSamples_lfdma22+ sqrt(noisePower/Q)*complexNoise;

%%%%%%%%%%%%%%%%%%%%%%%%%%%%%%%%%%%%%%%%%%%%%%%%%%%%%
%%%%%%%%%%%%%%%%Remove Cyclic Prefix%%%%%%%%%%%%%%%%%%%%
%%%%%%%%%%%%%%%%%%%%%%%%%%%%%%%%%%%%%%%%%%%%%%%%%%%%%

   rx1_ = rx1(SP.CPsize+1:numSymbols+SP.CPsize);
   rx2_ = rx2(SP.CPsize+1:numSymbols+SP.CPsize);

%%%%%%%%%%%%%%%%%%%%%%%%%%%%%%%%%%%%%%%%%%%%%%%%%%%%%
%%%%%%%%%%%%%%%%%%% SC-FDMA demodulation and FDE%%%%%%
%%%%%%%%%%%%%%%%%%%%%%%%%%%%%%%%%%%%%%%%%%%%%%%%%%%%%

   Y_lfdma11 = fft(rx1_, SP.FFTsize);
   Y_lfdma22 = fft(rx2_, SP.FFTsize);

if SP.sucariermapping = ='IFDMA'
   Y_lfdma1 = Y_lfdma11 (1+SP.subband:Q:numSymbols);
   Y_lfdma2 = Y_lfdma22 (1+SP.subband:Q:numSymbols);
   H_eff11 = H_channel11(1+SP.subband:Q:numSymbols);
   H_eff12 = H_channel12(1+SP.subband:Q:numSymbols);
   H_eff21 = H_channel21(1+SP.subband:Q:numSymbols);
   H_eff22 = H_channel22(1+SP.subband:Q:numSymbols);

else
   Y_lfdma1 = Y_lfdma11([1:SP.inputBlockSize]);
   Y_lfdma2 = Y_lfdma22([1:SP.inputBlockSize]);
   H_eff11 = H_channel11([1:SP.inputBlockSize]);
   H_eff12 = H_channel12([1:SP.inputBlockSize]);
   H_eff21 = H_channel21([1:SP.inputBlockSize]);
   H_eff22 = H_channel22([1:SP.inputBlockSize]);
end

A1 = diag(H_eff11);
A2 = diag(H_eff12);
A3 = diag(H_eff21);
A4 = diag(H_eff22);
AT = [A1 A2;A3 A4];
alpha_1 = 0.1;
alpha_2 = 0.1;

c1 = A3*inv(A1);
c2 = A2*inv(A4);

%%%%%%%%%%%%%%%%%%%%%%%%%%%%%%%%%%%%%%%%%%%%%%%%%%%%%
%%%%%%%%%%%%%%%%%% ZF Detector %%%%%%%%%%%%%%%%%%%%%%%%%
%%%%%%%%%%%%%%%%%%%%%%%%%%%%%%%%%%%%%%%%%%%%%%%%%%%%%

if SP.detection_type = ='ZFE'

C = inv(AT);
Y_lfdma33 = (C*[Y_lfdma1,Y_lfdma2].').';
```

```
%%%%%%%%%%%%%%%%%%%%%%%%%%%%%%%%%%%%%%%%%%%%%%%%%%%
%%%%%%%%%%%%%%%% RZF Detector %%%%%%%%%%%%%%%%%%%%
%%%%%%%%%%%%%%%%%%%%%%%%%%%%%%%%%%%%%%%%%%%%%%%%%%%

elseif SP.detection_type = ='RZF'

C = inv(AT'*AT+1/10^(SP.SNR(n)/10)*eye(2*SP.inputBlockSize))*AT';
Y_lfdma33 = (C*[Y_lfdma1,Y_lfdma2].').';
%
%%%%%%%%%%%%%%%%%%%%%%%%%%%%%%%%%%%%%%%%%%%%%%%%%%%
%%%%%%%%%%%%%%%%%%Low complexity ZF Detector (Propose 1)
%%%%%%%%%%%%%%%%%%%%%%%%%%%%%%%%%%%%%%%%%%%%%%%%%%%

elseif SP.detection_type = ='LZF'

% c1 = A3*inv(A1);
% c2 = A2*inv(A4);
Witr = [eye(SP.inputBlockSize) -c2;-c1 eye(SP.inputBlockSize)];
d1 = zeros(SP.inputBlockSize,SP.inputBlockSize);
Wira = [inv(A1-c2*A3) d1;d1 inv(A4-c1*A2)];
Y_lfdma33 = (Wira*Witr*[Y_lfdma1,Y_lfdma2].').';

%%%%%%%%%%%%%%%%%%%%%%%%%%%%%%%%%%%%%%%%%%%%%%%%%%%
%%%%%%%%%%%%% Low complexity RZF Detector (Propose 2)
%%%%%%%%%%%%%%%%%%%%%%%%%%%%%%%%%%%%%%%%%%%%%%%%%%%

elseif SP.detection_type = ='LRZ'

  c1 = A3*inv(A1);
  c2 = A2*inv(A4);
  Witr = [eye(SP.inputBlockSize) -c2;-c1 eye(SP.inputBlockSize)];
  d1 = zeros(SP.inputBlockSize,SP.inputBlockSize);
  a1 = inv((A1-c2*A3)'*(A1-c2*A3)+alpha_1*eye(SP.inputBlockSize))*(A1-c2*A3)';
  a2 = inv((A4-c1*A2)'*(A4-c1*A2)+alpha_2*eye(SP.inputBlockSize))*(A4-c1*A2)';
  Wira = [a1 d1;d1 a2];
  Y_lfdma33 = (Wira*Witr*[Y_lfdma1,Y_lfdma2].').';
end

%%%%%%%%%%%%%%%%%%%%%%%%%%%%%%%%%%%%%%%%%%%%%%%%%%%
%%%%%%%%%%%%%%% Demultiplexing
%%%%%%%%%%%%%%%%%%%%%%%%%%%%%%%%%%%%%%%%%%%%%%%%%%%

  EstSymbols_lfdma = zeros(1,2*SP.inputBlockSize);
  EstSymbols_lfdma(1:2:end) = ifft(Y_lfdma33(1:end/2));
  EstSymbols_lfdma(2:2:end) = ifft(Y_lfdma33(1+end/2:end));

%%%%%%%%%%%%%%%%%%%%%%%%%%%%%%%%%%%%%%%%%%%%%%%%%%%
%%%%%%%%%%%%%%%%%% Demodulation and Decoding%%%%%%%%
%%%%%%%%%%%%%%%%%%%%%%%%%%%%%%%%%%%%%%%%%%%%%%%%%%%

if modtyp = ='16QAM'

  switch (SP.cod_rate)
  case '1/2'
    demod_symbol1_conv_lfdma = EstSymbols_lfdma/scalin_fact;
    demodulated_symbol1_conv_lfdma = qamdemod(demod_symbol1_conv_lfdma,16);

  for y_conv = 1:size(demodulated_symbol1_conv_lfdma,1)

    demodulated_bit_conv_lfdma = de2bi(demodulated_symbol1_conv_lfdma(y_
    conv,:),symbol_size,'left-msb')';
    demodulated_bit1_conv_lfdma(y_conv,:) = demodulated_bit_conv_lfdma(:);
  end
  demodata_lfdma = demodulated_bit1_conv_lfdma;
  demodata_lfdma = matdeintrlv(demodata_lfdma,32,16);
  decoded_data_lfdma = vitdec(demodata_lfdma',trel,6,'cont','hard');% Decode.
    noe_lfdma(k) = sum(abs(data1(1:end-6)-decoded_data_lfdma(7:end).'));

  case'1'

    demod_symbol1_conv_lfdma = EstSymbols_lfdma/scalin_fact;
    demodulated_symbol1_conv_lfdma = qamdemod(demod_symbol1_conv_lfdma,16);

    for y_conv = 1:size(demodulated_symbol1_conv_lfdma,1)
```

```
    demodulated_bit_conv_lfdma = de2bi(demodulated_symbol1_conv_lfdma(y_
    conv,:),symbol_size,'left-msb')';
    demodulated_bit1_conv_lfdma(y_conv,:) = demodulated_bit_conv_lfdma(:);
    end
    demodata_lfdma = demodulated_bit1_conv_lfdma;
% demodata_lfdma = matdeintrlv(demodata_lfdma,32,16);
% decoded_data_lfdma = vitdec(demodata_lfdma',trel,6,'cont','hard');% Decode.
    noe_lfdma(k) = sum(abs(data1-demodata_lfdma));

    end

  elseif modtyp = ='QPSK1'
  switch (SP.cod_rate)
    case'1'
    EstSymbols_lfdma = sign(real(EstSymbols_lfdma)) + 1i*sign(imag(EstSymbols_
    lfdma));
    EstSymbols_lfdma = EstSymbols_lfdma/sqrt(2);
    temp_est_lfdma = QPSKDEMOD(EstSymbols_lfdma);
    noe_lfdma(k) = sum(abs(tmp-temp_est_lfdma));

    case '1/2'
    EstSymbols_lfdma = sign(real(EstSymbols_lfdma)) + 1i*sign(imag(EstSymbols_
    lfdma));
    EstSymbols_lfdma = EstSymbols_lfdma/sqrt(2);
    temp_est_lfdma = QPSKDEMOD(EstSymbols_lfdma);
    s1_est_lfdma = matdeintrlv(temp_est_lfdma,16,8);

    decoded_lfdma = vitdec(s1_est_lfdma',trel,6,'cont','hard');% Decode
    noe_lfdma(k) = sum(abs(tmp(1:end-6)-decoded_lfdma(7:end).'));

  end

    end

end

%%%%%%%%%%%%%%%%%%%%%%%%%%%%%%%%%%%%%%%%%%%%%%%%%%%%
%%%%%%%%%%%%%%%%%%%%%%%%% Calculate the BER %%%%%%%%%%%%
%%%%%%%%%%%%%%%%%%%%%%%%%%%%%%%%%%%%%%%%%%%%%%%%%%%%

if modtyp = ='16QAM'
  switch (SP.cod_rate)
  case '1/2'
    BER_(n,:) = sum(noe_lfdma)/(SP.numRun*(2*m1*SP.inputBlockSize/2-6));
%
  case '1'
    BER_(n,:) = sum(noe_lfdma)/(SP.numRun*(2*m1*SP.inputBlockSize));
  end
elseif modtyp = ='QPSK1'
  switch (SP.cod_rate)
  case '1/2'
    BER_(n,:) = sum(noe_lfdma)/(SP.numRun*(2*SP.inputBlockSize-6));
  case '1'
    BER_(n,:) = sum(noe_lfdma)/(SP.numRun*(2*2*SP.inputBlockSize));
  end
end

end

% = = = = = = = = = = = = = = = = = = = = = = = = = = = = = = = = = =
%This is the main run routine to determine the Regularization%%Parameter for MIMO
SC-FDMA.
% The parameters for the simulation are set here and the%%simulation functions are
called.
% = = = = = = = = = = = = = = = = = = = = = = = = = = = = = = = = = =

% function runSimSCFDMA()
clear all
tic;

% = = = = = = = = = Select the DFT and the IDFT length = = = = = = = = = = = = = = = = =
SP.FFTsize = 512;
SP.inputBlockSize = 64;
SP.subband = 0;
```

```
% = = = = = = = = = Number of iterations = = = = = = = = = = = = = = = = = = = =
SP.numRun = 10;%10^4;

SP.detection_type = 'LRZ'
SP.sucariermapping = 'LFDMA'

% = = = = = = = = = = = = = Select the coding rate and the Modulation Format = = = =
SP.cod_rate = '1/2';
SP.modtype = 'QPSK1';
% SP.modtype = '16QAM';

SP.nd = SP.FFTsize;
SP.CPsize = 20;

snr_no = [6 9 12 15];
for ii = 1:length(snr_no)
SP.SNR = snr_no(ii);

BER_lfdma = MULTI_CODE(SP);
s_lfdma(ii,:) = BER_lfdma;
clear BER_lfdma
end
alfa_ = [0.0001 0.001 0.01 0.1 1];

%% = = = = = = = = = = = = = = Plot the Results = = = = = = = = = = = = = = = =
loglog(alfa_,s_lfdma(1,:),'k-*',alfa_,s_lfdma(2,:),'k- *',alfa_,s_lfdma(3,:),
'r-*',alfa_,s_lfdma(4,:),'r- *');

toc
legend('SNR = 6 dB','SNR = 9 dB','SNR = 12 dB','SNR = 15 dB')
xlabel('Regularization Parameter'); ylabel('BER');
axis([1e-4 1 1e-4 1])
grid

% = = = = = = = = = = = = = = = = = = = = = = = = = = = = = = = = = = = = = =
% This is the simulator for to determine the Regularization%Parameter.
% = = = = = = = = = = = = = = = = = = = = = = = = = = = = = = = = = = = = = =

function [BER_lfdma] = MULTI_CODE(SP)

numSymbols = SP.FFTsize;
Q = numSymbols/SP.inputBlockSize;
modtyp = SP.modtype;%'16QAM';
alfa_ = [0.0001 0.001 0.01 0.1 1];

for n = 1:length(alfa_),

rand('state',0)
randn('state',0)
noe_ifdma = zeros(1,SP.numRun);
noe_lfdma = zeros(1,SP.numRun);
for k = 1:SP.numRun,
  k

%%%%%%%%%%%%%%%%%%%%%%%%%%%%%%%%%%%%%%%%%%%%%%%%%%%%%%%%%%%%%%%%%
%%%%%%%%%%%%%%%%%%%%%% Channels Generation%%%%%%%%%%%%%%%%%%%%%%%%%%
%%%%%%%%%%%%%%%%%%%%%%%%%%%%%%%%%%%%%%%%%%%%%%%%%%%%%%%%%%%%%%%%%

  % Uniform channel
  SP.channel = (1/sqrt(7))*(randn(4,7)+sqrt(-1)*randn(4,7))/sqrt(2);

% = = = = = = = = = = = = = = = = = = = = = = = = = = = = = = = = = = = = = =
  h1 = SP.channel(1,:);

%% = = = = = = = = = = = = = = = = = = = = = = = = = = = = = = = = = = = = =
  h2 = SP.channel(2,:);

%% = = = = = = = = = = = = = = = = = = = = = = = = = = = = = = = = = = = = =
  h3 = SP.channel(3,:);

%% = = = = = = = = = = = = = = = = = = = = = = = = = = = = = = = = = = = = =
  h4 = SP.channel(4,:);

%% = = = = = = = = = = = = = = = = = = = = = = = = = = = = = = = = = = = = =

    H_channel11 = fft(h1,SP.FFTsize);
    H_channel12 = fft(h2,SP.FFTsize);
```

```
    H_channel21 = fft(h3,SP.FFTsize);
    H_channel22 = fft(h4,SP.FFTsize);


%%%%%%%%%%%%%%%%%%%%%%%%%%%%%%%%%%%%%%%%%%%%%%%%%%%%%%%%%%%%%%%%%%%%%
%%%%%%%%%%%%%%%%%%%%%%%%%% Encoder and Modulation%%%%%%%%%%%%%%%%%%%%
%%%%%%%%%%%%%%%%%%%%%%%%%%%%%%%%%%%%%%%%%%%%%%%%%%%%%%%%%%%%%%%%%%%%%

    if modtyp = ='16QAM'
    m1 = 4;

    switch (SP.cod_rate)
    case '1'
       data1 = round(rand(1,2*SP.inputBlockSize*m1));
       inputSymbols = mod_(tmp_coded,'16QAM');
    case '1/2'
       m1 = 4;
       trel = poly2trellis(7,[171 133]);% Trellis
       data1 = round(rand(1,2*(SP.inputBlockSize*m1)/2-6));
       data1 = [data1 0 0 0 0 0 0];
       tmp_coded = convenc(data1,trel);% Encode the message.
       tmp_coded = matintrlv(tmp_coded.',32,16);%Matrix Interleaving
       tmp_coded = tmp_coded.';
       inputSymbols = mod_(tmp_coded,'16QAM');
    end
    elseif modtyp = ='QPSK1'

    switch (SP.cod_rate)
    case '1'
       m = 2;
       tmp = round(rand(1,2*m*SP.inputBlockSize));
       inputSymbols = mod_(tmp,'QPSK');
    case '1/2'
       tmp = round(rand(1,2*SP.inputBlockSize-6));
       trel = poly2trellis(7,[171 133]);% Trellis
       tmp = [tmp zeros(1,6)];
       tmp_coded = convenc(tmp,trel);% Encode the message.
% tmp_coded(2,:) = convenc(tmp(2,:),trel);%
       tmp_coded = matintrlv(tmp_coded.',16,16);%Matrix Interleaving
% tmp_coded(2,:) = matintrlv(tmp_coded(2,:).',16,8);%Matrix Interleaving
       tmp_coded = tmp_coded.';
       inputSymbols = mod_(tmp_coded,'QPSK');
    end

    end

%%%%%%%%%%%%%%%%%%%%%%%%%%%%%%%%%%%%%%%%%%%%%%%%%%%%%%%%
%%%%%%%%%%%%%%%%%%% SM %%%%%%%%%%%%%%%%%%%%%%%%%%%%%%%%%
%%%%%%%%%%%%%%%%%%%%%%%%%%%%%%%%%%%%%%%%%%%%%%%%%%%%%%%%

    inputSymbols_freq1 = fft(inputSymbols(1:2:end));
    inputSymbols_freq2 = fft(inputSymbols(2:2:end));
    inputSamples_ifdma1 = zeros(1,numSymbols);
    inputSamples_ifdma2 = zeros(1,numSymbols);
    inputSamples_lfdma1 = zeros(1,numSymbols);
    inputSamples_lfdma2 = zeros(1,numSymbols);

%%%%%%%%%%%%%%%%%%%%%%%%%%%%%%%%%%%%%%%%%%%%%%%%%%%%%%%%
%%%%%%%%%%%%%%%%%%% SC-FDMA Modulation %%%%%%%%%%%%%%%%%
%%%%%%%%%%%%%%%%%%%%%%%%%%%%%%%%%%%%%%%%%%%%%%%%%%%%%%%%

    if SP.sucariermapping = ='IFDMA'
    inputSamples_ifdma1(1+SP.subband:Q:numSymbols) = inputSymbols_freq1;
    inputSamples_ifdma2(1+SP.subband:Q:numSymbols) = inputSymbols_freq2;
    inputSamples_lfdma3 = ifft(inputSamples_ifdma1);
    inputSamples_lfdma4 = ifft(inputSamples_ifdma2);
    else
    inputSamples_lfdma1([1:SP.inputBlockSize]) = inputSymbols_freq1;
    inputSamples_lfdma2([1:SP.inputBlockSize]) = inputSymbols_freq2;
    inputSamples_lfdma3 = ifft(inputSamples_lfdma1);
    inputSamples_lfdma4 = ifft(inputSamples_lfdma2);
    end
```

```
%%%%%%%%%%%%%%%%%%%%%%%%%%%%%%%%%%%%%%%%%%%%%%%%%%%%%%%
%%%%%%%%%%%%%%%%%%%%%% Add Cyclic Prefix %%%%%%%%%%%%%%%%
%%%%%%%%%%%%%%%%%%%%%%%%%%%%%%%%%%%%%%%%%%%%%%%%%%%%%%%
    TxSamples_lfdma1 = [inputSamples_lfdma3(numSymbols-SP.CPsize+1:numSymbols)
    inputSamples_lfdma3];
    TxSamples_lfdma2 = [inputSamples_lfdma4(numSymbols-SP.CPsize+1:numSymbols)
    inputSamples_lfdma4];

%%%%%%%%%%%%%%%%%%%%%%%%%%%%%%%%%%%%%%%%%%%%%%%%%%%%%%%
%%%%%%%%%%%%%%%%% Rayleigh channels%%%%%%%%%%%%%%%%%%%%
%%%%%%%%%%%%%%%%%%%%%%%%%%%%%%%%%%%%%%%%%%%%%%%%%%%%%%%
    RxSamples_lfdma11 = filter(h1, 1, TxSamples_lfdma1);%
    RxSamples_lfdma12 = filter(h2, 1, TxSamples_lfdma2);%
    RxSamples_lfdma21 = filter(h3, 1, TxSamples_lfdma1);%
    RxSamples_lfdma22 = filter(h4, 1, TxSamples_lfdma2);%

    tmpn = randn(2, numSymbols+SP.CPsize);
    complexNoise = (tmpn(1,:) + i*tmpn(2,:))/sqrt(2);
    noisePower = 2*10^(-SP.SNR/10);
    rx1 = RxSamples_lfdma11+RxSamples_lfdma12 + sqrt(noisePower/Q)*complexNoise;
    rx2 = RxSamples_lfdma21+RxSamples_lfdma22+ sqrt(noisePower/Q)*complexNoise;

%%%%%%%%%%%%%%%%%%%%%%%%%%%%%%%%%%%%%%%%%%%%%%%%%%%%%%%
%%%%%%%%%%%%%%%%%%%%%%%%%Remove Cyclic Prefix%%%%%%%%%%%%%
%%%%%%%%%%%%%%%%%%%%%%%%%%%%%%%%%%%%%%%%%%%%%%%%%%%%%%%
    rx1_ = rx1(SP.CPsize+1:numSymbols+SP.CPsize);
    rx2_ = rx2(SP.CPsize+1:numSymbols+SP.CPsize);

%%%%%%%%%%%%%%%%%%%%%%%%%%%%%%%%%%%%%%%%%%%%%%%%%%%%%%%
%%%%%%%%%%%%%%%%%% SC-FDMA demodulation and FDE%%%%%%%%%
%%%%%%%%%%%%%%%%%%%%%%%%%%%%%%%%%%%%%%%%%%%%%%%%%%%%%%%
    Y_lfdma11 = fft(rx1_, SP.FFTsize);
    Y_lfdma22 = fft(rx2_, SP.FFTsize);

if SP.sucariermapping = ='IFDMA'
    Y_lfdma1 = Y_lfdma11 (1+SP.subband:Q:numSymbols);
    Y_lfdma2 = Y_lfdma22 (1+SP.subband:Q:numSymbols);
    H_eff11 = H_channel11(1+SP.subband:Q:numSymbols);
    H_eff12 = H_channel12(1+SP.subband:Q:numSymbols);
    H_eff21 = H_channel21(1+SP.subband:Q:numSymbols);
    H_eff22 = H_channel22(1+SP.subband:Q:numSymbols);

else
    Y_lfdma1 = Y_lfdma11([1:SP.inputBlockSize]);
    Y_lfdma2 = Y_lfdma22([1:SP.inputBlockSize]);
    H_eff11 = H_channel11([1:SP.inputBlockSize]);
    H_eff12 = H_channel12([1:SP.inputBlockSize]);
    H_eff21 = H_channel21([1:SP.inputBlockSize]);
    H_eff22 = H_channel22([1:SP.inputBlockSize]);
end

A1 = diag(H_eff11);
A2 = diag(H_eff12);
A3 = diag(H_eff21);
A4 = diag(H_eff22);
AT = [A1 A2;A3 A4];

%%%%%%%%%%%%%%%%%%%%%%%%%%%%%%%%%%%%%%%%%%%%%%%%%%%%%%%
%%%%%%%%%%%%%%%%% ZF Detector %%%%%%%%%%%%%%%%%%%%%%%%%
%%%%%%%%%%%%%%%%%%%%%%%%%%%%%%%%%%%%%%%%%%%%%%%%%%%%%%%

if SP.detection_type = ='ZFE'

C = inv(AT);
Y_lfdma33 = (C*[Y_lfdma1,Y_lfdma2].').';

%%%%%%%%%%%%%%%%%%%%%%%%%%%%%%%%%%%%%%%%%%%%%%%%%%%%%%%
%%%%%%%%%%%%%%%%%%% RZF Detector %%%%%%%%%%%%%%%%%%%%%%
%%%%%%%%%%%%%%%%%%%%%%%%%%%%%%%%%%%%%%%%%%%%%%%%%%%%%%%

elseif SP.detection_type = ='RZF'

C = inv(AT'*AT+alfa_(n)*eye(2*SP.inputBlockSize))*AT';
Y_lfdma33 = (C*[Y_lfdma1,Y_lfdma2].').';
%
```

```
%%%%%%%%%%%%%%%%%%%%%%%%%%%%%%%%%%%%%%%%%%%%%%%%%
%%%%%%%%%%%Low complexity ZF Detector (Propose 1) %%%%
%%%%%%%%%%%%%%%%%%%%%%%%%%%%%%%%%%%%%%%%%%%%%%%%%

elseif SP.detection_type = ='LZF'

c1 = A3*inv(A1);
c2 = A2*inv(A4);
Witr = [eye(SP.inputBlockSize) -c2;-c1 eye(SP.inputBlockSize)];
d1 = zeros(SP.inputBlockSize,SP.inputBlockSize);
Wira = [inv(A1-c2*A3) d1;d1 inv(A4-c1*A2)];
Y_lfdma33 = (Wira*Witr*[Y_lfdma1,Y_lfdma2].').';

%%%%%%%%%%%%%%%%%%%%%%%%%%%%%%%%%%%%%%%%%%%%%%%%%
%%%%%%%%%%%%%%% Low complexity RZF Detector %%%%%%%%%%%
%%%%%%%%%%%%%%%%%%%%%%%%%%%%%%%%%%%%%%%%%%%%%%%%%

elseif SP.detection_type = ='LRZ'

  c1 = A3*inv(A1);
  c2 = A2*inv(A4);
  Witr = [eye(SP.inputBlockSize) -c2;-c1 eye(SP.inputBlockSize)];
  d1 = zeros(SP.inputBlockSize,SP.inputBlockSize);
  a1 = inv((A1-c2*A3)'*(A1-c2*A3)+alfa_(n)*eye(SP.inputBlockSize))*(A1-c2*A3)';
  a2 = inv((A4-c1*A2)'*(A4-c1*A2)+alfa_(n)*eye(SP.inputBlockSize))*(A4-c1*A2)';
  Wira = [a1 d1;d1 a2];
  Y_lfdma33 = (Wira*Witr*[Y_lfdma1,Y_lfdma2].').';
end

%%%%%%%%%%%%%%%%%%%%%%%%%%%%%%%%%%%%%%%%%%%%%%%%%%%
%%%%%%%%%%%%%%%%%%%%%%%%% Demultiplexing%%%%%%%%%%%%%%%%%%%%%
%%%%%%%%%%%%%%%%%%%%%%%%%%%%%%%%%%%%%%%%%%%%%%%%%%%

  EstSymbols_lfdma = zeros(1,2*SP.inputBlockSize);
  EstSymbols_lfdma(1:2:end) = ifft(Y_lfdma33(1:end/2));
  EstSymbols_lfdma(2:2:end) = ifft(Y_lfdma33(1+end/2:end));

%%%%%%%%%%%%%%%%%%%%%%%%%%%%%%%%%%%%%%%%%%%%%%%%%%%%%%%%%%%%
%%%%%%%%%%%%%%%%%%%%%%%%%%%%% Demodulation and Decoding %%%%%%%%%%%%%%%%%%%%%%%
%%%%%%%%%%%%%%%%%%%%%%%%%%%%%%%%%%%%%%%%%%%%%%%%%%%%%%%%%%%%

if modtyp = ='16QAM'
  switch (SP.cod_rate)
  case '1/2'
    demod_symbol1_conv_lfdma = EstSymbols_lfdma/scalin_fact;
    demodulated_symbol1_conv_lfdma = qamdemod(demod_symbol1_conv_lfdma,16);
    for y_conv = 1:size(demodulated_symbol1_conv_lfdma,1)
      demodulated_bit_conv_lfdma = de2bi(demodulated_symbol1_conv_lfdma(y_
      conv,:),symbol_size,'left-msb')';
      demodulated_bit1_conv_lfdma(y_conv,:) = demodulated_bit_conv_lfdma(:);
    end
    demodata_lfdma = demodulated_bit1_conv_lfdma;
    demodata_lfdma = matdeintrlv(demodata_lfdma,32,16);
    decoded_data_lfdma = vitdec(demodata_lfdma',trel,6,'cont','hard');% Decode.
    noe_lfdma(k) = sum(abs(data1(1:end-6)-decoded_data_lfdma(7:end).'));

  case'1'

    demod_symbol1_conv_lfdma = EstSymbols_lfdma/scalin_fact;
    demodulated_symbol1_conv_lfdma = qamdemod(demod_symbol1_conv_lfdma,16);

    for y_conv = 1:size(demodulated_symbol1_conv_lfdma,1)

    demodulated_bit_conv_lfdma = de2bi(demodulated_symbol1_conv_lfdma(y_
    conv,:),symbol_size,'left-msb')';
    demodulated_bit1_conv_lfdma(y_conv,:) = demodulated_bit_conv_lfdma(:);
    end
    demodata_lfdma = demodulated_bit1_conv_lfdma;
% demodata_lfdma = matdeintrlv(demodata_lfdma,32,16);
% decoded_data_lfdma = vitdec(demodata_lfdma',trel,6,'cont','hard');% Decode.
    noe_lfdma(k) = sum(abs(data1-demodata_lfdma));

  end
```

```
  elseif modtyp = ='QPSK1'
  switch (SP.cod_rate)
    case'1'
    EstSymbols_lfdma = sign(real(EstSymbols_lfdma)) + 1i*sign(imag(EstSymbols_
    lfdma));
    EstSymbols_lfdma = EstSymbols_lfdma/sqrt(2);
    temp_est_lfdma = QPSKDEMOD(EstSymbols_lfdma);
    noe_lfdma(k) = sum(abs(tmp-temp_est_lfdma));

    case '1/2'
    EstSymbols_lfdma = sign(real(EstSymbols_lfdma)) + 1i*sign(imag(EstSymbols_
    lfdma));
    EstSymbols_lfdma = EstSymbols_lfdma/sqrt(2);
    temp_est_lfdma = QPSKDEMOD(EstSymbols_lfdma);
    s1_est_lfdma = matdeintrlv(temp_est_lfdma,16,16);
    decoded_lfdma = vitdec(s1_est_lfdma',trel,6,'cont','hard');% Decode
    noe_lfdma(k) = sum(abs(tmp(1:end-6)-decoded_lfdma(7:end).'));

  end

  end

end

%%%%%%%%%%%%%%%%%%%%%%%%%%%%%%%%%%%%%%%%%%%%%%%%%%%%%%%%%%%%%%%%%%%%%%%%%%%%%%%
%%%%%%%%%%%%%%%%%%%%%%%%%%% Calculate the BER %%%%%%%%%%%%%%%%%%%%%%%%%%%%%%%%%
%%%%%%%%%%%%%%%%%%%%%%%%%%%%%%%%%%%%%%%%%%%%%%%%%%%%%%%%%%%%%%%%%%%%%%%%%%%%%%%
if modtyp = ='16QAM'
  switch (SP.cod_rate)
  case '1/2'
    BER_lfdma(n,:) = sum(noe_lfdma)/(SP.numRun*(2*m1*SP.inputBlockSize/2-6));
%
  case '1'
    BER_lfdma(n,:) = sum(noe_lfdma)/(SP.numRun*(2*m1*SP.inputBlockSize));
  end
elseif modtyp = ='QPSK1'
  switch (SP.cod_rate)
  case '1/2'
    BER_lfdma(n,:) = sum(noe_lfdma)/(SP.numRun*(2*SP.inputBlockSize-6));
  case '1'
    BER_lfdma(n,:) = sum(noe_lfdma)/(SP.numRun*(2*2*SP.inputBlockSize));
  end
end

end
```

Appendix G: MATLAB® Simulation Codes for Chapters 7 through 9

G.1 MATLAB® Codes for DSTC

```
%this program perform AAF technique using Linear Dispersion STC and 1 relay

function BER = test_1r()

Q = 2; %consallatin size
L = 4; %no. of symbols per frame
modulation = 'BPSK';
SNR = 0:5:25;
K = 1; %no. of relays
N = [10^3 10^3 10^4 10^4 10^4 10^4]; %total no. of transmitted symbols

% = = = = = = = = = = = = = = = = = = = = = = = = = = = = = = = = = = = = = = = = = = =

for j = 1:length(SNR)
N_0 = 1;
P = N_0*10^(SNR(j)/10);
P1 = 0.5*P; %source transmitted power
P2 = P-P1; %all relays transmitted power
SER = 0;

for i1 = 1:N(j)
  hsr1 = sqrt(1/2)*(randn(1,1)+i*randn(1,1));
%zero mean and unit variance channel realization
  hr1d = sqrt(1/2)*(randn(1,1)+i*randn(1,1));

% Data generation
  tx_bits = randint(1,L*log2(Q));
%bits required for two symbols
  [mod_symbols,table] = tx_modulate(tx_bits,modulation);
  x = mod_symbols.';

%received signals at relays
  ysr1 = sqrt(P1)*hsr1*x+sqrt(N_0/2)*(randn(L,1)+i*randn(L,1));

%relay1 processing
  V1 = sqrt(1/2)*(randn(L,L)+i*randn(L,L));
  [Q1 R1] = qr(V1);
  yr1 = sqrt((P2/(K))/(P1+1))*(Q1*ysr1);
```

```
%the destination signal
   yd = hr1d'yr1+sqrt(N_0/2)'(randn(L,1)+i'randn(L,1));

%Exhaustive MLD
% = = = = = = = = = = = = = = = = =

   dist = inf;
   for hh = 1:Q^L
   kk = hh;
   for ii = 1:L
      num(ii,1) = rem(kk,Q);
      kk = floor(kk/Q);
   end
   %form one possible codeword

   for jj = 1:L
      D_poss(jj,1) = table(num(jj)+1);
   end

   c_1 = (Q1'D_poss);
%one of the possible codewords

distance = norm(sqrt(P1'P2/(P1+1))'hsr1'hr1d'c_1-yd);

   if distance<dist
      dist = distance;
      yrec = D_poss;
   end
   end
%calculate SER
   SER = SER+sum(yrec~ = x);
end
SERR(j) = SER/(L'N(j));
BER(j) = SERR(j)/log2(Q)
end

% = = = = = = = = = = = = = = = = = = = = = = = = = = = = = = = = = = = = = = = = =
% = = = = = = = = = = = = = = = = = = = = = = = = = = = = = = = = = = = = = = = = =

%this program perform AAF technique using Linear Dispersion STC and 1relays when the
%source node works as a virtual relay
function BER = test_1rs()

Q = 2; %consallatin size
L = 4; %no. of symbols per frame
modulation = 'BPSK';
SNR = 0:5:25;
K = 1; %no. of relays
N = [10^4 10^4 10^4 10^4 10^5 10^6]; %total no. of transmitted symbols

% = = = = = = = = = = = = = = = = = = = = = = = = = = = = = = = = = = = = = = = = =

for j = 1:length(SNR)
N_0 = 1;
P = N_0'10^(SNR(j)/10);
P1 = 0.3333'P;
%source transmitted power (optimum power allocation)
P2 =.2'P; % relay transmitted power
P3 =.4667'P;
SER = 0;

for i1 = 1:N(j)

   hsr1 = sqrt(1/2)'(randn(1,1)+i'randn(1,1));
%zero mean and unit variance channel realization

   hr1d = sqrt(1/2)'(randn(1,1)+i'randn(1,1));
   hsd = sqrt(1/2)'(randn(1,1)+i'randn(1,1));

% Data generation
   tx_bits = randint(1,L'log2(Q));
%bits required for two symbols
   [mod_symbols,table] = tx_modulate(tx_bits,modulation);
```

```
    x = mod_symbols.';
%recieved signals at relays
    ysr1 = sqrt(P1)'hsr1'x+sqrt(N_0/2)'(randn(L,1)+i'randn(L,1));
%relay1 processing
    Q1 = [1 0 0 0; 0 -1 0 0;0 0 -1 0;0 0 0 -1];
    yr1 = sqrt((P2)/(P1+1))'(Q1'ysr1);

%source processing phase 2

    Q2 = [0 1 0 0;1 0 0 0;0 0 0 1; 0 0 -1 0];
    ysd = sqrt(P3)'Q2'(x);

%the destination signal
    yd = hr1d'yr1+hsd'ysd+sqrt(N_0/2)'(randn(L,1)+i'randn(L,1));


%Exhaustive MLD
% = = = = = = = = = = = = = = = = = = = = = = = = = = = = = = = = = =

    dist = inf;
    for hh = 1:Q^L
    kk = hh;
    for ii = 1:L
        num(ii,1) = rem(kk,Q);
        kk = floor(kk/Q);
    end

    %form one possible codeword

    for jj = 1:L
        D_poss(jj,1) = table(num(jj)+1);
    end

    c_1 = (Q1'D_poss);
%one of the possible codewords
    c_2 = (Q2'(D_poss));
distance = norm(sqrt(P3)'c_2'hsd+sqrt(P1'P2/(P1+1))'hsr1'hr1d'c_1-yd);

    if distance<dist
        dist = distance;
        yrec = D_poss;
    end
    end
%calculate SER
    SER = SER+sum(yrec~ = x);
end
SERR(j) = SER/(L'N(j));
BER(j) = SERR(j)/log2(Q)
end
% = = = = = = = = = = = = = = = = = = = == = = = = = = = = = = = = = = = = = =

function [mod_symbols table = tx_modulate(data_in,modulation)

% = = = = = = = = = = = = = = = = = = = = = = = = = = = = = = = = = = = = = =
%This function modulates the input binary data
%The inputs are:
%data_in = = = > the binary input bits (0,1)
%modulation = = = > choose one of 'BPSK', 'QPSK', '8PSK', '16QAM'
%The outputs are:
%mod_symbols = = = > output modulated symbols
%table = = = > used by receiver
%p = = = > number of constellation points (used by receiver)
% = = = = = = = = = = = = = = = = = = = = = = = = = = = = = = = = = = = = = =

full_length = length(data_in);

%PBSK Modulation
if ~isempty(findstr(modulation,'BPSK'))
%Angle [0 pi] corresponds to
%Grey code vector [1 0], respectively.
table = exp(j'[0 pi]); %generate BPSK symbols
table = table([2 1]); %only reverse the order for Grey code to be [-1 1]
input = data_in;
mod_symbols = table(input+1);%maps transmitted bits into BPSK symbols
p = 2;%Number of constellation points
```

```
%QPSK Modulation
elseif ~isempty(findstr(modulation,'QPSK'))
%Angle vector [pi/4 3`pi/4 5`pi/4 7`pi/4]
%corresponds to Grey code vector:
%[00 10 11 01],respectively
table = exp(j`[pi/4 3`pi/4 7`pi/4 5`pi/4]);%generate QPSK symbols ordered
            % this way to meet Grey code
input = reshape(data_in,2,full_length/2);
mod_symbols = table([1 2]`input+1);
p = 4;%4 constellation points

elseif ~isempty(findstr(modulation,'8PSK'))
%Angle vector [0 pi/4 pi/2 3`pi/4 pi 5`pi/4 3`pi/2 7`pi/4]
%corresponds to Grey code bvector:
%[000 001 011 010 111 110 100 101]

%generate 8PSK symbols
table = exp(j`[0 pi/4 3`pi/4 pi/2 3`pi/2 7`pi/4 5`pi/4 pi]);% Generate 8PSK
                %symbols
input = reshape(data_in,3,full_length/3);
mod_symbols = table([4 2 1]`input+1);
p = 8;%8 constellation points

%16QAM Modulation
elseif ~isempty(findstr(modulation,'16QAM'))
m = 1;
for k = -3:2:3
   for l = -3:2:3
   table(m) = (k+j`l)/sqrt(10);%power normalization
   m = m+1;
   end
end
table = table([0 1 3 2 4 5 7 6 12 13 15 14 8 9 11 10]+1);% Gray code mapping pattern
input = reshape(data_in,4,full_length/4);
mod_symbols = table([8 4 2 1]`input+1);%maps transmitted bits into 16QAM
p = 16;%16 constellation points
else
error('Unimplemented modulation');
end

% = = = = = = = = = = = = = = = = = = = = = = = = = = = = = = = = = = = = = = =

Function [yrec] = demodMLD(yd,Q,K,Q1,Q2,table,P1,P2,hsr1,hsr2,hr1d,hr2d)

dist = inf;
   for hh = 1:Q^K
   kk = hh;
   for ii = 1:K
      num(ii,1) = rem(kk,Q);
      kk = floor(kk/Q);
   end

   %form one possible codeword

   for jj = 1:K
      D_poss(jj,1) = table(num(jj)+1);
   end

   c_1 = (Q1`D_poss);
%one of the possible codewords
   c_2 = (Q2`D_poss);
   distance = norm(sqrt(P1`P2/(P1`abs(hsr1)^2+1))`hsr1`hr1d`c_1+sqrt(P1`P2/(P1`abs
   (hsr2)^2+1))`hsr2`hr2d`c_2-yd);

   if distance<dist
      dist = distance;
      yrec = D_poss;
   end
   end

% = = = = = = = = = = = = = = = = = = = = = = = = = = = = = = = = = = = = = = =

function [out] = demodqpsk(y,table, Q)

%this function perform MLD for input QPSK symbols
```

```
%y = received signal
%table = transmitted symbols
%Q = no. of constellation points

for m = 1:Q
d(:,m) = (abs(y-table(m))).^2;
end

[i1 i2] = min(d,[],2);
out = table(i2).';

% = = = = = = = = = = = = = = = = = = = = = = = = = = = = = = = = = = = = = = = = = = =

%this program perform AAF technique using Random codes
%QPSK Modulation
%clear all;
%close all;
%clc

function BER = test_2r_rand()

Q = 4; %consallatin size
L = 4; %no. of transmitted symbols(coherent time)T
modulation = 'QPSK';
SNR = 0:5:30;
K = 2; %no of relays %no. of relays
N = [10^4 10^4 10^4 10^5 10^5 5*10^5 5*10^5]; %total no. of transmitted symbols

% = = = = = = = = = = = = = = = = = = = = = = = = = = = = = = = = = = = = = = = = = = =

for j = 1:length(SNR)
N_0 = 1;
P = N_0*10^(SNR(j)/10);
P1 = 0.5*P; %source transmitted power
P2 = P-P1; %all relays transmitted power
SER = 0;

for i1 = 1:N(j)

   hsr1 = sqrt(1/2)*(randn(1,1)+i*randn(1,1)); %zero mean and unit variance channel
   realization
   hsr2 = sqrt(1/2)*(randn(1,1)+i*randn(1,1));
   hr1d = sqrt(1/2)*(randn(1,1)+i*randn(1,1));
   hr2d = sqrt(1/2)*(randn(1,1)+i*randn(1,1));

% Data generation
   tx_bits = randint(1,L*log2(Q));
%bits required for two symbols
   [mod_symbols,table] = tx_modulate(tx_bits,modulation);

   x = mod_symbols.';
%recieved signals at relays
   ysr1 = sqrt(P1)*hsr1*x+sqrt(N_0/2)*(randn(L,1)+i*randn(L,1));
   ysr2 = sqrt(P1)*hsr2*x+sqrt(N_0/2)*(randn(L,1)+i*randn(L,1));

%relay1 processing
   V1 = sqrt(1/2)*(randn(L,L)+i*randn(L,L));
   [Q1 R1] = qr(V1);
%Q1 = diag([1 1 1 1]);
   yr1 = sqrt((P2/(K))/(P1+1))*(Q1*ysr1);

%relay2 processing
   V2 = sqrt(1/2)*(randn(L,L)+i*randn(L,L));
   [Q2 R2] = qr(V2);
%Q2 = [0 -1 0 0;1 0 0 0;0 0 0 -1; 0 0 1 0];
   yr2 = sqrt((P2/(K))/(P1+1))*(Q2*(ysr2));
      yd = hr1d*yr1+hr2d*yr2+sqrt(N_0/2)*(randn(L,1)+i*randn(L,1));

%Exhaustive MLD
% = = = = = = = = = =

   dist = inf;
   for hh = 1:Q^L
   kk = hh;
```

```
    for ii = 1:L
        num(ii,1) = rem(kk,Q);
        kk = floor(kk/Q);
    end

    %form one possible codeword

    for jj = 1:L
        D_poss(jj,1) = table(num(jj)+1);
    end
    c_1 = (Q1'D_poss);
%one of the possible codewords
    c_2 = (Q2'D_poss);
    distance = norm(sqrt(P1'(P2/K)/(P1+1))'hsr1'hr1d'c_1+sqrt(P1'(P2/K)/
    (P1+1))'(hsr2)'hr2d'c_2-yd);

    if distance<dist
        dist = distance;
        yrec = D_poss;
    end
    end
%calculate SER
    SER = SER+sum(yrec~ = x);
end
SERR(j) = SER/(L'N(j));
BER(j) = SERR(j)/log2(Q);
end

% = = = = = = = = = = = = = = = = = = = = = = = = = = = = = = = = = = = = = =
%this program perform AAF technique using Linear Dispersion STC and 2relays and the
%source node with random codes
%QPSK Modulation
%clear all;
%close all;
%clc

function BER = test_2rs_rand()

Q = 4; %consallatin size

modulation = 'QPSK';
SNR = 0:5:30;
K = 2; %no. of relays
L = 4;

N = [10^4 10^4 10^4 10^5 10^5 5'10^5 5'10^5]; %total no. of transmitted symbols

% = = = = = = = = = = = = = = = = = = = = = = = = = = = = = = = = = = = = = =
for j = 1:length(SNR)
N_0 = 1;
P = N_0'10^(SNR(j)/10);
P1 =.4'P; %source transmitted power
P2 =.1478'P; %all relays transmitted power
P3 =.3043'P;
SER = 0;

for i1 = 1:N(j)
    hsr1 = sqrt(1/2)'(randn(1,1)+i'randn(1,1)); %zero mean and unit variance channel
    realization
    hsr2 = sqrt(1/2)'(randn(1,1)+i'randn(1,1));
    hr1d = sqrt(1/2)'(randn(1,1)+i'randn(1,1));
    hr2d = sqrt(1/2)'(randn(1,1)+i'randn(1,1));
    hsd = sqrt(1/2)'(randn(1,1)+i'randn(1,1));

% Data generation
    tx_bits = randint(1,L'log2(Q)); %bits required for two symbols
    [mod_symbols,table] = tx_modulate(tx_bits,modulation);

    x = mod_symbols.';

%recieved signals the relays
    ysr1 = sqrt(P1)'hsr1'x+sqrt(N_0/2)'(randn(L,1)+i'randn(L,1));
    ysr2 = sqrt(P1)'hsr2'x+sqrt(N_0/2)'(randn(L,1)+i'randn(L,1));
```

```
%relay1 processing
   V1 = sqrt(1/2)'(randn(L,L)+i'randn(L,L));
   [Q1 R1] = qr(V1);
%Q1 = [0 -1 0 0;1 0 0 0;;0 0 0 -1; 0 0 1 0];
   yr1 = sqrt((P2)/(P1+1))'(Q1'(ysr1));

%relay2 processing
   V2 = sqrt(1/2)'(randn(L,L)+i'randn(L,L));
   [Q2 R2] = qr(V2);
%Q2 = [0 0 -1 0;0 0 0 -1;1 0 0 0;0 1 0 0];
   yr2 = sqrt((P2)/(P1+1))'(Q2'(ysr2));

%source node processing phase 2
   V3 = sqrt(1/2)'(randn(L,L)+i'randn(L,L));
   [Q3 R3] = qr(V3);
%Q3 = diag([1 1 1 1]);
   ysd = sqrt(P3)'Q3'x;

%the destination signal
   yd = hr1d'yr1+hr2d'yr2+hsd'ysd+sqrt(N_0/2)'(randn(L,1)+i'randn(L,1));

%Exhaustive MLD
% = = = = = = = = = = = = =
   dist = inf;
   for hh = 1:Q^L
   kk = hh;
   for ii = 1:L
      num(ii,1) = rem(kk,Q);
      kk = floor(kk/Q);
   end

   %form one possible codeword
   for jj = 1:L
      D_poss(jj,1) = table(num(jj)+1);
   end

   c_1 = (Q1'(D_poss)); %one of the possible codewords
   c_2 = (Q2'(D_poss));
   c_3 = (Q3'D_poss);
   distance = norm(sqrt(P3)'c_3'hsd+sqrt(P1'P2/(P1+N_0))'(hsr1)'hr1d'c_1+sqrt(P1'P2/
   (P1+N_0))'(hsr2)'hr2d'c_2-yd);

   if distance<dist
      dist = distance;
      yrec = D_poss;
   end
   end
%calculate SER
   SER = SER+sum(yrec~ = x);
end
SERR(j) = SER/(L'N(j));
BER(j) = SERR(j)/log2(Q)
end

% = = = = = = = = = = = = = = = = = = = = = = = = = = = = = = = = = = = = = = = =
Power allocation schemes
Scheme 1
% = = = = = = = = = = = = = = = = = = = = = = = = = = = = = = = = = = = = = = = =

function BER = test_2rs_1()

Q = 4; %consallatin size

modulation = 'QPSK';
SNR = 10:5:30;
K = 2; %no. of relays
L = 4;
N = [10^4 2'10^4 2'10^4 2'10^5 2'10^5]; %total no. of transmitted symbols
% = = = = = = = = = = = = = = = = = = = = = = = = = = = = = = = = = = = = = = = =
for j = 1:length(SNR)
N_0 = 1;
P = N_0'10^(SNR(j)/10);
```

```
P1 =.5*P; %source transmitted power
P2 =.5*P/3; %all relays transmitted power
P3 =.5*P/3;
SER = 0;

for i1 = 1:N(j)
  hsr1 = sqrt(1/2)*(randn(1,1)+i*randn(1,1)); %zero mean and unit variance channel
  %realization
  hsr2 = sqrt(1/2)*(randn(1,1)+i*randn(1,1));
  hr1d = sqrt(1/2)*(randn(1,1)+i*randn(1,1));
  hr2d = sqrt(1/2)*(randn(1,1)+i*randn(1,1));
  hsd = sqrt(1/2)*(randn(1,1)+i*randn(1,1));

% Data generation
  tx_bits = randint(1,L*log2(Q)); %bits required for two symbols
  [mod_symbols,table] = tx_modulate(tx_bits,modulation);

  x = mod_symbols.';

%recieved signals the relays
  ysr1 = sqrt(P1)*hsr1*x+sqrt(N_0/2)*(randn(L,1)+i*randn(L,1));
  ysr2 = sqrt(P1)*hsr2*x+sqrt(N_0/2)*(randn(L,1)+i*randn(L,1));

%relay1 processing
%V1 = sqrt(1/2)*(randn(L,L)+i*randn(L,L));
%[Q1 R1] = qr(V1);
  Q1 = diag([1 1 1 1]);
  yr1 = sqrt((P2)/(P1+1))*(Q1*ysr1);

%relay2 processing
%V2 = sqrt(1/2)*(randn(L,L)+i*randn(L,L));
%[Q2 R2] = qr(V2);
  Q2 = [0 0 0 1;0 0 -1 0;0 -1 0 0;1 0 0 0];
  yr2 = sqrt((P2)/(P1+1))*(Q2*ysr2);

%source node processing phase 2
%V3 = sqrt(1/2)*(randn(L,L)+i*randn(L,L));
%[Q3 R3] = qr(V3);
  Q3 = [0 -1 0 0;1 0 0 0;0 0 0 -1;0 0 1 0];
  ysd = sqrt(P3)*Q3*conj(x);

%the destination signal

  yd = hr1d*yr1+hr2d*yr2+hsd*ysd+sqrt(N_0/2)*(randn(L,1)+i*randn(L,1));

%Exhaustive MLD
% = = = = = = = = = = = = = =

  dist = inf;
  for hh = 1:Q^L
  kk = hh;
  for ii = 1:L
    num(ii,1) = rem(kk,Q);
    kk = floor(kk/Q);
  end

  %form one possible codeword
  for jj = 1:L
    D_poss(jj,1) = table(num(jj)+1);
  end
  c_1 = (Q1*D_poss); %one of the possible codewords
  c_2 = (Q2*D_poss);
  c_3 = (Q3*conj(D_poss));
  distance = norm(sqrt(P3)*c_3*hsd+sqrt(P1*P2/(P1+N_0))*hsr1*hr1d*c_1+sqrt(P1*P2/
  (P1+N_0))*hsr2*hr2d*c_2-yd);

  if distance<dist
    dist = distance;
    yrec = D_poss;
  end
  end
%calculate SER
  SER = SER+sum(yrec~ = x);
end
```

```
SERR(j) = SER/(L*N(j));
BER(j) = SERR(j)/log2(Q)
end

% = = = = = = = = = = = = = = = = = = = = = = = = = = = = = = = = = = = = =

Power allocation schemes
Scheme 2

function BER = test_2rs_2()

Q = 4; %consallatin size

modulation = 'QPSK';
SNR = 10:5:30;
K = 2; %no. of relays
L = 4;

N = [10^4 2*10^4 2*10^4 2*10^5 2*10^5]; %total no. of transmitted symbols

% = = = = = = = = = = = = = = = = = = = = = = = = = = = = = = = = = = = = =

for j = 1:length(SNR)
N_0 = 1;
P = N_0*10^(SNR(j)/10);
P1 =.25*P; %source transmitted power
P2 =.25*P; %all relays transmitted power
P3 =.25*P;
SER = 0;

for i1 = 1:N(j)
  hsr1 = sqrt(1/2)*(randn(1,1)+i*randn(1,1)); %zero mean and unit variance channel
  %realization
  hsr2 = sqrt(1/2)*(randn(1,1)+i*randn(1,1));
  hr1d = sqrt(1/2)*(randn(1,1)+i*randn(1,1));
  hr2d = sqrt(1/2)*(randn(1,1)+i*randn(1,1));
  hsd = sqrt(1/2)*(randn(1,1)+i*randn(1,1));

% Data generation
  tx_bits = randint(1,L*log2(Q)); %bits required for two symbols
  [mod_symbols,table] = tx_modulate(tx_bits,modulation);

  x = mod_symbols.';

%recieved signals the relays
  ysr1 = sqrt(P1)*hsr1'*x+sqrt(N_0/2)*(randn(L,1)+i*randn(L,1));
  ysr2 = sqrt(P1)*hsr2'*x+sqrt(N_0/2)*(randn(L,1)+i*randn(L,1));

%relay1 processing
%V1 = sqrt(1/2)*(randn(L,L)+i*randn(L,L));
%[Q1 R1] = qr(V1);
  Q1 = diag([1 1 1 1]);
  yr1 = sqrt((P2)/(P1+1))*(Q1*ysr1);

%relay2 processing
%V2 = sqrt(1/2)*(randn(L,L)+i*randn(L,L));
%[Q2 R2] = qr(V2);
  Q2 = [0 0 0 1;0 0 -1 0;0 -1 0 0;1 0 0 0];
  yr2 = sqrt((P2)/(P1+1))*(Q2*ysr2);

%source node processing phase 2
%V3 = sqrt(1/2)*(randn(L,L)+i*randn(L,L));
%[Q3 R3] = qr(V3);
  Q3 = [0 -1 0 0;1 0 0 0;0 0 0 -1;0 0 1 0];
  ysd = sqrt(P3)*Q3*conj(x);

%the destination signal
  yd = hr1d*yr1+hr2d*yr2+hsd*ysd+sqrt(N_0/2)*(randn(L,1)+i*randn(L,1));

%Exhaustive MLD
% = = = = = = = = = = = = = =
  dist = inf;
  for hh = 1:Q^L
  kk = hh;
```

```
    for ii = 1:L
       num(ii,1) = rem(kk,Q);
       kk = floor(kk/Q);
    end

    %form one possible codeword

    for jj = 1:L
       D_poss(jj,1) = table(num(jj)+1);
    end

    c_1 = (Q1'D_poss); %one of the possible codewords
    c_2 = (Q2'D_poss);
    c_3 = (Q3'conj(D_poss));
    distance = norm(sqrt(P3)'c_3'hsd+sqrt(P1'P2/(P1+N_0))'hsr1'hr1d'c_1+sqrt(P1'P2/
    (P1+N_0))'hsr2'hr2d'c_2-yd);

    if distance<dist
       dist = distance;
       yrec = D_poss;
    end
    end
%calculate SER
    SER = SER+sum(yrec~ = x);
end
SERR(j) = SER/(L'N(j));
BER(j) = SERR(j)/log2(Q);
end

% = = = = = = = = = = = = = = = = = = = = = = = = = = = = = = = = = = = = = = = = =

Power allocation schemes
Optimum power allocation
Scheme3
% = = = = = = = = = = = = = = = = = = = = = = = = = = = = = = = = = = = = = = = = =

function BER = test_2rs_op()

Q = 4; %consallatin size

modulation = 'QPSK';
SNR = 10:5:30;
K = 2; %no. of relays
L = 4;

N = [10^4 2'10^4 2'10^4 2'10^5 2'10^5]; %total no. of transmitted symbols

% = = = = = = = = = = = = =  = = = = = = = = = = = = = = =  = == = = = = = = = = = = = = = =

for j = 1:length(SNR)
N_0 = 1;
P = N_0'10^(SNR(j)/10);
P1 =.4'P; %source transmitted power
P2 =.1478'P; %all relays transmitted power
P3 =.3043'P;
SER = 0;

for i1 = 1:N(j)

   hsr1 = sqrt(1/2)'(randn(1,1)+i'randn(1,1)); %zero mean and unit variance channel
   realization
   hsr2 = sqrt(1/2)'(randn(1,1)+i'randn(1,1));
   hr1d = sqrt(1/2)'(randn(1,1)+i'randn(1,1));
   hr2d = sqrt(1/2)'(randn(1,1)+i'randn(1,1));
   hsd = sqrt(1/2)'(randn(1,1)+i'randn(1,1));

% Data generation
   tx_bits = randint(1,L'log2(Q)); %bits required for two symbols
   [mod_symbols,table] = tx_modulate(tx_bits,modulation);

   x = mod_symbols.';

%recieved signals the relays
   ysr1 = sqrt(P1)'hsr1'x+sqrt(N_0/2)'(randn(L,1)+i'randn(L,1));
   ysr2 = sqrt(P1)'hsr2'x+sqrt(N_0/2)'(randn(L,1)+i'randn(L,1));
```

```
%relay1 processing
%V1 = sqrt(1/2)'(randn(L,L)+i'randn(L,L));
%[Q1 R1] = qr(V1);
  Q1 = diag([1 1 1 1]);
  yr1 = sqrt((P2)/(P1+1))'(Q1'ysr1);

%relay2 processing
  Q2 = [0 0 0 1;0 0 -1 0;0 -1 0 0;1 0 0 0];
  yr2 = sqrt((P2)/(P1+1))'(Q2'ysr2);

  Q3 = [0 -1 0 0;1 0 0 0;0 0 0 -1;0 0 1 0];
  ysd = sqrt(P3)'Q3'conj(x);

%the destination signal
  yd = hr1d'yr1+hr2d'yr2+hsd'ysd+sqrt(N_0/2)'(randn(L,1)+i'randn(L,1));

%Exhaustive MLD
% = = = = = = = = = = = = =

  dist = inf;
  for hh = 1:Q^L
  kk = hh;
  for ii = 1:L
    num(ii,1) = rem(kk,Q);
    kk = floor(kk/Q);
  end

  %form one possible codeword

  for jj = 1:L
    D_poss(jj,1) = table(num(jj)+1);
  end

  c_1 = (Q1'D_poss); %one of the possible codewords
  c_2 = (Q2'D_poss);
  c_3 = (Q3'conj(D_poss));
  distance = norm(sqrt(P3)'c_3'hsd+sqrt(P1'P2/(P1+N_0))'hsr1'hr1d'c_1+sqrt(P1'P2/
  (P1+N_0))'hsr2'hr2d'c_2-yd);

  if distance<dist
    dist = distance;
    yrec = D_poss;
  end
  end
%calculate SER
  SER = SER+sum(yrec~ = x);
end
SERR(j) = SER/(L'N(j));
BER(j) = SERR(j)/log2(Q)
end

% = = = = = = = = = = = = = = = = = = = = = = = = = = =

A = [-1 -2 -2; 1 2 2];
b = [0 ;72];
%options = optimset('LargeScale','off');
x0 = [1; 1; 10];
[x,fval] = fmincon(@myfun1,x0,A,b)

% = = = = = = = = = = = = = = = = = = = = = = = = = = = =
% = = = = = = = = = = = = = = = = = = = = = = = = = = = =

function f = myfun1(x)
f = -x(1)'x(2)'x(3);

% = = = = = = = = = = = = = = = = = = = = = = = = = = = =
% = = = = = = = = = = = = = = = = = = = = = = = = = = = =
```

G.2 MATLAB® Codes for DSFC

```
%SER for DSFC,BPSK,two relays, and one channel path
%clear all;
%clc
function SEER = test2r1p()
```

```
% = = = = = = = = = = = = = = = = = = = = = = = = = =

N = 2; %number of relays
L = 1; %number of paths
delay = [0]; %2 rays
% ^ = = = = = = = = = = = = = = = = = = = = = = = = = =

Q = 2; %constellation size
modulation = 'BPSK';
BW = 20*10^6; %system BW
no_pkts = [5*10^4 5*10^4 5*10^4 5*10^4 5*10^4 5*10^5 5*10^5]; %number of simulated packets
K = 64; %number of subcarriers
df = BW/K; %the subcarrier frequency spacing

M = K/L; %number of source subblocks
G = K/(L*N); %number of groups at the relay

SNR = 0:4:24;

V2 = sqrt(1/2)*vandermonde(2) % vandermonde matrix at the relay nodes

P = 2; %the total transmitted power
P1 = 0.5*P; %source transmitted power
P2 = P-P1; %all relays transmitted power

for j = 1:length(SNR)

N_0 = P*10^(-SNR(j)/10);
SER = 0;

    clear D_poss
    clear C_poss

for i1 = 1:no_pkts(j)

%the data generation
    tx_bits = randint(1,2*log2(Q)); %bits required to transmit 2 symbols
    [mod_symbols table] = tx_modulate(tx_bits,modulation);
    x = mod_symbols.'; %column vector of the generated symbols(K*1)

%channel coefficients (S- >R)
%.........................................
    hsr1 = multiray(0);%sqrt(1/4)*(randn(1,L)+i*randn(1,L)); %zero mean and unit variance
    hsr2 = multiray(0);%sqrt(1/4)*(randn(1,L)+i*randn(1,L));

    for k = 1:2
    H1(k,1) = 0; %source to relay1
    H2(k,1) = 0;
    for d = 1:length(delay)
       H1(k,1) = H1(k,1)+hsr1(d)*exp(-i*2*pi*(k-1)*df*delay(d)); %K*1
       H2(k,1) = H2(k,1)+hsr2(d)*exp(-i*2*pi*(k-1)*df*delay(d)); %K*1
    end
    end

%channel coefficients (R- >D)
%.........................................
    hr1d = multiray(0);%sqrt(1/4)*(randn(1,L)+i*randn(1,L)); %zero mean and unit variance
    hr2d = multiray(0);%sqrt(1/4)*(randn(1,L)+i*randn(1,L));
    for k = 1:2
    H3(k,1) = 0;
    H4(k,1) = 0;
    for d = 1:length(delay)
       H3(k,1) = H3(k,1)+hr1d(d)*exp(-i*2*pi*(k-1)*df*delay(d)); %K*1
       H4(k,1) = H4(k,1)+hr2d(d)*exp(-i*2*pi*(k-1)*df*delay(d)); %K*1
    end
    end

%the received signal at the relays ysr1 = sqrt(P1)*x.*H1+sqrt(N_0/2)*(randn(2,1)+i*ra
ndn(2,1));
    %detection

        ysr2 = sqrt(P1)*x.*H2+sqrt(N_0/2)*(randn(2,1)+i*randn(2,1));
    %detection

    yr1_dem = demodMLD(ysr1,table,Q,H1);
    yr2_dem = demodMLD(ysr2,table,Q,H2);
```

```
%verify which relay can decode
%............................................

   if yr1_dem = =x %both relays decodes correctly
   if yr2_dem = = x

     Cr = (x.''V2).';
     H = zeros(L'N,1);
     H(1,1) = H3(1,1);
     H(2,1) = H4(2,1);
     yd = sqrt(P2)'Cr.'H+sqrt(N_0/2)'(randn(L'N,1)+i'randn(L'N,1));

   %detection

     dist = inf;
     for hh = 1:Q^2
     kk = hh;
     for ii = 1:2
     num(ii,1) = rem(kk,Q);
     kk = floor(kk/Q);
     end
     %form one possible codeword
     for jj = 1:2
     D_poss(jj,1) = table(num(jj)+1);
     end

     C_poss = ((D_poss.')'V2).';
     R_poss = sqrt(P2)'H.'C_poss;

     distance = norm(yd-R_poss,2);

     if distance<dist
     dist = distance;
     decision = D_poss;
     end
     end
   SER = SER+sum(decision~ = x);

   else %relay 1 decode correctly and R2 not
     Cr = (x.''V2).';

yd = sqrt(P2)'Cr(1,1).'H3(1,1)+sqrt(N_0/2)'(randn(1,1)+i'randn(1,1));

   %detection

     dist = inf;
     for hh = 1:Q^2
     kk = hh;
     for ii = 1:2
     num(ii,1) = rem(kk,Q);
     kk = floor(kk/Q);
     end
     %form one possible codeword
     for jj = 1:2
     D_poss(jj,1) = table(num(jj)+1);
     end

     C_poss = ((D_poss.')'V2).';
     R_poss = sqrt(P2)'H3(1,1).'C_poss(1,1);

     distance = norm(yd-R_poss,2);

     if distance<dist
     dist = distance;
     decision = D_poss;
     end
     end
   SER = SER+sum(decision~ = x);

   end

   else
   if yr2_dem = =x %relay 2 decodes correctly and the first not
     Cr = (x.''V2).';

     yd = sqrt(P2)'Cr(2,1).'H4(2,1)+sqrt(N_0/2)'(randn(1,1)+i'randn(1,1));
```

```
%detection
   dist = inf;
   for hh = 1:Q^2
   kk = hh;
   for ii = 1:2
   num(ii,1) = rem(kk,Q);
   kk = floor(kk/Q);
   end
   %form one possible codeword
   for jj = 1:2
   D_poss(jj,1) = table(num(jj)+1);
   end

   C_poss = ((D_poss.')'V2).';
   R_poss = sqrt(P2)'H4(2,1).'C_poss(2,1);

   distance = norm(yd-R_poss,2);

   if distance<dist
   dist = distance;
   decision = D_poss;
   end
   end
   SER = SER+sum(decision~ = x);

   else
      SER = SER+2;
   end
   end
%end
end
SEER(j) = SER/(no_pkts(j)'2)
end
semilogy(SNR,SEER)

% = = = = = = = = = = = = = = = = = = = = = = = = = =
% = = = = = = = = = = = = = = = = = = = = = = = = = =
% = = = = = = = = = = = = = = = = = = = = = = = = = =
% vandermonde Matrix
% = = = = = = = = = = = = = = = = = = = = = = = = = =

function V=vandermonde(L)
V=zeros(L);

for column=1:L
   for row=1:L
     V(row,coloumn)=(exp(i*(4*column-3)/(2*L)*pi))^(row-1);
   end
end

function [out] = demodMLD(y,table,Q,H)

%this function perform MLD for input QPSK symbols
%y = received signal
%table = transmitted symbols
%Q = no. of constellation points

for m = 1:Q
d(:,m) = (abs(y-table(m).'H)).^2;
end

[i1 i2] = min(d,[],2);
out = table(i2).';

% = = = = = = = = = = = = = = = = = = = = = = = = =
% = = = = = = = = = = = = = = = = = = = = = = = = =

   %SER for DSFC,BPSK,two relays,
%example of frequency selective channel of 2 relays and 2 paths
%clear all;
%clc
function SEER = test2r2p()
% = = = = = = = = = = = = = = = = = = = = = = = = =
N = 2; %number of relays
L = 2; %number of paths
delay = [0 .5'10^-6]; %2 rays
% ^ = = = = = = = = = = = = = = = = = = = = = = = = =
```

```
Q = 2; %constellation size
modulation = 'BPSK';
BW = 20*10^6; %system BW
no_pkts = [10^3 10^3 10^4 10^4 10^5 10^5 10^6]; %number of simulated packets
K = 64; %number of subcarriers
df = BW/K; %the subcarrier frequency spacing

M = K/L; %number of source subblocks
G = K/(L*N); %number of groups at the relay

SNR = 0:4:24;

V1 = sqrt(1/L)*vandermonde(L); % vandermonde matrix at the source node
V2 = sqrt(1/4)*vandermonde(4) % vandermonde matrix at the relay nodes

P = 2; %the total transmitted power
P1 = 0.5*P; %source transmitted power
P2 = P-P1; %all relays transmitted power
%·············································

for j = 1:length(SNR)

N_0 = P*10^(-SNR(j)/10);
SER = 0;

for i1 = 1:no_pkts(j)
  clear D_poss
  clear C_poss

%the data generation

   tx_bits = randint(1,L*N*log2(Q)); %bits required to transmit 2 symbols
   [mod_symbols table] = tx_modulate(tx_bits,modulation);
   x = mod_symbols.'; %column vector of the generated symbols(K*1)

%channel coefficients (S- >R)
%·····································

   hsr1 = multiray(1); hsr2 = multiray(1);%sqrt(1/4)*(randn(1,L)+i*randn(1,L));

   H1 = zeros(4,1);
   H2 = zeros(4,1);

   for k = 1:4

   for d = 1:length(delay)
      H1(k,1) = H1(k,1)+hsr1(d)*exp(-i*2*pi*(k-1)*df*delay(d)); %K*1
      H2(k,1) = H2(k,1)+hsr2(d)*exp(-i*2*pi*(k-1)*df*delay(d)); %K*1
   end
   end

%channel coefficients (R- >D)
%·····································
% hr1d = multiray(1); sqrt(1/4)*(randn(1,L)+i*randn(1,L)); %zero mean and unit variance
%hr2d = multiray(1); sqrt(1/4)*(randn(1,L)+i*randn(1,L));

   H3 = zeros(4,1);
   H4 = zeros(4,1);

   for k = 1:4

   for d = 1:length(delay)
      H3(k,1) = H3(k,1)+hr1d(d)*exp(-i*2*pi*(k-1)*df*delay(d)); %K*1
      H4(k,1) = H4(k,1)+hr2d(d)*exp(-i*2*pi*(k-1)*df*delay(d)); %K*1
   end
   end

%code construction at the source node
%·······················································
   for m = 1:L
   C(2*m-1:2*m,1) = ((x(2*m-1:2*m,1).')*V1).';
   end

%the received signal at the relays
   ysr1 = sqrt(P1)*C.*H1+sqrt(N_0/2)*(randn(L*N,1)+i*randn(L*N,1));
   yr1_dem = ML_s_2r2p(ysr1,Q,L,H1,V1,M,table,P1);

   ysr2 = sqrt(P1)*C.*H2+sqrt(N_0/2)*(randn(L*N,1)+i*randn(L*N,1));
   yr2_dem = ML_s_2r2p(ysr2,Q,L,H2,V2,M,table,P1);
```

```
%verify which relay can decode
%•••••••••••••••••••••••••••••••••••••••••••••••••••
    if yr1_dem = =x
    if yr2_dem = =x %bot relays decode correctly

        Cr = (x.''V2).';
        H = zeros(L'N,1);
        H(1:2:4) = H3(1:2:4);
        H(2:2:4) = H4(2:2:4);
        yd = sqrt(P2)'Cr.'H+sqrt(N_0/2)'(randn(L'N,1)+i'randn(L'N,1));

    %detection

        dist = inf;
        for hh = 1:Q^4
        kk = hh;
        for ii = 1:4
        num(ii,1) = rem(kk,Q);
        kk = floor(kk/Q);
        end
        %form one possible codeword
        for jj = 1:4
        D_poss(jj,1) = table(num(jj)+1);
        end

        C_poss = ((D_poss.')'V2).';
        R_poss = sqrt(P2)'H.'C_poss;

        distance = norm(R_poss-yd);

        if distance<dist
        dist = distance;
        decision = D_poss;
        end
        end
    SER = SER+sum(decision~ = x);

    else %relay 1 decodes correctly and the second not

        Cr = (x.''V2).';
        H = zeros(L'N,1);
        H(1:2:end) = H3(1:2:end);
        H(2:2:end) = H4(2:2:end);
        yd = sqrt(P2)'Cr(1:2:4).'H(1:2:4)+sqrt(N_0/2)'(randn(L,1)+i'randn(L,1));

    %detection

        dist = inf;
        for hh = 1:Q^4
        kk = hh;
        for ii = 1:4
        num(ii,1) = rem(kk,Q);
        kk = floor(kk/Q);
        end
        %form one possible codeword
        for jj = 1:4
        D_poss(jj,1) = table(num(jj)+1);
        end

        C_poss = ((D_poss.')'V2).';
        R_poss = sqrt(P2)'H(1:2:4).'C_poss(1:2:4);

        distance = norm(R_poss-yd);

        if distance<dist
        dist = distance;
        decision = D_poss;
        end
        end
    SER = SER+sum(decision~ = x);

    end

    else
    if yr2_dem = =x %second relay decodes correctly and first not
        Cr = (x.''V2).';
        H = zeros(L'N,1);
        H(1:2:end) = H3(1:2:end);
```

```
    H(2:2:end) = H4(2:2:end);
    yd = sqrt(P2)*Cr(2:2:4).*H(2:2:4)+sqrt(N_0/2)*(randn(L,1)+i*randn(L,1));

%detection

    dist = inf;
    for hh = 1:Q^4
    kk = hh;
    for ii = 1:4
    num(ii,1) = rem(kk,Q);
    kk = floor(kk/Q);
    end
    %form one possible codeword
    for jj = 1:4
    D_poss(jj,1) = table(num(jj)+1);
    end

    C_poss = ((D_poss.')*V2).';
    R_poss = sqrt(P2)*H(2:2:4).*C_poss(2:2:4);

    distance = norm(R_poss-yd);

    if distance<dist
    dist = distance;
    decision = D_poss;
    end
    end
    SER = SER+sum(decision~ = x);

    else
      SER = SER+4;
    end
    end

end

SEER(j) = SER/(no_pkts(j)*L*N);
end
%semilogy(SNR,SEER)

% = = = = = = = = = = = = = = = = = = = = = = = = = = = =
% = = = = = = = = = = = = = = = = = = = = = = = = = = = =
```

G.3   MATLAB® Code for SC–FDMA Chapter

```
%DAF DSTC frequency domain
%4 cases at according to the states of the relays
%all channel variance = 1

function SEER = DAF()

%clear all;
%clc;

sim_options = read_options;

%map simulation options
modulation = sim_options.modulation;
channel_model = sim_options.channel_model;
fft_length = sim_options.fft_length;

Nt = sim_options.Tx_antenna;
Nr = sim_options.Rx_antenna;
CP = sim_options.CP;
M = sim_options.block_size;
Q1 = fft_length/M;

%no_pkts = sim_options.no_pkts;
no_pkts = [10^4 10^4 10^4 10^4 5*10^4 5*10^4 10^5];

% SNR up to 30 dBs
SNR = 0:4:24;
```

```
% = = = = = = = = = = = = = = = = = = =
%Main simulation Loop
% = = = = = = = = = = = = = = = = = = =

switch (modulation)
   case 'BPSK'
      BITS = 1;
      Q = 2;
   case 'QPSK'
      BITS = 2;
      Q = 4;
   case '8PSK'
      BITS = 3;
      Q = 8;
   case '16QAM'
      BITS = 4;
      Q = 16;
   case '64QAM'
      BITS = 16;
      Q = 64
end

for j = 1:length(SNR)

SER = 0;
P = 10^(SNR(j)/(10));
P1 = P/2;
P2 = P-P1;
N_0 = 1;

for packet_count = 1:no_pkts(j)

%·····················································
%Transmitter
%·····················································

%generate the information bits

   inf_bits = randint(2,M*BITS);

   %Modulation
   temp = [];
   for k = 1:Nt
   [mod_symbols table Q] = tx_modulate(inf_bits(k,:),modulation);
   temp = [temp; mod_symbols];
   end
   x = temp.'; %M*2 matrix

   X1 = fft(x);

   X2 = [-conj(X1(:,2)) conj(X1(:,1))];
%·····················································
%received signals at the relay node
%·····················································
cir12 = sqrt(1)*get_channel_ir(sim_options);
cir21 = sqrt(1)*get_channel_ir(sim_options);

hF12 = fft(cir12,fft_length);
hF21 = fft(cir21,fft_length);

H12 = hF12(1:M).';
H21 = hF21(1:M).';

Rr1 = sqrt(P1)*H12.*X1(:,1)+fft(sqrt(N_0*.5)*(randn(M,1)+i*randn(M,1)));
Rr2 = sqrt(P1)*H21.*X1(:,2)+fft(sqrt(N_0*.5)*(randn(M,1)+i*randn(M,1)));

%Equalization
C1 = (conj(H12)./(conj(H12).*H12+N_0/P1));
C2 = (conj(H21)./(conj(H21).*H21+N_0/P1));

%C = 1./H2;

Rr1 = Rr1.*C1;
Rr2 = Rr2.*C2;
```

```
rr1 = ifft(Rr1);
rr2 = ifft(Rr2);

%demodulation
r1_dem = demodMLD(rr1,table,Q);
r2_dem = demodMLD(rr2,table,Q);

I1 = 0;
I2 = 0;
if r1_dem = =x(:,1)
    I1 = 1;
end

if r2_dem = =x(:,2)
    I2 = 1;
end

%⋯⋯⋯⋯⋯⋯⋯⋯⋯⋯⋯⋯⋯⋯⋯⋯⋯⋯⋯⋯⋯⋯
%channel
%⋯⋯⋯⋯⋯⋯⋯⋯⋯⋯⋯⋯⋯⋯⋯⋯⋯⋯⋯⋯⋯⋯
    cir1 = get_channel_ir(sim_options); %S- ->D
    cir3 = get_channel_ir(sim_options); %R- ->D

%find fft for the channel and store it for the receiver
    hF1 = fft(cir1,fft_length);
    hF3 = fft(cir3,fft_length);

    H1 = hF1(1:M).';
    H3 = hF3(1:M).';

%⋯⋯⋯⋯⋯⋯⋯⋯⋯⋯⋯⋯⋯⋯⋯⋯⋯⋯⋯⋯⋯⋯
%Recieved signal at destination
%⋯⋯⋯⋯⋯⋯⋯⋯⋯⋯⋯⋯⋯⋯⋯⋯⋯⋯⋯⋯⋯⋯

%recieved at destination

    if I1 = =1 && I2 = =1
    Rd1 = sqrt(P1)*X1(:,1).*H1+sqrt(P1)*X1(:,2).*H3+fft(sqrt(N_0*.5)*(randn(M,1)+i*randn
    (M,1)))+fft(sqrt(N_0*.5)*(randn(M,1)+i*randn(M,1)));
    Rd2 = sqrt(P1)*X2(:,1).*H1+sqrt(P1)*X2(:,2).*H3+fft(sqrt(N_0*.5)*(randn(M,1)+i*rand
    n(M,1)));

    Z1 = (Rd1.*conj(H1)+conj(Rd2).*H3).';%s0
    Z2 = (Rd1.*conj(H3)-conj(Rd2).*H1).';%s1
    end

    if I1 = =1 && I2 = =0 %X1(:,1) received correct at the relay and X1(:,2) not
    Rd1 = sqrt(P1)*X1(:,1).*H1+fft(sqrt(N_0*.5)*(randn(M,1)+i*randn(M,1)));
    Rd2 = sqrt(P1)*X1(:,2).*H3+fft(sqrt(N_0*.5)*(randn(M,1)+i*randn(M,1)));
    Rd3 = sqrt(P1)*conj(X1(:,1)).*H3+fft(sqrt(N_0*.5)*(randn(M,1)+i*randn(M,1)));

    Z1 = (Rd1.*conj(H1)+conj(Rd3).*H3).';%s0
    Z2 = (Rd2.*conj(H3)).'; %s1
    end

    if I1 = =0 && I2 = =1 %X1(:,2) received correct at the relay and X1(:,1) not
    Rd1 = sqrt(P1)*X1(:,1).*H1+fft(sqrt(N_0*.5)*(randn(M,1)+i*randn(M,1)));
    Rd2 = sqrt(P1)*X1(:,2).*H3+fft(sqrt(N_0*.5)*(randn(M,1)+i*randn(M,1)));
    Rd3 = -sqrt(P1)*conj(X1(:,2)).*H1+fft(sqrt(N_0*.5)*(randn(M,1)+i*randn(M,1)));

    Z1 = (Rd1.*conj(H1)).'; %s0
    Z2 = (Rd2.*conj(H3)-conj(Rd3).*H1).'; %s1
    end

    if I1 = =0 && I2 = =0 %X1(:,2) received correct at the relay and X1(:,1) not
    Rd1 = sqrt(P1)*X1(:,1).*H1+fft(sqrt(N_0*.5)*(randn(M,1)+i*randn(M,1)));
    Rd2 = sqrt(P1)*X1(:,2).*H3+fft(sqrt(N_0*.5)*(randn(M,1)+i*randn(M,1)));

    Z1 = (Rd1.*conj(H1)).'; %s0
    Z2 = (Rd2.*conj(H3)).'; %s1
    end

z1 = ifft(Z1);
z2 = ifft(Z2);
```

```
% MLD
  for m = 1:Q
  d1(:,m) = abs(z1-table(m)).^2;
  d2(:,m) = abs(z2-table(m)).^2;
  end

%decision for detecting s1 (coded)
[y1 i1] = min(d1,[],2);
s1d = table(i1).';
clear d1

%decision for detecting s2 (coded)
[y2 i2] = min(d2,[],2);
s2d = table(i2).';
clear d2;

xd = [s1d s2d];
%demodulation

  SER = SER+sum(sum(xd~ = x));

end
  SEER(j) = SER/(2*M*no_pkts(j))
  BER(j) = SEER(j)/log2(Q);
end

%semilogy(SNR,SEER,'b');

% = = = = = = = = = = = = = = = = = = = = = = = =
% = = = = = = = = = = = = = = = = = = = = = = = =

%DAF DSTC frequency domain
%Classical scheme

function SEER = DAF_class()

%clear all;
%clc;

sim_options = read_options;

%map simulation options
modulation = sim_options.modulation;
channel_model = sim_options.channel_model;
fft_length = sim_options.fft_length;

Nt = sim_options.Tx_antenna;
Nr = sim_options.Rx_antenna;
CP = sim_options.CP;
M = sim_options.block_size;
Q1 = fft_length/M;

%no_pkts = sim_options.no_pkts;
no_pkts = [10^4 10^4 10^4 10^4 5*10^4 5*10^4 10^5];

% SNR up to 30 dBs
SNR = 0:4:24;

% = = = = = = = = = = = = = = = = = = = = = = = =
%Main simulation Loop
% = = = = = = = = = = = = = = = = = = = = = = = =

switch (modulation)
  case 'BPSK'
    BITS = 1;
    Q = 2;
  case 'QPSK'
    BITS = 2;
    Q = 4;
  case '8PSK'
    BITS = 3;
    Q = 8;
  case '16QAM'
    BITS = 4;
    Q = 16;
```

```
   case '64QAM'
      BITS = 16;
      Q = 64
end

for j = 1:length(SNR)

SER = 0;
P = 10^(SNR(j)/(10));
P1 = P/2;
P2 = P-P1;
N_0 = 1;

for packet_count = 1:no_pkts(j)

%••••••••••••••••••••••••••••••••••••••••••••••••••••••••
%Transmitter
%••••••••••••••••••••••••••••••••••••••••••••••••••••••••

%generate the information bits

   inf_bits = randint(2,M*BITS);
   %Modulation
   temp = [];
   for k = 1:Nt
   [mod_symbols table Q] = tx_modulate(inf_bits(k,:),modulation);
   temp = [temp; mod_symbols];
   end
   x = temp.'; %M*2 matrix

   % fft- >mapping- >ifft- >add CP

   X1 = fft(x);

   X2 = [-conj(X1(:,2)) conj(X1(:,1))];

%••••••••••••••••••••••••••••••••••••••••••••••••••••••••
%received signals at the relay node
%••••••••••••••••••••••••••••••••••••••••••••••••••••••••
cir12 = sqrt(1)*get_channel_ir(sim_options);
cir21 = sqrt(1)*get_channel_ir(sim_options);

hF12 = fft(cir12,fft_length);
hF21 = fft(cir21,fft_length);

H12 = hF12(1:M).';
H21 = hF21(1:M).';

Rr1 = sqrt(P1)*H12.*X1(:,1)+fft(sqrt(N_0*.5)*(randn(M,1)+i*randn(M,1)));
Rr2 = sqrt(P1)*H21.*X1(:,2)+fft(sqrt(N_0*.5)*(randn(M,1)+i*randn(M,1)));

%Equalization
C1 = (conj(H12)./(conj(H12).*H12+N_0/P1));
C2 = (conj(H21)./(conj(H21).*H21+N_0/P1));

%C = 1./H2;

Rr1 = Rr1.*C1;
Rr2 = Rr2.*C2;

rr1 = ifft(Rr1);
rr2 = ifft(Rr2);

%demodulation
r1_dem = demodMLD(rr1,table,Q);
r2_dem = demodMLD(rr2,table,Q);

I1 = 0;
I2 = 0;
if r1_dem = =x(:,1)
   I1 = 1;
end

if r2_dem = =x(:,2)
   I2 = 1;
end
```

```
%·····················································
%channel S- >D, R- >D
%·····················································

   cir1 = get_channel_ir(sim_options); %S- ->D
   cir3 = get_channel_ir(sim_options); %R- ->D

%find fft for the channel and store it for the receiver
   hF1 = fft(cir1,fft_length);
   hF3 = fft(cir3,fft_length);

   H1 = hF1(1:M).';
   H3 = hF3(1:M).';

%·····················································
%Recieved signal at destination
%·····················································

%recieved at destination

   if I1 = =1 && I2 = =1
   Rd1 = sqrt(P1)'X1(:,1).'H1+sqrt(P1)'X1(:,2).'H3+fft(sqrt(N_0'.5)'(randn(M,1)+i'rand
   n(M,1)));
   Rd2 = -sqrt(P1)'conj(X1(:,2)).'H1+sqrt(P1)'conj(X1(:,1)).'H3+fft(sqrt(N_0'.5)'(randn(M
   ,1)+i'randn(M,1)));
   Z1 = (Rd1.'conj(H1)+conj(Rd2).'H3).';%s0
   Z2 = (Rd1.'conj(H3)-conj(Rd2).'H1).';%s1
   end

   if I1 = =0 | I2 = =0 %X1(:,2) received correct at the relay and X1(:,1) not
   Rd1 = sqrt(P1)'X1(:,1).'H1+fft(sqrt(N_0'.5)'(randn(M,1)+i'randn(M,1)));
   Rd2 = sqrt(P1)'X1(:,2).'H3+fft(sqrt(N_0'.5)'(randn(M,1)+i'randn(M,1)));

   Z1 = (Rd1.'conj(H1)).'; %s0
   Z2 = (Rd2.'conj(H3)).'; %s1
   end

z1 = ifft(Z1);
z2 = ifft(Z2);
% MLD
   for m = 1:Q
   d1(:,m) = abs(z1-table(m)).^2;
   d2(:,m) = abs(z2-table(m)).^2;
   end

%decision for detecting s1 (coded)
[y1 i1] = min(d1,[],2);
s1d = table(i1).';
clear d1

%decision for detecting s2 (coded)
[y2 i2] = min(d2,[],2);
s2d = table(i2).';
clear d2;

xd = [s1d s2d];

%demodulation

   SER = SER+sum(sum(xd~ = x));
end
   SEER(j) = SER/(2'M'no_pkts(j))
   BER(j) = SEER(j)/log2(Q);
end

%semilogy(SNR,SEER,'b');

% = = = = = = = = = = = = = = = = = = = = = = = = =
% = = = = = = = = = = = = = = = = = = = = = = = = =

   function cir = get_channel_ir(sim_options)

td_max = sim_options.td_max;
f_samp = sim_options.sampling_freq;
K = sim_options.kfactor;
```

```
channel_model = sim_options.channel_model;

if ~isempty(findstr(channel_model,'Rayleigh'))
if td_max = =0
  Kmax = 0;
  vark = 1;
else
%calculate the exponential decay envelope
  Kmax = ceil(10*td_max*f_samp); %no of channel paths
  var0 = (1-exp(-1/(f_samp*td_max)))/(1-exp(-1*((Kmax+1)*f_samp)/td_max));
  k = 0:Kmax;
  env = var0*exp(-k/(f_samp*td_max));
end
std_dev = sqrt(env/2);
cir = std_dev.*(randn(1,Kmax+1)+i*randn(1,Kmax+1));
end

% = = = = = = = = = = = = = = = = = = = = = = = = = = = =
% = = = = = = = = = = = = = = = = = = = = = = = = = = = =

%MIMO 2*2 frequency domain
%clear all;
%clc;
function SEER = MIMO()
sim_options = read_options;

%map simulation options
modulation = sim_options.modulation;
channel_model = sim_options.channel_model;
fft_length = sim_options.fft_length;
Nt = sim_options.Tx_antenna;
Nr = sim_options.Rx_antenna;
CP = sim_options.CP;
M = sim_options.block_size;
Q1 = fft_length/M;

%no_pkts = sim_options.no_pkts;
no_pkts = [10^4 10^4 10^4 10^4 10^4 10^5 10^5];

SNR = 0:4:24;

% = = = = = = = = = = = = = = = = = = = = = = = = = = = =
%Main simulation Loop
% = = = = = = = = = = = = = = = = = = = = = = = = = = = =

switch (modulation)
  case 'BPSK'
    BITS = 1;
    Q = 2;
  case 'QPSK'
    BITS = 2;
    Q = 4;
  case '8PSK'
    BITS = 3;
    Q = 8;
  case '16QAM'
    BITS = 4;
    Q = 16;
  case '64QAM'
    BITS = 16;
    Q = 64
end

P = 2;
P1 = P/2;
P2 = P-P1;

for j = 1:length(SNR)

SER = 0;
N_0 = 10^(-SNR(j)/(10));

for packet_count = 1:no_pkts(j)
```

```
%**********************************************************
%Transmitter
%**********************************************************

%generate the information bits
   inf_bits = randint(2,M*BITS);

   %Modulation
   temp = [];
   for k = 1:Nt
   [mod_symbols table Q] = tx_modulate(inf_bits(k,:),modulation);
   temp = [temp; mod_symbols];
   end
   x = temp.'; %M*2 matrix

   X1 = fft(x);

   X2 = [-conj(X1(:,2)) conj(X1(:,1))];

%**********************************************************
%recieved signals at the relay
%**********************************************************

   cir2 = get_channel_ir(sim_options);
   hF2 = fft(cir2,fft_length);
   H2 = hF2(1:M).';

%**********************************************************
%channel S- >D, R- >D
%**********************************************************

   cir1 = get_channel_ir(sim_options); %S- ->D
   cir3 = get_channel_ir(sim_options); %R- ->D
   cir4 = get_channel_ir(sim_options); %S- ->D
   cir5 = get_channel_ir(sim_options); %R- ->D

%find fft for the channel and store it for the receiver
   hF1 = fft(cir1,fft_length);
   hF3 = fft(cir3,fft_length);
   hF4 = fft(cir4,fft_length);
   hF5 = fft(cir5,fft_length);

   H1 = hF1(1:M).';
   H3 = hF3(1:M).';
   H4 = hF4(1:M).';
   H5 = hF5(1:M).';

%**********************************************************
%Recieved signal at destination
%**********************************************************

%recieved at destination

R1 = sqrt(1/2).*X1(:,1).*H1+sqrt(1/2).*X1(:,2).*H3+fft(sqrt(N_0*.5)*(randn(M,1)+i*rand
n(M,1)));
   R2 = sqrt(1/2).*X2(:,1).*H1+sqrt(1/2).*X2(:,2).*H3+fft(sqrt(N_0*.5)*(randn(M,1)+i*rand
   n(M,1)));

   R3 = sqrt(1/2).*X1(:,1).*H4+sqrt(1/2).*X1(:,2).*H5+fft(sqrt(N_0*.5)*(randn(M,1)+i*rand
   n(M,1)));
   R4 = sqrt(1/2).*X2(:,1).*H4+sqrt(1/2).*X2(:,2).*H5+fft(sqrt(N_0*.5)*(randn(M,1)+i*rand
   n(M,1)));

%Alamouti Combining

Z1 = (R1.*conj(H1)+conj(R2).*H3+R3.*conj(H4)+conj(R4).*H5).';%s0
Z2 = (R1.*conj(H3)-conj(R2).*H1+R3.*conj(H5)-conj(R4).*H4).';%s1

z1 = ifft(Z1);
z2 = ifft(Z2);

% MLD
   for m = 1:Q
   d1(:,m) = abs(z1-table(m)).^2;
   d2(:,m) = abs(z2-table(m)).^2;
   end
```

```
%decision for detecting s1 (coded)
[y1 i1] = min(d1,[],2);
s1d = table(i1).';
clear d1

%decision for detecting s2 (coded)
[y2 i2] = min(d2,[],2);
s2d = table(i2).';
clear d2;

xd = [s1d s2d];
%demodulation

   SER = SER+sum(sum(xd~ = x));

end
   SEER(j) = SER/(2*M*no_pkts(j))
   BER(j) = SEER(j)/log2(Q);
end

%semilogy(SNR,SEER,'b');

% = = = = = = = = = = = = = = = = = = = = = = = = =

%This is the main program. SC_FDE transmitter and receiver (1*1)
%It starts reading the simulation options from the function 'read_options'.
%clear all;
%clc;
function SEER = SISO()
sim_options = read_options;

%map simulation options
modulation = sim_options.modulation;
channel_model = sim_options.channel_model;
fft_length = sim_options.fft_length;

Nt = sim_options.Tx_antenna;
Nr = sim_options.Rx_antenna;
CP = sim_options.CP;
M = sim_options.block_size;
Q1 = fft_length/M;

%no_pkts = sim_options.no_pkts;
no_pkts = [10^4 10^4 10^4 10^4 10^4 10^4 5*10^4];

% SNR up to 30 dBs
SNR = 0:4:24;

% = = = = = = = = = = = = = = = = = = = = = = = = =
%Main simulation Loop
% = = = = = = = = = = = = = = = = = = = = = = = = =
switch (modulation)
   case 'BPSK'
     BITS = 1;
     Q = 2;
   case 'QPSK'
     BITS = 2;
     Q = 4;
   case '8PSK'
     BITS = 3;
     Q = 8;
   case '16QAM'
     BITS = 4;
     Q = 16;
   case '64QAM'
     BITS = 16;
     Q = 64
end

for j = 1:length(SNR)
% N_0 = 1/10^(SNR/10);
SER = 0;

for packet_count = 1:no_pkts(j)
```

```
%''''''''''''''''''''''''''''''''''''''''''''''''''''''''''
%Transmitter
%''''''''''''''''''''''''''''''''''''''''''''''''''''''''''

%generate the information bits

    inf_bits = randint(1,M*BITS);

    %Modulation
    [mod_symbols table Q] = tx_modulate(inf_bits,modulation);
    x = mod_symbols.';
    % fft- >mapping- >ifft- >add CP
    X1 = fft(x);

%''''''''''''''''''''''''''''''''''''''''''''''''''''''''''
%channel
%''''''''''''''''''''''''''''''''''''''''''''''''''''''''''

    cir = get_channel_ir(sim_options);

%find fft for the channel and store it for the receiver
    hF = fft(cir,fft_length);
    H1 = hF(1:M).';

%additive noise
    N_0 = 10^(-SNR(j)/(10));
    noise = fft(sqrt(N_0*.5)*(randn(M,1)+i*randn(M,1)));

    %''''''''''''''''''''''''''''''''''''''''''''''''''''''''''''''''''''

    R = H1.*X1+noise;
%Y = Y./hF;

    H_eff = hF(1:M);
    C = (conj(H1)./(conj(H1).*H1+10^(-SNR(j)/10)));

    R = R.*C;
%Y1 = Y1./H_eff;

    r = ifft(R);

%demodulation
    y_dem = demodMLD(r,table,Q);
    SER = SER+sum(y_dem~ = x);

end
    SEER(j) = SER/(M*no_pkts(j))
    BER(j) = SEER(j)/log2(Q);
end

%semilogy(SNR,SEER,'b');

% = = = = = = = = = = = = = = = = = = = = = = = = = = =

function sim_options = read_options

sim_options = struct('no_pkts',{10000},…
        'modulation', {'QPSK'},…
        'channel_model',{'Rayleigh'},…
        'Tx_antenna',{2},…
        'Rx_antenna',{1},…
          'fft_length',{512},…
          'block_size',{16},…
          'CP',{16},… %if zeropad = 0 = = > datacarriers = fftpoints
        'zeropad',{0},…
        'DataCarriers',{64},…
        'VBLAST',{0},…
          'TxDiv',{0},…
          'sampling_freq',{20*10^6},…
        'td_max',{50*10^-9},…
          'kfactor',{[]})

% = = = = = = = = = = = = = = = = = = = = = = = = = = =
% = = = = = = = = = = = = = = = = = = = = = = = = = = =
```

```
%MISO 2'1 frequency domain
%clear all;

%clc;
function SEER = MISO()
sim_options = read_options;

%map simulation options
modulation = sim_options.modulation;
channel_model = sim_options.channel_model;
fft_length = sim_options.fft_length;

Nt = sim_options.Tx_antenna;
Nr = sim_options.Rx_antenna;
CP = sim_options.CP;
M = sim_options.block_size;
Q1 = fft_length/M;

%no_pkts = sim_options.no_pkts;
no_pkts = [10^4 10^4 10^4 10^4 5'10^4 5'10^4 5'10^5];

% SNR up to 30 dBs
SNR = 0:4:24;

% = = = = = = = = = = = = = = = = = = = = = = = = = = =
%Main simulation Loop
% = = = = = = = = = = = = = = = = = = = = = = = = = = =

switch (modulation)
  case 'BPSK'
    BITS = 1;
    Q = 2;
  case 'QPSK'
    BITS = 2;
    Q = 4;
  case '8PSK'
    BITS = 3;
    Q = 8;
  case '16QAM'
    BITS = 4;
    Q = 16;
  case '64QAM'
    BITS = 16;
    Q = 64
end

for j = 1:length(SNR)

SER = 0;
P = 10^(SNR(j)/(10));
P1 = P/2;
P2 = P-P1;
N_0 = 1;

for packet_count = 1:no_pkts(j)

%••••••••••••••••••••••••••••••••••••••••••••••••••••••••••••
%Transmitter
%••••••••••••••••••••••••••••••••••••••••••••••••••••••••••••

%generate the information bits

  inf_bits = randint(2,M'BITS);

  %Modulation
  temp = [];
  for k = 1:Nt
  [mod_symbols table Q] = tx_modulate(inf_bits(k,:),modulation);
  temp = [temp; mod_symbols];
  end
  x = temp.'; %M'2 matrix

  X1 = fft(x);

  X2 = [-conj(X1(:,2)) conj(X1(:,1))];
```

```
%·····················································
%received signals at the relay
%·····················································

  cir2 = get_channel_ir(sim_options);
  hF2 = fft(cir2,fft_length);
  H2 = hF2(1:M).';
%·····················································
%channel S- >D, R- >D
%·····················································

  cir1 = get_channel_ir(sim_options); %S- ->D
  cir3 = get_channel_ir(sim_options); %R- ->D

%find fft for the channel and store it for the receiver
  hF1 = fft(cir1,fft_length);
  hF3 = fft(cir3,fft_length);

  H1 = hF1(1:M).';
  H3 = hF3(1:M).';

%·····················································
%Received signal at destination
%·····················································

%received at destination
  R1 = sqrt(P1).*X1(:,1).*H1+sqrt(P1).*X1(:,2).*H3+fft(sqrt(N_0*.5)*(randn(M,1)+i*rand
  n(M,1)));
  R2 = sqrt(P1).*X2(:,1).*H1+sqrt(P1).*X2(:,2).*H3+fft(sqrt(N_0*.5)*(randn(M,1)+i*rand
  n(M,1)));

%Alamouti Combining

Z1 = (R1.*conj(H1)+conj(R2).*H3).';%s0
Z2 = (R1.*conj(H3)-conj(R2).*H1).';%s1

z1 = ifft(Z1);
z2 = ifft(Z2);
% MLD
  for m = 1:Q

  d1(:,m) = abs(z1-table(m)).^2;
  d2(:,m) = abs(z2-table(m)).^2;
  end

%decision for detecting s1 (coded)
[y1 i1] = min(d1,[],2);
s1d = table(i1).';
clear d1

%decision for detecting s2 (coded)
[y2 i2] = min(d2,[],2);
s2d = table(i2).';
clear d2;

xd = [s1d s2d];

%demodulation

  SER = SER+sum(sum(xd~ = x));

end
  SEER(j) = SER/(2*M*no_pkts(j))
  BER(j) = SEER(j)/log2(Q);
end

%semilogy(SNR,SEER,'b');

% = = = = = = = = = = = = = = = = = = = = = = = = = =
% = = = = = = = = = = = = = = = = = = = = = = = = = =

%example of different number of subcarriers comparison

%clear all;
%clc;
```

```
function SEER = DAF10_MLD_16()
sim_options = read_options;

%map simulation options
modulation = sim_options.modulation;
channel_model = sim_options.channel_model;
fft_length = sim_options.fft_length;

Nt = sim_options.Tx_antenna;
Nr = sim_options.Rx_antenna;
CP = sim_options.CP;
M = sim_options.block_size;
M = 16;
Q1 = fft_length/M;

%no_pkts = sim_options.no_pkts;
no_pkts = [10^4 10^4 10^4 10^4 10^4 5*10^4 5*10^4];

% SNR up to 30 dBs
SNR = 0:4:24;

% = = = = = = = = = = = = = = = = = = = = = = = = =
%Main simulation Loop
% = = = = = = = = = = = = = = = = = = = = = = = = =

switch (modulation)
   case 'BPSK'
      BITS = 1;
      Q = 2;
   case 'QPSK'
      BITS = 2;
      Q = 4;
   case '8PSK'
      BITS = 3;
      Q = 8;
   case '16QAM'
      BITS = 4;
      Q = 16;
   case '64QAM'
      BITS = 16;
      Q = 64
end

for j = 1:length(SNR)

SER = 0;
P = 10^(SNR(j)/(10));
P1 = P/2;
P2 = P-P1;
N_0 = 1;

for packet_count = 1:no_pkts(j)

%•••••••••••••••••••••••••••••••••••••••••••••••••••••••
%Transmitter
%•••••••••••••••••••••••••••••••••••••••••••••••••••••••

%generate the information bits

   inf_bits = randint(2,M*BITS);

   %Modulation
   temp = [];
   for k = 1:1
   [mod_symbols table Q] = tx_modulate(inf_bits(k,:),modulation);
   temp = [temp; mod_symbols];
   end
   x = temp.'; %M*2 matrix

   % fft- >mapping- >ifft- >add CP

   X1 = fft(x);
```

```
%.........................................................
%received signals at the relay node
%.........................................................

cir2 = sqrt(10)'get_channel_ir(sim_options);
hF2 = fft(cir2,fft_length);
H2 = hF2(1:M).';

Rr1 = sqrt(P1)'H2.'X1(:,1)+fft(sqrt(N_0'.5)'(randn(M,1)+i'randn(M,1)));

H2 = sqrt(P1)'H2;
%Equalization
C = (conj(H2)./(conj(H2).'H2+P1/N_0));
%C = 1./H2;

Rr1 = Rr1.'C;

rr1 = ifft(Rr1);

%demodulation
r1_dem = demodMLD(rr1,table,Q);

I = 0; %relay state
if r1_dem = =x
   I = 1;
end

%...........................................................
%channel S- >D, R- >D
%...........................................................

   cir1 = get_channel_ir(sim_options); %S- ->D
   cir3 = get_channel_ir(sim_options); %R- ->D
%find fft for the channel and store it for the receiver
   hF1 = fft(cir1,fft_length);
   hF3 = fft(cir3,fft_length);

   H1 = hF1(1:M).';
   H3 = hF3(1:M).';

%...........................................................
%Received signal at destination
%...........................................................

   Rd1 = sqrt(P1)'X1(1:M/2).'H1(1:M/2)-sqrt(P1)'conj(X1(M/2+1:M)).'H3(1:M/2).'I+fft(sqrt(
   N_0'.5)'(randn(M/2,1)+i'randn(M/2,1)));
   Rd2 = sqrt(P1)'X1(M/2+1:M).'H1(M/2+1:M)+sqrt(P1)'conj(X1(1:M/2)).'H3(M/2+1:M).'I+fft(s
   qrt(N_0'.5)'(randn(M/2,1)+i'randn(M/2,1)));

%Alamouti Combining

Z1 = (Rd1.'conj(H1(1:M/2))+conj(Rd2).'H3(M/2+1:M).'I).'; %s0

Z2 = (-conj(Rd1).'H3(1:M/2).'I+Rd2.'conj(H1(M/2+1:M))).';%s1

z1 = ifft([Z1 Z2]);
% MLD
   for m = 1:Q
   d1(:,m) = abs(z1-table(m)).^2;

   end
%decision for detecting s1 (coded)
[y1 i1] = min(d1,[],2);
s1d = table(i1).';
clear d1

xd = [s1d];
%demodulation

   SER = SER+sum(sum(xd~ = x));

end
   SEER(j) = SER/(M'no_pkts(j))
   BER(j) = SEER(j)/log2(Q);
end

%semilogy(SNR,SEER,'b');
```

```
% = = = = = = = = = = = = = = = = = = = = = = = = = =
% = = = = = = = = = = = = = = = = = = = = = = = = = =

%Matlab code for protocol II
%clear all;
%clc;
function SEER = DAF5_mmse_protocol2()
sim_options = read_options;

%map simulation options
modulation = sim_options.modulation;
channel_model = sim_options.channel_model;
fft_length = sim_options.fft_length;

Nt = sim_options.Tx_antenna;
Nr = sim_options.Rx_antenna;
CP = sim_options.CP;
M = sim_options.block_size;
Q1 = fft_length/M;

%no_pkts = sim_options.no_pkts;
no_pkts = [10^4 10^4 10^4 10^4 5'10^4 10^5 5'10^5];

% SNR up to 30 dBs
SNR = 0:4:24;

% = = = = = = = = = = = = = = = = = = = = = = = = = =
%Main simulation Loop
% = = = = = = = = = = = = = = = = = = = = = = = = = =
switch (modulation)
    case 'BPSK'
        BITS = 1;
        Q = 2;
    case 'QPSK'
        BITS = 2;
        Q = 4;
    case '8PSK'
        BITS = 3;
        Q = 8;
    case '16QAM'
        BITS = 4;
        Q = 16;
    case '64QAM'
        BITS = 16;
        Q = 64
end

for j = 1:length(SNR)
SER = 0;
P = 10^(SNR(j)/(10));
P1 = P/2;
P2 = P-P1;
N_0 = 1;

for packet_count = 1:no_pkts(j)

%··········································································
%Transmitter
%··········································································

%generate the information bits
    inf_bits = randint(2,M'BITS);

    %Modulation
    temp = [];
    for k = 1:1
    [mod_symbols table Q] = tx_modulate(inf_bits(k,:),modulation);
    temp = [temp; mod_symbols];
    end
    x = temp.'; %M'2 matrix

    % fft- >mapping- >ifft- >add CP

    X1 = (1/sqrt(M))'fft(x);
```

```
%··················································································
%received signal at the destination node in the first phase
%··················································································

cir_sd = sqrt(1)'get_channel_ir(sim_options);
hFsd = fft(cir_sd,fft_length);
Hsd = hFsd(1:M).';

Rd = sqrt(P1)'Hsd.'X1(:,1)+(sqrt(N_0'.5)'(randn(M,1)+i'randn(M,1)));

   Rd = conj((Hsd)).'Rd;

%·····················································
%received signals at the relay node
%·····················································

cir2 = sqrt(5)'get_channel_ir(sim_options);
hF2 = fft(cir2,fft_length);
H2 = hF2(1:M).';

Rr1 = sqrt(P1)'H2.'X1(:,1)+(sqrt(N_0'.5)'(randn(M,1)+i'randn(M,1)));
   H2 = H2'sqrt(P1);

%Equalization
%C = (conj(H2)./(conj(H2).'H2+10^(-SNR(j)/10)));
C = (conj(H2)./(conj(H2).'H2+N_0/P1));
%C = 1./H2;

Rr1 = Rr1.'C;
rr1 = ifft(Rr1);

%demodulation
r1_dem = demodMLD(rr1,table,Q);

I = 0; %relay state
if r1_dem = =x

   I = 1;
end

%·····················································
%channel S- >D, R- >D
%·····················································

   cir1 = get_channel_ir(sim_options); %S- ->D
   cir3 = get_channel_ir(sim_options); %R- ->D

%find fft for the channel and store it for the receiver
   hF1 = fft(cir1,fft_length);
   hF3 = fft(cir3,fft_length);

   H1 = hF1(1:M).';
   H3 = hF3(1:M).';
%·····················································
%Received signal at destination
%·····················································

%received at destination

   X11 = X1(1:M/2);
   X22 = X1(M/2+1:M);

   H11 = diag(H1(1:M/2));
   H12 = diag(H1(M/2+1:M));
   H21 = diag(H3(1:M/2));
   H22 = diag(H3(M/2+1:M));

   yy = conj(Hsd).'Hsd;
   yy(M/2+1:M) = conj(yy(M/2+1:M));

   g1 = [H11 -H21'I; I'conj(H22) conj(H12)];
   g1 = conj(transpose(g1));

   H = g1'[H11 -H21'I; I'conj(H22) conj(H12)]+diag(yy);

   Rd(M/2+1:M) = conj(Rd(M/2+1:M)); %added
```

```
R = sqrt(P1)'[H11 -H21'I; I'conj(H22) conj(H12)]'[X11;conj(X22)]+(sqrt(N_0'.5)'(rand
n(M,1)+i'randn(M,1)));

R = g1'R;

R = R+Rd; %added

w = ones(1,M);
II = diag(w);
RR = inv(conj(transpose(H))'H+(N_0/P)'II)'conj(transpose(H))'R;

RR(M/2+1:M) = conj(RR(M/2+1:M));

z1 = ifft(RR);
% MLD
  for m = 1:Q

  d1(:,m) = abs(z1-table(m)).^2;
  end

%decision for detecting s1 (coded)
[y1 i1] = min(d1,[],2);
s1d = table(i1).';
clear d1

xd = [s1d];
%demodulation

  SER = SER+sum(sum(xd~ = x));

end
  SEER(j) = SER/(M'no_pkts(j))
  BER(j) = SEER(j)/log2(Q);
end

%semilogy(SNR,SEER,'b');

% = = = = = = = = = = = = = = = = = = = = = = = = = = = = = = = = = = = = =
% = = = = = = = = = = = = = = = = = = = = = = = = = = = = = = = = = = = = =

%PAPR
%proposed scheme

function SEER = papr_DAF_prop_relay()

% = = = = = = = = = = = = = = = = = = = = = = = = = = = = = = = = = = = = =

N = 2; %number of relays
L = 2; %number of paths
delay = [0 .5'10^-6]; %2 rays

%^ = = = = = = = = = = = = = = = = = = = = = = = = = =  = = = = = = = = = = =

Q = 4; %constellation size
modulation = 'QPSK';
BW = 20'10^6; %system BW
no_pkts = [10^5]; %number of simulated packets
K = 64; %number of subcarriers
df = BW/K; %the subcarrier frequency spacing
M = K/L; %number of source subblocks
G = K/(L'N); %number of groups at the relay

SNR = 0:4:24;
P = 2; %the total transmitted power
P1 = 0.5'P; %source transmitted power
P2 = P-P1; %all relays transmitted power

%·················································································································

M = 64;
fft_length = 512;

papr1 = zeros(1,no_pkts);
for i1 = 1:no_pkts

%the data generation
```

```
tx_bits = randint(1,M'log2(Q)); %bits required to transmit 2 symbols
[mod_symbols table] = tx_modulate(tx_bits,modulation);
x = mod_symbols; %column vector of the generated symbols(K'1)
X1 = fft(x).';
X = zeros(1,M);
X(1:M/2) = -conj(X1(M/2+1:M));
X(M/2+1:M) = conj(X1(1:M/2));

% = = = = = = = = = = = = = = = = = = = = = = = = =
Z = zeros(1,fft_length);
Z(1:M) = X;
t = ifft(Z);
papr1(i1) = 10'log10(max(abs(t).^2)/mean(abs(t).^2));
% = = = = = = = = = = = = = = = = = = = = = = = = =

end

% = = = = = = = = = = = = = = = = = = = = = = = = =
[kkk,mm] = hist(papr1,20);
semilogy(mm,1-cumsum(kkk)/max(cumsum(kkk)),'b')
% = = = = = = = = = = = = = = = = = = = = = = = = =

hold on
```

F.4 Relay Selection for Security Enhancement

```
% = = = = = = = = = = = = = = = = = = = = = = = = =
% = = = = = = = = = = = = = = = = = = = = = = = = =

%Simulation of relay selection of Serecy outage probability vs. SNR (P)
%Non-cooperative eavesdroppers without jamming
%N = 4
%M = 2

clear all
clc

% = = = = = = = = = = = = = = = = = = = = = = = = =
%system parameters
% = = = = = = = = = = = = = = = = = = = = = = = = =

N = 4; %number of relay nodes
M = 2; %number of eavesdroppers
R0 = 2; %transmission rate (to check which relay can decode correctly)
Rs = 0.1; %target secrecy rate (BPCU)-for outage probability calculation
P_dB = 0:5:50 %Transmitted power (dB)
Num_Trials = 3'1000; %number of trials
PL_exp = 3; %path los exponent
N0 = 1; %noise variance

% = = = = = = = = = = = = = = = = = = = = = = = = =
loc_s = [0.5];
loc_d = [1.5];
loc_r = [.5.2;.5.4;.5.6;.5.8];
loc_e = [1.3; 1.7];

var_sd = (sqrt((loc_s(1)-loc_d(1)).^2+(loc_s(2)-loc_d(2)).^2)).^(-PL_exp);

z1 = (loc_s(1)-loc_e(:,1)).^2;
z2 = (loc_s(2)-loc_e(:,2)).^2;
z = sqrt(z1+z2);
var_ses = z.^(-PL_exp);

z1 = (loc_s(1)-loc_r(:,1)).^2;
z2 = (loc_s(2)-loc_r(:,2)).^2;
z = sqrt(z1+z2);
var_srs = z.^(-PL_exp);

z1 = (loc_d(1)-loc_r(:,1)).^2;
z2 = (loc_d(2)-loc_r(:,2)).^2;
z = sqrt(z1+z2);
var_rsd = z.^(-PL_exp);
```

```
z1 = (loc_r(:,1)-loc_e(1,1)).^2;
z2 = (loc_r(:,2)-loc_e(1,2)).^2;
z11 = sqrt(z1+z2);
z3 = (loc_r(:,1)-loc_e(2,1)).^2;
z4 = (loc_r(:,2)-loc_e(2,2)).^2;
z22 = sqrt(z3+z4);
z = [z11 z22];
var_rses = z.^(-PL_exp);

% = = = = = = = = = = = = = = = = = = = = = = = = = =
%simulation start
% = = = = = = = = = = = = = = = = = = = = = = = = = =

for ii = 1:length(P_dB)
Ps = 10^(P_dB(ii)/10);
Pr = Ps;
Pj = Ps/100; %jammer power less than source node power

%Cs_ncj = zeros(1,Num_Trials) ;

outage_nc = 0;
outage_ncs = 0;
outage_ncj = 0;
outage_ncjs = 0;
outage_ncj1 = 0;
outage_ncj2 = 0;
outage_nccj = 0;
outage_nccjs = 0;

outage_c = 0;
outage_cj = 0;
outage_cj2 = 0;
outage_ccj = 0;

for jj = 1:Num_Trials

   %Channel coefficients
   % = = = = = = = = = = = = = = = = = = = =
   %broadcasting phase

   hsd = sqrt(var_sd/2).*(randn(1,1)+i*randn(1,1)); %S- >destination
   hsrs = sqrt(var_srs/2).*(randn(N,1)+i*randn(N,1)); % S- >relays
   hses = sqrt(var_ses/2).*(randn(M,1)+i*randn(M,1)); % S- >eavesdroppers

%cooperative links
   hrsd = sqrt(var_rsd/2).*(randn(N,1)+i*randn(N,1)); %R- >destination

   hrses = sqrt(var_rses/2).*(randn(N,M)+i*randn(N,M)); % R- >eavesdroppers

%The case of symmetric
% = = = = = = = = = = = = = = = = = = = = = = = = = = =
% = = = = = = = = = = = = = = = = = =

   %instantaneous SNR calculations
% = = = = = = = = = = = = = = = = = =
   snr_sd = Ps*abs(hsd).^2;
   snr_srs = Ps*abs(hsrs).^2;
   snr_ses = Ps*abs(hses).^2;

   snr_rsd = Pr*abs(hrsd).^2;
   snr_rses = Pr*abs(hrses).^2;

   %checking for correct decoding
   Csrs =.5*log2(1+snr_srs); %capacity f source-relays links
   y = Csrs>R0; %all valid N relays

% = = = = = = = = = = = = = = = = = = = = = = = = = =
%Case 1: Non-cooperative without gamming (nc)
% = = = = = = = = = = = = = = = = = = = = = = = = = =

F = [];
for k1 = 1:N
C = 0.5*max(0,log2(1+snr_sd+y(k1)*snr_rsd(k1))-log2(max(1+snr_ses+y(k1)
*snr_rses(k1,:)')));
```

```
F = [F C];
end

Cs_nc(1,jj) = max(F);

outage_nc = outage_nc+(Rs>Cs_nc(1,jj));

% = = = = = = = = = = = = = = = = = = = = = = = = = = = = = = = =
%Case 1: Non-cooperative without gamming: Asymptotic (at high SNR)
% = = = = = = = = = = = = = = = = = = = = = = = = = = = = = = = =

y1 = ones(N,1); %all the relays decode correctly
F = [];
for k1 = 1:N

C = 0.5*max(0,log2(snr_sd+y1(k1)*snr_rsd(k1))-log2(max(snr_ses+y1(k1)*snr_
rses(k1,:)')));

F = [F C];

end
Cs_ncs(1,jj) = max(F);

outage_ncs = outage_ncs+(Rs>Cs_ncs(1,jj));

% = = = = = = = = = = = = = = = = = = = = = = = = = = = = = = =
%Case 2: Non-Cooperative with jamming (ncj)-Optimal case
% = = = = = = = = = = = = = = = = = = = = = = = = = = = = = = =
F = [];
for j1 = 1:N
for r = 1:N
%if y(r) = =1
for j2 = 1:N
  if (r~ = j2)

C = max(0,0.5*log2(1+(snr_sd./(1+snr_rsd(j1)/100))+(y(r)*snr_rsd(r)./(1+snr_
rsd(j2)/100)))…
     -0.5*log2(max(1+(snr_ses./(1+snr_rses(j1,:).'/100))+(y(r)*snr_rses(r,:).'./
     (1+snr_rses(j2,:).'/100)))));

F = [F C];

  end
end

%end
end
end
Cs_ncj(1,jj) = max(F);
outage_ncj = outage_ncj+(Rs>Cs_ncj(1,jj));

% = = = = = = = = = = = = = = = = = = = = = = = = = = = = = = = =
%Case : Non-Cooperative with jamming (ncj)-Asymptotic (high SNR)
% = = = = = = = = = = = = = = = = = = = = = = = = = = = = = = =
F = [];
for j1 = 1:N
for r = 1:N
%if y(r) = =1
for j2 = 1:N
  if (r~ = j2)
C = max(0,0.5*log2((snr_sd./(snr_rsd(j1)/100))+(y1(r)*snr_rsd(r)./(snr_rsd(j2)/100)))…
     -0.5*log2(max((snr_ses./(snr_rses(j1,:).'/100))+(y1(r)*snr_rses(r,:).'./(snr_
     rses(j2,:).'/100)))));

F = [F C];

  end
end

%end
end
end
```

```
Cs_ncjs(1,jj) = max(F);

outage_ncjs = outage_ncjs+(Rs>Cs_ncjs(1,jj));

% = = = = = = = = = = = = = = = = = = = = = = = = = = = = = = = = = =
%Case : Non-Cooperative with jamming (ncj)- suboptimal (individual selection)
% = = = = = = = = = = = = = = = = = = = = = = = = = = = = = = = = = =

%selection the jammer in the first phase
[q1 q2] = max(min(snr_rses,[],2)./snr_rsd);

[q3 q4] = max(y.'(snr_rsd./max(snr_rses,[],2)));

Q1 = (q3 ==0); %Q1 = 1 = => all the relays fail to decode
temp = ones(N,1);
temp(q4,1) = 0;

%selecting the jammer in the second phase
[q7 q8] = max(temp.'min(snr_rses,[],2)./snr_rsd);

Cs_ncj1(1,jj) = 0.5'max(0,log2(1+(snr_sd./(1+snr_rsd(q2)/100))+(y(q4))'(snr_rsd(q4)./
(1+snr_rsd(q8)/100)))...
      -log2(max(1+(snr_ses./(1+snr_rses(q2,:).'/100))+(y(q4))'(snr_rses(q4,:).'./
      (1+snr_rses(q8,:).'/100)))));

outage_ncj1 = outage_ncj1+(Rs>Cs_ncj1(1,jj));

% = = = = = = = = = = = = = = = = = = = = = = = = = = = = = = = = = =
%Case : Non-Cooperative with jamming (ncj)- joint selection in phase II
% = = = = = = = = = = = = = = = = = = = = = = = = = = = = = = = = = =

%selection the jammer in the first phase
[q1 q2] = max(min(snr_rses,[],2)./snr_rsd);
%joint selection in the second phase

temp2 = [];
for r = 1:N
  temp1 = [];
  for j2 = 1:N

  if (r~ = j2)
    F = max(y(r)'(snr_rsd(r)/(snr_rsd(j2)/100))./(max(snr_rses(r,:)./(snr_
    rses(j2,:)/100))));
    temp1 = [temp1 F];
  end

  end
  [q3 q4] = max(temp1);
  temp2 = [temp2 q3];
end
[q5 q6] = max(temp2); %q6 is the relay number

%search for the jammer
  temp3 = [];
  for j2 = 1:N
  if (q6~ = j2)
F = max((y(q6)'snr_rsd(q6)/(snr_rsd(j2)/100))./(max(snr_rses(q6,:)./(snr_
rses(j2,:)/100))));
  temp3 = [temp3 F];
  else
  temp3 = [temp3 0];
  end

  end
  [q7 q8] = max(temp3);

Cs_ncj2(1,jj) = 0.5'max(0,log2(1+(snr_sd./(1+snr_rsd(q2)/100))+(y(q6))'(snr_rsd(q6)./
(1+snr_rsd(q8)/100)))...
      -log2(max(1+(snr_ses./(1+snr_rses(q2,:).'/100))+(y(q6))'(snr_rses(q6,:).'./
      (1+snr_rses(q8,:).'/100)))));

outage_ncj2 = outage_ncj2+(Rs>Cs_ncj2(1,jj));
```

```
% = = = = = = = = = = = = = = = = = = = = = = = = = =
%Case : Non-Cooperative with controlled jamming (NCCJ)
% = = = = = = = = = = = = = = = = = = = = = = = = = =

%selection the jammer in the first phase
[q1 q2] = max(min(snr_rses,[],2));

%selecting the relay in the second phase
[q3 q4] = max(y.*(snr_rsd./max(snr_rses,[],2)));
Q1 = (q3 = =0); %Q1 = 1 = => all the relays fail to decode
temp = ones(N,1);
temp(q4,1) = 0;

%selecting the jammer in the second phase
[q7 q8] = max(temp.*min(snr_rses,[],2));

Cs_nccj(1,jj) = 0.5*max(0,log2(1+(snr_sd)+(y(q4))'*(snr_rsd(q4)))…
        -log2(max(1+(snr_ses./(1+snr_rses(q2,:).'/100))+(y(q4))'*(snr_rses(q4,:).'./
        (1+snr_rses(q8,:).'/100)))));

outage_nccj = outage_nccj+(Rs>Cs_nccj(1,jj));

% = = = = = = = = = = = = = = = = = = = = = = = = = = = = = = = =
%Case : Non-Cooperative with controlled jamming- Asymptotic (high SNR)
% = = = = = = = = = = = = = = = = = = = = = = = = = = = = = = = =

%selection the jammer in the first phase
[q1 q2] = max(min(snr_rses,[],2)); % q2 is j1

%selecting the relay in the second phase
[q3 q4] = max(y.*(snr_rsd./max(snr_rses,[],2)));

Q1 = (q3 = =0); %Q1 = 1 = => all the relays fail to decode
temp = ones(N,1);
temp(q4,1) = 0;

%selecting the jammer in the second phase
[q7 q8] = max(temp.*min(snr_rses,[],2));

Cs_nccjs(1,jj) = 0.5*max(0,log2((snr_sd)+(y1(q4))'*(snr_rsd(q4)))…
        -log2(max((snr_ses./(snr_rses(q2,:).'/100))+(y1(q4))'*(snr_rses(q4,:).'./
        (snr_rses(q8,:).'/100)))));
outage_nccjs = outage_nccjs+(Rs>Cs_nccjs(1,jj));

% = = = = = = = = = = = = = = = = = = = = = = = =
%Case 3: Cooperative without gamming (Cw/oJ)
% = = = = = = = = = = = = = = = = = = = = = = = =

   F = [];
for k1 = 1:N

C = 0.5*max(0,log2(1+snr_sd+y(k1)*snr_rsd(k1))-log2(sum(1+snr_ses+y(k1)*snr_
rses(k1,:)')));

F = [F C];

end

Cs_c(1,jj) = max(F);

outage_c = outage_c+(Rs>Cs_c(1,jj));

% = = = = = = = = = = = = = = = = = = = = = = = =
%Case 4: Cooperative with jamming- Optimal
% = = = = = = = = = = = = = = = = = = = = = = = =

F = [];
for j1 = 1:N
  for r = 1:N
   for j2 = 1:N
     if (r~ = j2)

C = 0.5*max(0,log2(1+(snr_sd./(1+snr_rsd(j1)/100))+y(r)'*(snr_rsd(r)./(1+snr_
rsd(j2)/100)))-…
```

```
        log2(sum(1+(snr_ses./(1+snr_rses(j1,:).'/100))+y(r)'(snr_rses(r,:).'./
        (1+snr_rses(j2,:).'/100)))));
      F = [F C];
    end
  end
end
end

Cs_cj(1,jj) = max(F);
outage_cj = outage_cj+(Rs>Cs_cj(1,jj));

% = = = = = = = = = = = = = = = = = = = = = = = = = =
%Case : Cooperative with jamming- SubOptimal
% = = = = = = = = = = = = = = = = = = = = = = = = =

%selection the jammer in the first phase
[q1 q2] = min(snr_rsd./sum(snr_rses,2)); % q2 is j1

%selecting the relay in the second phase
[q3 q4] = max(y.'(snr_rsd./sum(snr_rses,2))); % q4 is R
Q1 = (q3 ==0); %Q1 = 1 = => all the relays fail to decode
temp = ones(N,1);
temp(q4,1) = 0;

%selecting the jammer in the second phase
[q7 q8] = max(temp.'(sum(snr_rses,2)./snr_rsd)); % q8 is

% = = = = = = = = = = = = = = = = = = = = = = = = = =  = = = = =
%Case : Cooperative with jamming- Joint selection in phase II
% = = = = = = = = = = = = = = = = = = = = = = = = = = = = = = = =

%selection the jammer in the first phase
[q1 q2] = min(snr_rsd./sum(snr_rses,2));

%joint selection in the second phase

temp2 = [];
for r = 1:N
  temp1 = [];
  for j2 = 1:N

  if (r~= j2)
    F = max(y(r)'(snr_rsd(r)/(snr_rsd(j2)/100))./(sum(snr_rses(r,:)./(snr_
    rses(j2,:)/100))));
    temp1 = [temp1 F];
  end

  end
  [q3 q4] = max(temp1);
  temp2 = [temp2 q3];
end
[q5 q6] = max(temp2); %q6 is the relay number

%search for the jammer
  temp3 = [];
  for j2 = 1:N
  if (q6~= j2)
  F = max((y(q6)'snr_rsd(q6)/(snr_rsd(j2)/100))./(sum(snr_rses(q6,:)./(snr_
  rses(j2,:)/100))));
  temp3 = [temp3 F];
  else
  temp3 = [temp3 0];
  end

  end

  [q7 q8] = max(temp3); %q8 is the jammer in the second phase
Cs_cj2(1,jj) = 0.5*max(0,log2(1+(snr_sd./(1+snr_rsd(q2)/100))+y(q6)'(snr_rsd(q6)./
(1+snr_rsd(q8)/100)))-…
       log2(sum(1+(snr_ses./(1+snr_rses(q2,:).'/100))+y(q6)'(snr_rses(q6,:).'./
       (1+snr_rses(q8,:).'/100)))));

outage_cj2 = outage_cj2+(Rs>Cs_cj2(1,jj));
```

```
% = = = = = = = = = = = = = = = = = = = = = = = = =
%Case : Cooperative with controlled jamming (CCJ)
% = = = = = = = = = = = = = = = = = = = = = = = = =

%selection the jammer in the first phase
[q1 q2] = max(sum(snr_rses,2)); % q2 is j1

%selecting the relay in the second phase
[q3 q4] = max(y.'(snr_rsd./sum(snr_rses,2))); % q4 is R
Q1 = (q3 = =0); %Q1 = 1 = => all the relays fail to decode
temp = ones(N,1);
temp(q4,1) = 0;

%selecting the jammer in the second phase
[q7 q8] = max(temp.'(sum(snr_rses,2))); % q8 is j2

Cs_ccj(1,jj) = 0.5*max(0,log2(1+(snr_sd)+y(q4)'(snr_rsd(q4)))-…
        log2(sum(1+(snr_ses./(1+snr_rses(q2,:).'/100))+y(q4)'(snr_rses(q4,:).'./
        (1+snr_rses(q8,:).'/100)))));

outage_ccj = outage_ccj+(Rs>Cs_ccj(1,jj));

% = = = = = = = = = = = = = = = = = = = = = = = = = = = = = = = = = = = = = = = = = = =

end

out_nc(ii) = outage_nc/Num_Trials;
C_nc(ii) = mean(Cs_nc);

out_ncs(ii) = outage_ncs/Num_Trials; %NC without jamming, asymptotic case.
C_ncs(ii) = mean(Cs_ncs);

out_ncj(ii) = outage_ncj/Num_Trials;
C_ncj(ii) = mean(Cs_ncj)

out_ncjs(ii) = outage_ncjs/Num_Trials; %NCJ, asymptotic case.
C_ncjs(ii) = mean(Cs_ncjs)

out_ncj1(ii) = outage_ncj1/Num_Trials;
C_ncj1(ii) = mean(Cs_ncj1)
out_ncj2(ii) = outage_ncj2/Num_Trials;
C_ncj2(ii) = mean(Cs_ncj2)

out_nccj(ii) = outage_nccj/Num_Trials;
C_nccj(ii) = mean(Cs_nccj)

out_nccjs(ii) = outage_nccjs/Num_Trials; %NCCJ asymptotic
C_nccjs(ii) = mean(Cs_nccjs)

out_c(ii) = outage_c/Num_Trials;
C_c(ii) = mean(Cs_c);

out_cj(ii) = outage_cj/Num_Trials;
C_cj(ii) = mean(Cs_cj);

out_cj2(ii) = outage_cj2/Num_Trials;
C_cj2(ii) = mean(Cs_cj2);

out_ccj(ii) = outage_ccj/Num_Trials;
C_ccj(ii) = mean(Cs_ccj);

end
plot(P_dB,C_nc,'o-')

hold on
plot(P_dB,C_ncj,'o- ')

hold on
plot(P_dB,C_ncj1,'o- ')

hold on
plot(P_dB,C_ncj2,'o- ')

hold on
plot(P_dB,C_nccj,'o- ')
```

```
hold on
plot(P_dB,C_c,'s-')

hold on
plot(P_dB,C_cj,'s- ')

hold on
plot(P_dB,C_cj2,'s- ')

hold on
plot(P_dB,C_ccj,'s- ')

axis([0 50 0 4]);

figure

semilogy(P_dB,out_nc,'o-')
hold on

semilogy(P_dB,out_ncs,'o-')
hold on

semilogy(P_dB,out_ncj,'o- ')
hold on

semilogy(P_dB,out_ncjs,'o- ')
hold on

semilogy(P_dB,out_ncj1,'o- ')
hold on

semilogy(P_dB,out_ncj2,'o- ')
hold on

semilogy(P_dB,out_nccj,'o- ')
hold on

%semilogy(P_dB,out_nccjs,'o- ')
%hold on

semilogy(P_dB,out_c,'s-')
hold on

semilogy(P_dB,out_cj,'s- ')
hold on

semilogy(P_dB,out_cj2,'s- ')
hold on

semilogy(P_dB,out_ccj,'s- ')

axis([0 50 10^-5 1]);

% = = = = = = = = = = = = = = = = = = = = = = = = = = = = = = = = = = = = = = = = = = = = = = =
```

References

1. F. Khan, *LTE for 4G Mobile Broadband Air Interface Technologies and Performance*, Cambridge University Press, Cambridge, U.K., 2009.
2. D. Falconer, S. Ariyavisitakul, A. Benyamin-Seeyar, and B. Eidson, Frequency domain equalization for single-carrier broadband wireless systems, *IEEE Commun. Mag.*, 40(4), 58–66, April 2002.
3. J. Coon, S. Armour, M. Beach, and J. McGeehan, Adaptive frequency-domain equalization for single-carrier multiple-input multiple-output wireless transmissions, *IEEE Trans. Signal Process.*, 53(8), 3247–3256, August 2005.
4. F. S. Al-kamali, M. I. Dessouky, B. M. Sallam, and F. E. Abd El-Samie, Performance evaluation of Cyclic Prefix CDMA systems with frequency domain interference cancellation, *Digital Signal Process.*, 19(1), 2–13, January 2009.
5. A. Gusmao, P. Torres, R. Dinis, and N. Esteves, A reduced-CP approach to SC/FDE block transmission for broadband wireless communications, *IEEE Trans. Commun.*, 55(4), 801–809, April 2007.
6. Y. Yoshida, K. Hayashi, H. Sakai, and W. Bocquet, Analysis and compensation of transmitter IQ imbalances in OFDMA and SC-FDMA systems, *IEEE Trans. Signal Process.*, 57(8), 3119–3129, August 2009.
7. I. Koffman and V. Roman, Broadband wireless access solutions based on OFDM access in IEEEE 802.16, *IEEE Commun. Mag.*, 40(4), 96–103, April 2002.
8. H. Schulze and C. Luders, *Theory and Applications of OFDM and CDMA*, John Wiley & Sons, Ltd., Chichester, U.K., 2005.
9. H. G. Myung and D. J. Goodman, *Single Carrier FDMA: A New Air Interface for Long Term Evaluation*, John Wiley & Sons, Ltd., Chichester, U.K., 2008.

10. Y. Wu and W. Y. Zou, Orthogonal frequency division multiplexing: A multicarrier modulation scheme, *IEEE Trans. Consum. Electron.*, 41(3), 392–399, August 1995.

11. N. Parasad, S. Wang, and X. Wang, Efficient receiver algorithms for DFT-spread OFDM systems, *IEEE Trans. Wireless Commun.*, 8(6), 3216–3225, June 2009.

12. Y. Zhu and K. B. Lataief, CFO estimation and compensation in single carrier interleaved FDMA systems, in *Proceedings of the IEEE GLOBECOM 2009*, Honolulu, HI, pp. 1–5, 2009.

13. Z. Cao, U. Tureli, and Y. Yao, Low-complexity orthogonal spectral signal construction for generalized OFDMA uplink with frequency synchronization errors, *IEEE Trans. Vehic. Technol.*, 56(3), 1143–1154, May 2007.

14. H. G. Myung, J. Lim, and D. J. Goodman, Single carrier FDMA for uplink wireless transmission, *IEEE Vehic. Technol. Mag.*, 1(3), 30–38, September 2006.

15. Z. Lin, P. Xiao, and B. Vucetic, Analysis of receiver algorithms for LTE SC-FDMA based uplink MIMO systems, *IEEE Trans. Wireless Commun.*, 9(1), 60–65, January 2010.

16. 3rd Generation Partnership Project, TR 25.814—Technical Specification Group Radio Access Network; Physical layer aspects for evolved Universal Terrestrial Radio Access (UTRA) (Release 7), Section 9.1, 2006.

17. A. Ghosh, R. Ratasuk, B, Mondal, N. Mangalvedhe, and T. Thomas, LTE-advanced: Next-generation wireless broadband technology, *IEEE Wireless Commun.*, 17(3), 10–22, June 2010.

18. A. Wilzeck, Q. Cai, M. Schiewer, and T. Kaiser, Effect of multiple carrier frequency offsets in MIMO SC-FDMA systems, in *Proceedings of the International ITG/IEEE Workshop on Smart Antennas*, Vienna, Austria, February 2007.

19. A. Goldsmith, *Wireless Communications*, Cambridge University Press, Cambridge, U.K., 2005.

20. D. M. Sacristan, J. F. Monserrat, J. C. Penuelas, D. Calabuig, S. Garrigas, and N. Cardona, On the way towards fourth-generation mobile: 3GPP LTE and LTE-advanced, *EURASIP J. Wireless Commun. Network.*, 2009, Article ID 354089, 1–10, 2009.

21. J. D. Gibson, *Mobile Communications Handbook*, 2nd edn., Springer-Verlag New York, Secaucus, NJ, 1999.

22. T. S. Rappaport, *Wireless Communications Principles and Practice*, 2nd edn., Pearson Education, Indianapolis, IN, 2002.

23. J. Proakis, *Digital Communications*, 4th edn., McGraw-Hill, New York, 2001.

24. B. Lo and K. Ben Letaief, Adaptive equalization and interference cancellation for wireless communication systems, *IEEE Trans. Commun.*, 47(4), 538–545, April 1999.

25. C. Laot, A. Glavieux, and J. Labat, Turbo equalization: Adaptive equalization and channel decoding jointly optimized, *IEEE J. Sel. Areas Commun.*, 47(9), 1744–1752, April 2001.

26. E. Larsson, Y. Selen, and P. Stoica, Adaptive equalization for frequency selective channels of unknown length, *IEEE Trans. Vehic. Technol.*, 54(2), 568–579, March 2005.
27. F. S. Al-kamali, M. I. Dessouky, B. M. Sallam, and F. E. Abd El-Samie, Low complexity frequency domain equalization for Cyclic Prefix CDMA systems, in *Proceedings of the URSI National Radio Science Conference (NRSC)*, Tanta, Egypt, March 2008.
28. R. Chang, Synthesis of band-limited orthogonal signals for multi-channel data transmission, *Bell Labs Tech. J.*, 1775–1796, December 1966.
29. S. Weinstein and P. Ebert, Data transmission by frequency division multiplexing using discrete Fourier transform, *IEEE Trans. Commun. Technol.*, 19(5), 628–634, October 1971.
30. K. Fazel and S. Kaiser, *Multi-Carrier and Spread Spectrum Systems*, John Wiley & Sons, Ltd., Chichester, U.K., 2003.
31. C. Y. Wong, R. S. Cheng, K. B. Letaief, and R. D. Murch, Multiuser OFDM with adaptive subcarrier, bit, and power allocation, *IEEE J. Sel. Areas Commun.*, 17(10), 1747–1757, October 1999.
32. R. Nogueroles, M. Bossert, A. Donder, and V. Zyablov, Improved performance of a random OFDMA mobile communication system, in *Proceedings of the IEEE VTC*, vol. 3, Ottawa, ON, pp. 2502–2506, May 1998.
33. A. V. Oppenheim, R. W. Schafer, and J. R. Buck, *Discrete-Time Signal Processing*, 2nd edn., Prentice Hall, Upper Saddle River, NJ, 1999.
34. N. Arrue, I. Velez, J. Sevillano, and L. Fontan, Two coarse frequency acquisition algorithms for OFDM based IEEE 802.11 Standards, *IEEE Consum. Electron.*, 53(1), 33–38, February 2007.
35. H. G. Myung, J. Lim, and D. J. Goodman, Peak-to-average power ratio of single carrier FDMA signals with pulse shaping, in *Proceedings of the IEEE PIMRC 2006*, Helsinki, Finland, pp. 1–5, September 2006.
36. G. Huang, A. Nix, and S. Armour, Impact of radio resource allocation and pulse shaping on PAPR of SC-FDMA signals, in *Proceedings of the IEEE PIMRC 2007*, Athens, Greece, pp. 1–5, September 3–7, 2007.
37. M. Rumney, 3GPP LTE: Introducing single-carrier FDMA, *Agilent Meas. J.*, 1–10, January 2008.
38. J. Gazda, Multicarrier based transmission systems undergoing nonlinear amplification, PhD thesis, Technical University of Kosice, Kosice, Slovak, August 2010.
39. E. Costa, M. Midrio, and S. Pupolin, Impact of amplifier nonlinearities on OFDM transmission system performance, *IEEE Commun. Lett.*, 3(2), 37–39, February 1999.
40. S. L. Miller and R. J. ODea, Peak power and bandwidth efficient linear modulation, *IEEE Trans. Commun.*, 46(12), 1639–1648, December 1998.
41. H. Myung, Single carrier orthogonal multiple access technique for broadband wireless communications, PhD thesis, Polytechnic University, Brooklyn, NY, January 2007.
42. D. Wulich and L. Goldfeld, Bound of the distribution of instantaneous power in single carrier modulation, *IEEE Trans. Wireless Commun.*, 4(4), 1773–1778, July 2005.

43. 3rd Generation Partnership Project, 3GPP TS 25.101—Technical Specification Group Radio Access Network; User Equipment (UE) Radio Transmission and Reception (FDD) (Release 7), Section B.2.2, September 2007.

44. F. S. Al-kamali, M. I. Dessouky, B. M. Sallam, F. Shawki, and F. E. Abd El-Samie, Impact of the power amplifier on the performance of the single carrier frequency division multiple access system, Accepted for publication in *J. Telecommun. Syst.*, DOI 10.1007/s11235-011-9439-y.

45. G. D. Mandyam, Sinusoidal transforms in OFDMA system, *IEEE Trans. Broadcasting*, 50(2), 172–184, June 2004.

46. N. Al-Dhahir and H. Minn, A new multicarrier transceiver based on the discrete cosine transform, in *Proceedings of the IEEE Wireless Communications and Networking Conference*, vol. 1, New Orleans, LA, pp. 45–50, March 13–17, 2005.

47. P. Tan and N. C. Beaulieu, A comparison of DCT-based OFDM and DFT-based OFDM in frequency offset and fading channels, *IEEE Trans. Commun.*, 54(11), 2113–2125, November 2006.

48. Y. Han and J. Leou, Detection and correction of transmission errors in JPEG images, *IEEE Trans. Circ. Syst. Video Technol.*, 8(2), 221–231, April 1998.

49. Z. D. Wang, Fast algorithms for the discrete W transform and the discrete Fourier transform, *IEEE Trans. Acoust. Speech Signal Process.*, 32(4), 803–816, August 1984.

50. G. D. Mandyam, On the discrete cosine transform and OFDM systems, in *Proceedings of the IEEE International Conference Acoustics, Speech, Signal Processing*, vol. 4, Hong Kong, China, pp. 544–547, April 2003.

51. F. S. Al-kamali, M. I. Dessouky, B. M. Sallam, F. E. Abd El-Samie, and F. Shawki, A new single carrier FDMA system based on the discrete cosine transform, in *Proceedings of the ICCES'9 Conference*, Cairo, Egypt, pp. 555–560, December 14–16, 2009.

52. X. Li and L. J. Cimini, Effects of clipping and filtering on the performance of OFDM, *IEEE Commun. Lett.*, 2(5), 131–133, June 1998.

53. X. Wang, T. T. Tjhung, and C. S. Ng, Reduction of peak-to-average power ratio of OFDM system using a companding technique, *IEEE Trans. Broadcasting*, 45(3), 303–307, September 1999.

54. X. Huang, J. Lu, J. Zheng, J. Chuang, and J. Gu, Reduction of peak to average-power-ratio of OFDM signals with companding transform, *IEEE Electron. Lett.*, 37(8), 506–507, April 2001.

55. J. Armstrong, Peak-to-average power reduction for OFDM by repeated clipping and frequency domain filtering, *Electron. Lett.*, 38, 246–247, February 2002.

56. S. H. Han and J. H. Lee, An overview of peak-to-average power ratio reduction techniques for multicarrier transmission, *IEEE Wireless Commun.*, 12(2), 56–65, April 2005.

57. N. Chaudhary and L. Cao, Comparison of compand-filter schemes for reducing PAPR in OFDM, in *Proceedings of the IEEE WCNC 2006*, vol. 4, Las Vegas, NV, pp. 2070–2075, 2006.

58. J. Kim and Y. Shin, An effective clipped companding scheme for PAPR reduction of OFDM signals, in *Proceedings of the IEEE ICC 2008*, Beijing, China, pp. 668–672, 2008.

59. J. Kim, S. Han, and Y. Shin, A robust companding scheme against nonlinear distortion of high power amplifiers in OFDM systems, in *Proceedings of the IEEE VTC 2008*, Singapore, pp. 1697–1701, 2008.

60. H. G. Myung, K. J. Pan, R. Olesen, and D. Grie, Peak power characteristics of single carrier FDMA MIMO precoding system, in *Proceedings of the IEEE VTC 2007*, Baltimore, MD, pp. 477–481, 2007.

61. F. E. Abd El-Samie, Super resolution reconstruction of images, PhD thesis, Minoufiya University, Shibin el Kom, Egypt, 2005.

62. B. G. Negash and H. Nikookar, Wavelet based OFDM for wireless channels, in *Proceedings of the IEEE VTC 2001*, 1, Rhodes, Greece, 688–691, 2001.

63. A. H. Kattoush, W. A. Mahmoud, and S. Nihad, The performance of multiwavelets based OFDM system under different channel conditions, *Digital Signal Process.*, 20, 472–482, 2010.

64. V. Erceg, K. V. Hari, M. S. Smith, D. S. Baum, K. P. Sheikh, C. Tappenden, J. M. Costa, C. Bushue, A. Sarajedini, R. Schwartz, D. Branlund, T. Kaitz, and D. Trinkwon, Channel models for fixed wireless applications, IEEE 802.16a cont. IEEE 802.16.3c-01/29r1, February 2001.

65. F. S. Al-kamali, M. I. Dessouky, B. M. Sallam, F. E. Abd El-Samie, and F. Shawki, Transceiver scheme for single-carrier frequency division multiple access implementing the wavelet transform and the PAPR reduction methods, *IET Commun.*, 4(1), 69–79, January 2010.

66. F. E. Abd El-Samie, F. S. Al-kamali, M. I. Dessouky, B. M. Sallam, and F. Shawki, Performance enhancement of SC-FDMA system using a companding technique, *Ann. Telecommun.*, 65(5), 293–300, May 2010.

67. H. Cheon, Frequency offset estimation for high speed users in E-UTRA uplink, in *Proceedings of the IEEE PIMRC 2007*, Athens, Greece, pp. 1–5, September 3–7, 2007.

68. P. H. Moose, A technique for orthogonal frequency division multiplexing frequency offset correction, *IEEE Trans. Commun.*, 42(10), 2908–2914, October 1994.

69. J. V. Beek, P. O. Borjesson, M. L. Boucheret, D. Landstrom, J. M. Arenas, P. Odling, C. Ostberg, M. Wahlqvist, and S. K. Wilson, A time and frequency synchronization scheme for multiuser OFDM, *IEEE J. Sel. Areas Commun.*, 17(11), 1900–1914, November 1999.

70. J. Choi, C. Lee, H. W. Jung, and Y. H. Lee, Carrier frequency offset compensation for uplink of OFDM-FDMA systems, *IEEE Commun. Lett.*, 4(12), 414–416, December 2000.

71. K. Sathananthan, R. M. Rajatheva, and S. B. Slimane, Cancellation technique to reduce intercarrier interference in OFDM, *Electrons Lett.*, 36(25), 2078–2079, December 2000.

72. K. Sathananthan and C. Tellambura, partial transmit sequence and selected mapping schemes to reduce ICI in OFDM systems, *IEEE Commun. Lett.*, 6(8), 313–315, August 2002.

73. D. Huang and K. B. Letaief, An interference-cancellation scheme for carrier frequency offsets correction in OFDMA systems, *IEEE Trans. Commun.*, 53(7), 1155–1165, July 2005.

74. D. Yan, W. Bai, Y. Xiao, and S. Li, Multiuser interference suppression for uplink interleaved FDMA with carrier frequency offset, in *Proceedings of the International Conference on Wireless Communications and Signal Processing 2009*, Nanjing, China, pp. 1–5, November 13–15, 2009.

75. X. Zhang, H. G. Ryu, and Y. Li, Joint suppression of phase noise and CFO by block type pilots, in *Proceedings of the Ninth International Symposium on Communications and Information Technology*, Icheon, South Korea, pp. 466–469, September 28–30, 2009.

76. G. Chen, Y, Zhu, and K. B. Letaief, Combined MMSE-FDE and interference cancellation for uplink SC-FDMA with carrier frequency offsets, in *Proceedings of the IEEE ICC 2010 Conference*, Cape Town, South Africa, pp. 1–5, 2010.

77. M. Ma, X. Huang, and Y. J. Guo, An interference self-cancellation technique for SC-FDMA systems, *IEEE Commun. Lett.*, 14(6), 3546–3551, June 2010.

78. P. Sun and L. Zhang, Low complexity iterative interference cancelation for OFDMA uplink with carrier frequency offsets, in *Proceedings of the APCC*, Shanghai, China, pp. 390–393, 2009.

79. F. S. Al-kamali, M. I. Dessouky, B. M. Sallam, F. E. Abd El-Samie, and F. Shawki, Carrier frequency offset problem in DCT-SC-FDMA system: Investigation and compensation, *ISRN Commun. Network. J.*, 2011, 1–7, 2011.

80. G. H. Golub, and C. F. Van Loan, *Matrix Computations*, 3rd edn., The Johns Hopkins University Press, Baltimore, MD, 1996.

81. T. Yucek and H. Arslan, Carrier frequency offset compensation with successive cancellation in uplink OFDMA systems, *IEEE Trans. Wireless Commun.*, 6(10), 3546–3551, October 2007.

82. F. S. Al-kamali, M. I. Dessouky, B. M. Sallam, F. E. Abd El-Samie, and F. Shawki, Uplink single-carrier frequency division multiple access system with joint equalization and carrier frequency offsets compensation, *IET Commun.*, 5(4), 425–433, March 2011.

83. R. D. Murch and K. B. Letaief, Antenna systems for broadband wireless access, *IEEE Commun. Mag.*, 40(4), 76–83, April 2002.

84. L. Zheng and D. N. C. Tse, Diversity and multiplexing: A fundamental tradeoff in multiple-antenna channels, *IEEE Trans. Inf. Theory*, 49(5), 1073–1096, May 2003.

85. S. Catreux, L. J. Greenstein, and V. Erceg, Some results and insights on the performance gains of MIMO systems, *IEEE J. Sel. Areas Commun.*, 21(5), 839–847, June 2003.

86. S. Yoon and S.-k. Lee, A detection algorithm for multi-input multi-output (MIMO) transmission using poly-diagonalization and Trellis decoding, *IEEE J. Sel. Areas Commun.*, 26(6), 993–1002, August 2008.

87. V. Kuhn, *Wireless Communications over MIMO Channel Application to CDMA and Multiple Antenna Systems*, John Wiley & Sons, Ltd, Somerset, NJ, 2006.

88. C. Yoon, K. Song, M. Cheong, and S. Lee, Low-complexity ZF detection for double space-frequency transmit diversity based OFDM system in frequency selective fading channel, in *Proceedings of the IEEE WCNC2007*, Hong Kong, China, pp. 2022–2026, March 11–15, 2007.

89. N. Tavangaran, A. Wilzeck, and T. Kaiser, MIMO SC-FDMA system performance for space time/frequency coding and spatial multiplexing, in *Proceedings of the International ITG/IEEE Workshop on Smart Antennas*, Vienna, Austria, pp. 382–386, February 26–27, 2008.

90. D. Li, P. Wei, and X. Zhu, Novel space-time coding and mapping scheme in single-carrier FDMA systems, in *Proceedings of the International Symposium on Personal, Indoor, and Mobile Radio Communications (PIMRC)*, Athens, Greece, pp. 1–4, September 3–7, 2007.

91. Y. Wu, X. Zhu, and A. Nandi, MIMO single-carrier FDMA with adaptive turbo multiuser detection and co-channel interference suppression, in *Proceedings of the IEEE ICC 2008*, Beijing, China, pp. 4521–4525, May 19–23, 2008.

92. J. L. Pan, R. L. Olesen, D. Grieco, and C. P. Yen, Efficient feedback design for MIMO SC-FDMA systems, in *Proceedings of the IEEE VTC 2007 Spring*, Dublin, Ireland, pp. 2399–2403, April 2007.

93. N. Al-Dhahir and A. H. Sayed, The finite-length multi-input multi-output MSE-DFE, *IEEE Trans. Signal Process.*, 48(10), 2921–2936, October 2000.

94. A. Lozano and C. Papadias, Layered space–time receivers for frequency selective wireless channels, *IEEE Trans. Commun.*, 50(1), 65–73, January 2002.

95. R. Kalbasi, D. D. Falconer, A. H. Banihashemi, and R. Dinis, A comparison of frequency-domain block MIMO transmission systems, *IEEE Trans. Vehic. Technol.*, 58(1), 165–175, January 2009.

96. F. S. Al-kamali, M. I. Dessouky, B. M. Sallam, F. E. Abd. El-Samie, and F. Shawki, Low-complexity equalization scheme for MIMO uplink SC-FDMA systems, in *Proceedings of the NRSC 2010*, Menouf, Egypt, 2010.

97. F. S. Al-kamali, M. I. Dessouky, B. M. Sallam, F. E. Abd El-Samie, and F. Shawki, A new equalization scheme for MIMO SC-FDMA system in the presence of CFO, in *Proceedings of the ECSE 2010 Conference*, Cairo, Egypt, November 1–3, 2010.

98. F. S. Al-kamali, M. I. Dessouky, B. M. Sallam, F. Shawki, W. Al-Hanafy, and F. E. Abd El-Samie, Joint low-complexity equalization and carrier frequency offsets compensation scheme for MIMO SC-FDMA systems, *IEEE Trans. Wireless Commun.*, 11(3), 869–873, 2012.

99. A. Sendonaris, E. Erkip, and B. Aazhang, User cooperation diversity—Part II: Implementations aspects and performance analysis, *IEEE Trans Commun.*, 51(11), 1927–1938, November 2003.

100. A. Sendonaris, E. Erkip, and B. Aazhang, User cooperation diversity—Part I: System description, *IEEE Trans Commun.*, 51(11), 1939–1948, November 2003.

101. J. Laneman, D. Tse, and G. W. Wornell, Cooperative diversity in wireless networks: Efficient protocols and outage behavior, *IEEE Trans. Inf. Theory*, 50(12), 3062–3080, December 2004.

102. ITU-R Rep. M.2134, Requirements related to technical performance for IMT-advanced radio interface(s), 2008.

103. S. W. Peter and R. W. Heath, The future of WiMAX: Multihop relaying with IEEE 802.16j, *IEEE Commun. Mag.*, 104–111, January 2009.

104. H. Hu, J. Xu, and G. Mao, Relay technologies for WiMAX and LTE-advanced mobile systems, *IEEE Commun. Mag.*, 104–111, 2009.

105. A. D. Wyner, The wire-tap channel, *Bell Syst. Technol. J.*, 54, 1355–1387, January 1975.

106. I. Csiszar and J. Korner, Broadcast channels with confidential messages, *IEEE Trans. Inf. Theory*, 24, 451–456, July 1978.

107. D. Tse and P. Viswanath, *Fundamentals of Wireless Communications*, Cambridge University Press, Cambridge, U.K., 2005.

108. W. C. Jacks, *Microwave Mobile Communications*, John Wiley & Sons, New York, 1974.

109. G. J. Foschini, Layered space-time architecture for wireless communication in a fading environment when using multi-element antennas, *Bell Labs Technol. J.*, 1(2), 41–59, Autumn 1996.

110. V. Tarokh, N. Seshadri, and A. R. Calderbank, Space-time codes for high data rate wireless communication: Performance criterion and code construction, *IEEE Trans. Inf. Theory*, 44(2), 744–765, March 1998.

111. V. Tarokh, H. Jafarkhani, and A. R. Calderbank, Space-time block coding for wireless communications: Performance results, *IEEE J. Sel. Areas Commun.*, 17(3), 451–460, March 1999.

112. V. Tarokh, H. Jafarkhani, and A. Calderbank, Space–time block codes from orthogonal designs, *IEEE Trans. Inf. Theory*, 45(5), 1456–1467, July 1999.

113. E. C. der Meulen, Three-terminal communication channels, *Adv. Appl. Prob.*, 3, 120–154, 1971.

114. T. Cover and A. Gamal, Capacity theorems for the relay channel, *IEEE Trans. Inf. Theory*, 25(5), 572–584, September 1979.

115. A. Host-Madsen, and J. Zhang, Capacity bounds and power allocation for wireless relay channels, *IEEE Trans. Inf. Theory*, 51(6), 2020–2040, June 2005.

116. G. Kramer, M. Gastpar, and P. Gupta, Cooperative strategies and capacity theorems for relay networks, *IEEE Trans. Inf. Theory*, 51, 3037–3063, September 2005.

117. R. Pabst, B. H. Walke, D. C. Schultz, P. Herhold, H. Yanikomeroglu, S. Mukherjee, H. Viswanathan, M. Lott, W. Zirwas, M. Dohler, H. Aghvami, D. Falconer, and G. P. Fettweis, Relay-based deployment concepts for wireless and mobile broadband radio, *IEEE Commun. Mag.*, 80–89, September 2004.

118. O. Muñoz, J. Vidal, and A. Agustin, Linear transceiver design in non-regenerative relays with channel state information, *IEEE Trans. Signal Process.*, 44(6), 2593–2604, June 2007.

119. A. Wyner and J. Ziv, The rate-distortion function for source coding with side information at the decoder, *IEEE Trans. Inf. Theory*, 22(1), 1–10, January 1976.

120. K. J. Ray Liu, A. K. Sadek, W. Su, and A. Kwasinski, *Cooperative Communications and Networking*, Cambridge University Press, Cambridge, U.K., 2009.

121. K. G. Seddik, A. K Sadek, and K. J. Ray Liu, Outage analysis and optimal power allocation for multimode relay networks, *IEEE Signal Process. Lett.*, 14(6), 377–380, June 2007.

122. J. Laneman and G. W. Wornell, Distributed space-time coded protocols for exploiting cooperative diversity in wireless networks, *IEEE Trans. Inf. Theory*, 50(12), 77–81, December 2004.

123. P. A. Anghel, G. Leus, and M. Kaveh, Multi-user space-time coding in cooperative networks, *International Conference on Acoustics, Speech and Signal Processing (ICASSP)*, Hong Kong, China, April 6–10, 2003.

124. Y. Jing and B. Hassibi, Distributed space-time coding in wireless relay networks, *IEEE Trans. Wireless Commun.*, 5(12), 3524–3536, December 2006.

125. B. Sirkeci-Mergen and B. Scaglione, Randomized space–time coding for distributed cooperative communications, *IEEE Trans. Signal Process.*, 55(10), 5003–5017, October 2007.

126. K. G. Seddik, A. K. Sadek, and K. J. Ray Liu, Design criteria and performance analysis for distributed space-time coding, *IEEE Trans. Vehic. Technol.*, 57(4), 2280–2292, July 2008.

127. H. Meheidat, M. Uysal, and N. Al-Dhahir, Equalizations techniques for distributed space-time block codes with amplify-and-forward relaying, *IEEE Trans. Signal Process.*, 55(5), 1839–1852, May 2007.

128. E. Beres and R. Adve, Selection cooperation in multi-source cooperative networks, *IEEE Trans. Wireless Commun.*, 7, 118–127, January 2008.

129. A. Bletsas, A. Khisti, D. Reed, and A. Lippman, A simple cooperative diversity method based on network path selection, *IEEE J. Sel. Areas Commun.*, 24, 659–672, March 2006.

130. A. Bletsas, A. Khisti, and M. Win, Cooperative communications with outage-optimal opportunistic relaying, *IEEE Trans. Wireless Commun.*, 6(9), 3450–3460, September 2007.

131. A. Adinoyi, Y. Fan, H. Yanikomeroglu, and V. Poor, On the performance of selective relaying, in *Proceedings of IEEE Vehicular Technology Conference (VTC) Fall*, Calgary, Alberta, September 2008.

132. A. S. Ibrahim, A. K. Sadek, W. Su, and K. J. Liu, Cooperative communications with relay selection: When to cooperate and whom to cooperate with, *IEEE Trans. Wireless Commun.*, 7(7), 2814–2827, July 2008.

133. D. S. Michalopoulos and G. K. Karagiannidis, Performance analysis of single relay selection in Rayleigh fading, *IEEE Trans. Wireless Commun.*, calgary, Alberta, 7(10), 3718–3724, October 2008.

134. T. E. Hunter and A. Nosratinia, Cooperative diversity through coding, in *Proceedings of IEEE International Symposium Information Theory (ISIT)*, Lausanne, Switzerland, p. 220, July 2002.

135. A. Stefanov and E. Erkip, Cooperative coding for wireless networks, *IEEE Trans. Commun.*, 52(9), 1470–1476, September 2004.

136. T. E. Hunter and A. Nosratinia, Diversity through coded cooperation, *IEEE Trans. Wireless Commun.*, 5(2), 283–289, February 2006.

137. H. Sari and G. Karam, Orthogonal division multiple access and its application to CATV networks, *Eur. Trans. Telecommun.*, 9(6), 507–516, November 1998.

138. R. Laroi, S. Uppala, and J. Li, Designing a mobile broadband wireless access network, *IEEE Signal Process. Mag.*, 25, 20–28, September 2004.

139. R. Dinis and D. Falconer, A multiple access scheme for the uplink of broadband wireless systems, *IEEE Global Telecommunication Conference (GLOBECOM)*, Dallas, TX, pp. 3808–3812, 2004.

140. 3GPP TS 36.201 V 8.1.0, LTE-physical layer-general description (Release 8), November 2007.

141. M. Noune and A. Nix, Frequency-domain transmit processing for MIMO SC-FDMA in wideband propagation channels, in *Proceedings of Wireless Communications and Networking Conference (WCNC)*, Budapest, Hungary, pp. 1–6, April 2009.

142. J. Niu and I. T. Lu, Coded cooperation on block-fading channels in single carrier FDMA systems, *IEEE Global Telecommunication Conference (GLOBECOM)*, Washington, DC, pp. 4339–4343, November 2007.

143. A. Y. Al-nahari, F. E. Abd El-Samie, and M. I. Dessouky, Cooperative diversity schemes for uplink single-carrier FDMA systems, *Digital Signal Process. J.*, 21(2), 320–331.

144. A. Y. Al-nahari, M. I. Dessouky, and F. E. Abd El-Samie, Cooperative space–time coding with amplify-and-forward relaying, *Signal Process. Syst.*, 67(2), 129–138, 2012.

145. S. M. Alamouti, A simple transmitter diversity scheme for wireless communications, *IEEE J. Sel. Areas Commun.*, 16(8), 1451–1458, October 1998.

146. R. Nabar, H. Bolkskei, and W. Kneubuhler, Fading relay channels: Performance limits and space–time signal design, *IEEE J. Sel. Areas Commun.*, 22(6), 1099–1109, 2004.

147. I. Krikidis, J. S. Thompson, and S. McLaughlin, On the diversity order of non-orthogonal amplify-and-forward over block fading channels, *IEEE Trans. Wireless Commun.*, 9(6), 1890–1900, June 2010.

148. P. Merkey and E. C. Posner, Optimal cyclic redundancy codes for noise channels, *IEEE Trans. Inf. Theory*, 30(3), 865–867, November 1984.

149. A. K. Sadek, W. Su, and K. G. R. Liu, Multi-node cooperative communications in wireless networks, *IEEE Trans. Signal Process.*, 55, 341–355, January 2007.

150. J. Laneman and G. W. Wornell, Distributed space-time coded protocols for exploiting cooperative diversity in wireless networks, *IEEE Trans. Inf. Theory*, 49(10), 2415–2425, October 2003.

151. K. F. Lee and D. B. Williams, A space-frequency transmitter diversity technique for OFDM systems, *IEEE Global Telecommunication Conference (GLOBECOM)*, Francisco, CA, 3, 1473–1477, November 2000.

152. D. Djenouri, L. Khelladi, and N. Badache, A survey of security issues in mobile ad hoc and sensor networks, *IEEE Commun. Surveys Tutorials*, 7, 2–28, 2005.

153. Y. Liang, H. V. Poor, and L. Ying, Wireless broadcast networks: Reliability, security, and stability, *IEEE Information Theory Workshop*, Porto, Portugal, pp. 249–255, February 2008.

154. J. Barros and M. R. D. Rodrigues, Secrecy capacity of wireless channels, in *Proceedings IEEE International Symposium Information Theory*, Seattle, WA, pp. 356–360, July 2006.

155. A. O. Hero, Secure space-time communication, *IEEE Trans. Inf. Theory*, 49(12), 3235–3249, December 2003.

156. F. Oggier and B. Hassibi, The secrecy capacity of the MIMO wiretap channel, in *IEEE International Symposium on Information Theory (ISIT)*, Toronto, Ontario, Canada, pp. 524–528, July 2008.

157. A. Khisti and G. W. Wornell, Secure transmission with multiple antennas I: The MISOME wiretap channel, *IEEE Trans. Inf. Theory*, 56(7), 3088–3104, July 2010.

158. Z. Li, W. Trappe, and R. Yates, Secret communication via multi-antenna transmission, in *Proceedings of 41st Conference Information Sciences Systems*, Baltimore, MD, March 2007.

159. P. Popovski and O. Simeone, Wireless secrecy in cellular systems with infrastructure-aided cooperation, *IEEE Trans. Inf. Foren. Sec.*, 4, 242–256, June 2009.

160. L. Dong, Z. Han, A. Petropulu, and H. V. Poor, Improving wireless physical layer security via cooperating relays, *IEEE Trans. Signal Process.*, 58(3), 1875–1888, March 2010.

161. J. Li, A. P. Petropulu, and S. Weber, Optimal cooperative relaying schemes for improving wireless physical layer security, *IEEE Trans. Signal Process.* (submitted for publication). Available online: http://arxiv.org/abs/1001.1389

162. L. Lai and H. El Gamal, The relay–eavesdropper channel: Cooperation for secrecy, *IEEE Trans. Inf. Theory*, 54, 4005–4019, September 2008.

163. I. Krikidis, Opportunistic relay selection for cooperative networks with secrecy constraints, *IET Commun.*, 4(15), 1787–1791, 2010.

164. I. Krikidis, Relay selection for secure cooperative networks with jamming, *IEEE Trans. Wireless Commun.*, 8(10), 5003–5011, October 2009.

165. P. Wang, G. Yu, and Z. Zhang, On the secrecy capacity of fading wireless channel with multiple eavesdroppers, in *Proceedings of the IEEE International Symposium on Information Theory (ISIT)*, Nice, France, June 2007.

166. A. Papoulis, *Probability, Random Variables, and Stochastic Processes*, 3rd edition, McGraw-Hill, New York, 1991.

167. J. Galambos, *The Asymptotic Theory of Extreme Order Statistics*, Krieger Publishing Company, Melbourne, FL, 1987.

168. I. S. Gradshteyn and I. M. Ryzhik, *Tables of Integrals, Series, and Products*, 7th edn., Elsevier, Amsterdam, the Netherlands, 2007.

169. A. Y. Al-nahari, I. Krikidis, A. S. Ibrahim, M. I. Dessouky, F. E. Abd El-Samie, Relaying techniques for enhancing the physical layer secrecy in cooperative networks with multiple eavesdroppers, Accepted for publication in *Trans. Emerg. Telecommun. Technol.*, DOI: 10.1002/ett.2581.

Index